# Fouling of
# Heat Exchangers

CHEMICAL ENGINEERING MONOGRAPHS

Advisory Editor: Professor S.W. CHURCHILL, Department of Chemical Engineering, University of Pennsylvania, Philadelphia, PA 19104, U.S.A.

Vol. 1 Polymer Engineering (Williams)
Vol. 2 Filtration Post-Treatment Processes (Wakeman)
Vol. 3 Multicomponent Diffusion (Cussler)
Vol. 4 Transport in Porous Catalysts (Jackson)
Vol. 5 Calculation of Properties Using Corresponding-State Methods (Štěrbáček et al.)
Vol. 6 Industrial Separators for Gas Cleaning (Štorch et al.)
Vol. 7 Twin Screw Extrusion (Janssen)
Vol. 8 Fault Detection and Diagnosis in Chemical and Petrochemical Processes (Himmelblau)
Vol. 9 Electrochemical Reactor Design (Pickett)
Vol. 10 Large Chemical Plants (Froment, editor)
Vol. 11 Design of Industrial Catalysts (Trimm)
Vol. 12 Steady-state Flow-sheeting of Chemical Plants (Benedek, editor)
Vol. 13 Chemical Reactor Design in Practice (Rose)
Vol. 14 Electrostatic Precipitators (Böhm)
Vol. 15 Toluene, the Xylenes and their Industrial Derivatives (Hancock, editor)
Vol. 16 Dense Gas Dispersion (Britter and Griffiths, editors)
Vol. 17 Gas Transport in Porous Media: The Dusty-Gas Model (Mason and Malinauskas)
Vol. 18 Principles of Electrochemical Reactor Analysis (Fahidy)
Vol. 19 The Kinetics of Industrial Crystallization (Nývlt et al.)
Vol. 20 Heat Pumps in Industry (Moser and Schnitzer)
Vol. 21 Electrochemical Engineering (Roušar et al.) in 2 volumes: Parts A-C and Parts D-F
Vol. 22 Heavy Gas Dispersion Trials at Thorney Island (McQuaid, editor)
Vol. 23 Advanced Design of Ventilation Systems for Contaminant Control (Goodfellow)
Vol. 24 Ventilation '85 (Goodfellow, editor)
Vol. 25 Adsorption Engineering (Suzuki)
Vol. 26 Fouling of Heat Exchangers (Bott)

# Fouling of
# Heat Exchangers

**T.R. BOTT**

*School of Chemical Engineering, University of Birmingham,
Birmingham B15 2TT, UK*

**1995**

**ELSEVIER**
**Amsterdam-Lausanne-New York-Oxford-Shannon-Tokyo**

ELSEVIER SCIENCE B.V.
Sara Burgerhartstraat 25
P.O. Box 211, 1000 AE Amsterdam, The Netherlands

Transferred to digital print on demand, 2005

ISBN:  0-444-82186-4

This book is printed on acid-free paper.

Printed and bound by Antony Rowe Ltd, Eastbourne

# PREFACE

There are many textbooks devoted to heat transfer and the design of heat exchangers ranging from the extreme theoretical to the very practical. The purpose of these publications is to provide improved understanding of the science and to give guidance on the design and operation of process heat exchangers. In many of these texts the problem of the accumulation of deposits on heat transfer surfaces is ignored or at best, dealt with through the traditional fouling resistance. It is common knowledge that this approach is severely limited and inaccurate and may lead to gross errors in design. Furthermore the very arbitrary choice of fouling resistance more than offsets the accuracy of correlations and sophisticated methods, for the application of fundamental heat transfer knowledge.

Little attention was paid to the heat exchanger fouling and the associated inefficiencies of heat exchanger operation till the so-called "oil crisis" of the 1970s, when it became vital to make efficient use of available energy. Heat exchanger fouling of course reduces the opportunity for heat recovery with its attendant effect on primary energy demands. Since the oil crisis there has been a modest interest in obtaining knowledge regarding all aspects of heat exchanger fouling, but the investment is nowhere near as large as in the field of heat transfer as a whole.

Although books have appeared from time to time since the 1970s, addressing the question of heat exchanger fouling, they are largely based on conferences and meetings so that there is a general lack of continuity. The purpose of this book therefore, is to present a comprehensive appraisal of current knowledge in all aspects of heat exchanger fouling including fundamental science, mathematical models such as they are, and aspects of the practical approach to deal with the problem of fouling through design and operation of heat exchangers. The techniques of on and off-line cleaning of heat exchangers to restore efficiency are also described in some detail.

The philosophy of the book is to provide a wide range of data in support of the basic concepts associated with heat exchanger fouling, but written in such a way that the non-mathematical novice as well as the expert, may find the text of interest.

T.R. Bott
December 1994

# ACKNOWLEDGEMENTS

The author wishes to record his sincere gratitude for the skill, dedication and persistence of Jayne Olden, without which this book would never have been completed.

All the diagrams and figures in this book were drawn by Pauline Hill and her considerable effort is acknowledged.

# NOMENCLATURE

Note: In the use of equations it will be necessary to use consistent units unless otherwise stated

| | |
|---|---|
| $A$ | Area or area for heat transfer |
| $A_1$ | Constant |
| $A_H$ | Hamaker constant |
| $a_I, a_C, a_D, a_T$ | Vector switches associated with the dimensionless deposition parameters $N_I, N_C, N_D$ and $N_T$ respectively |
| $B$ | Correction term Equation 12.28 |
| $C$ | Circulation rate |
| $C$ | Cunningham coefficient or a constant |
| $c$ | Concentration |
| $c_b$ | Concentration in bulk blowdown water |
| $c_m$ | Concentration in make up water |
| $c_p$ | Specific heat |
| $c_{ps}$ | Specific heat of solid |
| $c_x$ | Concentration of cells in suspension |
| $D$ | Diffusion coefficient or dimensionless grouping as described by Equation 10.50 |
| $D_c$ | Collector diameter |
| $D_p$ | Diffusion coefficient for particles |
| $d$ | Diameter |
| $E$ | Activation energy or dimensionless grouping as described by Equation 10.49 |

| | |
|---|---|
| $E_D$ | Eddy diffusivity |
| $F$ | Shear force |
| $F_A$ | Adhesion parameter |
| $F_R$ | Repulsion force |
| $FS$ | Slagging index |
| $F_w$ | Van der Waals force |
| $f$ | Friction factor |
| $f_k$ | Ball frequency (Balls/$h$) |
| $h$ | Lifshitz - van der Waals constant |
| $K$ | Transfer coefficient or Constant |
| $K_A$ | Mass transfer coefficient of species $A$ |
| $K_d$ | Deposition coefficient |
| $K_d$ | Mass transfer coefficient allowing for sticking probability |
| $K_m$ | Mass transfer coefficient |
| $K_M$ | Mass transfer coefficient of macro-molecules |
| $K_o$ | Constant in Equation 10.31 |
| $K_{so}$ | Solubility product |
| $K_t$ | Transport coefficient |
| $K_t^*$ | Dimensionless transport coefficient |
| $k_d, k_g, k_s$ | Rate constants Equation 12.10 |
| $k_r$ | Rate constant |
| $l$ | Length |
| $l_m$ | Characteristic length |
| $M$ | Mass flow rate |
| $M^*$ | Asymptotic deposit mass |

| | |
|---|---|
| $M_f$ | Mass of fouling deposit |
| $m$ | Mass |
| $N$ | Dimensionless deposition parameter (Equation 7.40) |
| $N_A$ | Mass flux of cells |
| $N_c$ | Dimensionless interception deposition parameter |
| $N_D$ | Dimensionless diffusion deposition parameter or mass flux away from reaction zone |
| $N_I$ | Dimensionless impaction deposition parameter |
| $N_M$ | Mass flux of macro-molecules |
| $N_r$ | Mass flux of reactants or precursors |
| $N_T$ | Dimensionless thermophoresis deposition parameter |
| $N_P$ | Particle number density |
| $n$ | Integer on concentration factor |
| $P$ | Sticking probability |
| $P_d$ | Probability of scale formation |
| $P_I$ | Sticking probability for impacting mechanisms |
| $P_O$ | Sticking probability for non-impacting mechanisms |
| $P_S$ | Overall sticking probability |
| $p$ | Pressure |
| $\Delta p$ | Pressure drop |
| $Q$ | Rate of heat transfer |
| $q$ | Heat flux |
| $R$ | Universal gas constant or parameter defined in Equation 9.14 |
| $R_f$ | Fouling resistance or Fouling potential (see Chapter 16) |
| $R_{fo}$ | Overall fouling resistance |
| $R_{ft}$ | Fouling resistance at time $t$ |

| | |
|---|---|
| $R_s$ | Slagging propensity |
| $R_T$ | Total resistance to heat transfer |
| $R_\infty$ | Asymptotic fouling resistance |
| $r$ | Radius |
| $r_{O_2}$ | Rate of oxygen supply |
| $r_c$ | Rate of corrosion |
| $r_{O_2}$ | Rate of oxygen supply |
| $S$ | Stopping distance or parameter defined by Equation 9.13 |
| $SR$ | Silica ratio |
| $T$ | Temperature |
| $T_c$ | Cloud point temperature |
| $T_{CV}$ | Temperature of critical viscosity |
| $T_f$ | Freezing temperature |
| $T_p$ | Pour point temperature |
| $t$ | Time |
| $t_d$ | Induction time |
| $t_u$ | Average ball circulation time |
| $U$ | Electrophoretic mobility of charged particles |
| $U_C$ | Overall heat transfer coefficient for clean conditions |
| $U_D$ | Overall heat transfer coefficient for fouled conditions |
| $u$ | Velocity |
| $u_o$ | Initial velocity or velocity in the absence of thermophoresis |
| $u_r$ | Radial velocity |
| $u_T$ | Stokes terminal velocity |
| $u_t$ | Velocity due to thermophoresis |

| | |
|---|---|
| $u^*$ | Friction velocity |
| $\bar{V}$ | Mean particle volume |
| $V$ | Volumetric flow |
| $V_R$ | Energy associated with double layers |
| $V_T$ | Total energy of adsorption |
| $V_W$ | Energy associated with van der Waals forces |
| $v_e$ | Electrophoretic mobility |
| $X$ | Number of cells per unit area |
| $X_T$ | Number of cells to cover completely unit area |
| $x$ | Thickness or distance |

## Subscripts

| | |
|---|---|
| $Av$ | Average |
| $b$ | Bulk |
| $Bio$ | Biomass |
| $C$ | Cold or clean |
| $c$ | Critical |
| $e$ | Crystal face |
| $f$ | Foulant, or freezing |
| $g$ | Gas or growth |
| $H$ | Hot |
| $I$ | Impact |
| $i$ | Induction, or initiation, or interference or inside |
| $in$ | Inhibitory |
| $irr$ | Irreversible |

| | |
|---|---|
| $m$ | Mean, or metal |
| $max$ | Maximum |
| $P$ | Particle or sticking probability |
| $p$ | Pressure |
| $rev$ | Reversible |
| $s$ | Scale, surface or solid, saturation |
| $t$ | Time |
| $w$ | Wall or surface |
| $x_{ads}$ | Adsorbed cells |
| $\infty$ | Asymptotic or infinite |

## Dimensionless numbers

| | |
|---|---|
| $Re = \dfrac{dv\rho}{\eta}$ | Reynolds number |
| $Pr = \dfrac{c_p\eta}{\lambda}$ | Prandtl number |
| $St = \dfrac{\alpha}{\rho v c_p}$ | Stanton number |
| $Nu = \dfrac{\alpha d}{\lambda}$ | Nusselt number |
| $Sc = \dfrac{\eta}{\rho D}$ | Schmidt number |
| $Sh = \dfrac{K l_m}{D}$ | Sherwood number |
| $Bi = \dfrac{\alpha l_m}{\lambda}$ | Biot number |

## Greek

| | |
|---|---|
| $\alpha$ | Heat transfer coefficient |
| $\beta$ | Time constant |
| $\gamma$ | Distance over which diffusion takes place |
| $\varepsilon$ | Induced EMF |
| $\eta$ | Viscosity |
| $\eta_i$ | Particle collection efficiency |
| $\eta_o$ | Combined collection efficiency for non-impacting mechanisms |
| $\eta_{tot}$ | Overall particle collection efficiency |
| $\theta$ | Fraction of surface |
| $\lambda$ | Thermal conductivity |
| $\lambda_f$ | Thermal conductivity of foulant deposit |
| $\lambda_s$ | Thermal conductivity of scale |
| $\mu$ | Kinematic viscosity |
| $\mu_c$ | Dimensionless group described by Equation 10.48 |
| $\rho$ | Density |
| $\rho_f$ | Foulant density |
| $\tau$ | Shear stress or particle relaxation time |
| $\tau^*$ | Dimensionless particle relaxation time |
| $\phi_D$ | Rate of deposition |
| $\phi_f$ | Particle flux |
| $\phi_p$ | Particle volume |
| $\phi_p$ | Particle volume function (see Equation 7.45) |
| $\phi_R$ | Rate of removal |
| $\psi$ | Scale strength factor |
| $\Omega$ | Water quality factor |

# CONTENTS

| | | PAGE |
|---|---|---|

**Preface**     v

**Acknowledgements**     vii

**Nomenclature**     ix

1. **Introduction**     1
   References     6

2. **Basic Principles**     7
   References     14

3. **The Cost of Fouling**     15
   3.1 Introduction     15
   3.2 Increased capital investment     15
   3.3 Additional operating costs     16
   3.4 Loss of production     18
   3.5 The cost of remedial action     19
   3.6 Financial incentives     20
   3.7 Concluding comments on the cost of fouling     21
   References     21

4. **General Models of Fouling**     23
   4.1 Introduction     23
   4.2 A simple general model     26
   4.3 Asymptotic fouling     26
   4.4 Falling rate fouling     30
   4.5 Concluding remarks     31
   References     32

5. **Fluid Flow and Mass Transfer**     33
   5.1 Introduction     33
   5.2 The flow of fluids     33
   5.3 Mass transfer     36
   5.4 Removal of deposits     37
   5.5 Concluding remarks     42
   References     43

6.  **Adhesion**                                                      45
    6.1 Introduction                                                 45
    6.2 Classification of interactions                               45
    6.3 Van der Waals forces                                         46
    6.4 Electrostatic double layer forces                            49
    6.5 Practical systems                                            52
    6.6 Concluding remarks                                           52
    References                                                       53

7.  **Particulate Deposition**                                       55
    7.1 Introduction                                                 55
    7.2 Fundamental concepts                                         56
    7.3 Particle transport - isothermal conditions                   58
        7.3.1 Diffusion regime                                       61
        7.3.2 Particles with inertia                                 63
        7.3.3 Impaction of particles                                 64
        7.3.4 The important variables                                64
    7.4 Non-isothermal particle deposition                           66
    7.5 Electromagnetic effects                                      69
    7.6 Particle deposition models                                   70
    7.7 Some experimental data on particulate deposition             78
        7.7.1 Liquid systems                                         78
        7.7.2 Gas systems                                            85
    References                                                       91

8.  **Crystallisation and Scale Formation**                          97
    8.1 Introduction                                                 97
    8.2 Theoretical background to crystallisation fouling           100
        8.2.1 Crystalline solids                                    100
        8.2.2 Solubility                                            103
    8.3 The mechanism of crystallisation                            106
        8.3.1 Supersaturation                                       107
        8.3.2 Formation of nuclei and crystallites                  110
        8.3.3 Crystal growth                                        110
    8.4 Models of crystallisation fouling                           114
    8.5 Crystallisation and scaling under nucleate boiling conditions 122
        8.5.1 Changes in heat transfer and boiling phenomena due to
              crystallisation fouling                               126
    8.6 Crystallisation fouling due to organic materials            127
    References                                                      133

9.  **Freezing Fouling or Liquid Solidification**                   137
    9.1 Introduction                                                137

9.2  Concepts and mathematical analysis                         137
    9.2.1  Static systems                                       137
    9.2.2  Flowing systems                                      141
References                                                      147

10.  **Fouling Due to Corrosion**                               149
    10.1  Introduction                                          149
    10.2  General liquid corrosion theory                       150
        10.2.1  Galvanic corrosion involving two dissimilar metals  156
        10.2.2  Crevice corrosion                               161
        10.2.3  Pitting corrosion                               162
        10.2.4  "Hot spot" corrosion                            162
        10.2.5  Stress corrosion cracking                       163
        10.2.6  Selective leaching                              163
        10.2.7  Corrosion fatigue cracking                      164
        10.2.8  Impingement attack                              164
    10.3  Corrosion fouling in gas systems                      164
        10.3.1  Low temperature corrosion                       164
        10.3.2  High temperature corrosion in gas systems       165
    10.4  Materials of construction to resist corrosion         170
    10.5  Factors affecting corrosion fouling                   171
        10.5.1  Liquid systems                                  171
        10.5.2  Gas systems                                     173
    10.6  Corrosion fouling models                              174
    References                                                  182

11.  **Chemical Reaction Fouling**                              185
    11.1  Introduction                                          185
    11.2  Basic chemistry                                       187
        11.2.1  Background                                      188
        11.2.2  Possible reactions leading to deposition        188
        11.2.3  Pyrolysis, cracking and dehydration             190
        11.2.4  Autoxidation                                    191
    11.3  Deposit composition                                   193
    11.4  The effects of system operating variables             197
        11.4.1  Temperature                                     197
        11.4.2  Pressure                                        200
        11.4.3  Flow rate or velocity                           200
        11.4.4  The effects of impurities                       201
    11.5  Models for chemical reaction fouling                  201
    References                                                  218

12.  **Biological Growth on Heat Exchanger Surfaces**                 223
        12.1  Introduction                                           223
        12.2  Basic biology                                          224
            12.2.1  Microbial biology                                224
            12.2.2  Micro-organisms on surfaces                      228
            12.2.3  Macrofouling                                     236
        12.3  Biofilm formation                                      239
            12.3.1  Velocity effects                                 242
            12.3.2  Temperature                                      247
            12.3.3  The effect of surface                            249
            12.3.4  The effects of suspended solids                  250
            12.3.5  *pH* changes                                     250
            12.3.6  Trace elements                                   253
        12.4  Models of biofouling                                   253
            12.4.1  Transport to the surface                         254
            12.4.2  Adhesion and growth                              257
        12.5  Concluding remarks                                     262
        References                                                   262

13.  **The Design, Installation, Commissioning and Operation
     of Heat Exchangers to Minimise Fouling**                        269
        13.1  Introduction                                           269
        13.2  Basic considerations                                   269
            13.2.1  Summary of practical preliminary design concepts 275
        13.3  The selection of heat exchangers                       275
        13.4  The choice of fouling resistance                       277
        13.5  Plant layout                                           281
        13.6  Heat exchanger control                                 282
        13.7  Acceptance, commissioning and start up of heat exchangers  283
        13.8  Plant operation                                        285
        References                                                   286

14.  **The Use of Additives to Mitigate Fouling**                    287
        14.1  Introduction
        14.2  Liquid systems                                         291
            14.2.1  Particle deposition                              292
            14.2.2  The prevention of scale formation                295
            14.2.3  Prevention of corrosion fouling using additives  306
            14.2.4  Additives to prevent chemical reaction           314
            14.2.5  The control of biofouling with additives         316
                14.2.5.1  The ideal biocide and factors affecting choice      319
                14.2.5.2  System factors that affect efficiency of biocides   321
                14.2.5.3  Biocidal activity and industrial biocides  324

14.2.5.4  The future use of biocides 346
14.3  Gas systems 347
References 352

15. **Heat Exchanger Cleaning** 357
   15.1  Introduction 357
   15.2  On-line cleaning 357
      15.2.1  Circulation of sponge rubber balls 357
      15.2.2  Brush and cage systems 362
      15.2.3  Air or gas injection 363
      15.2.4  Magnetic devices 363
      15.2.5  Soot blowers 365
      15.2.6  The use of sonic technology 367
      15.2.7  Water washing 368
      15.2.8  Shot cleaning 369
      15.2.9  Galvanic protection 369
      15.2.10  The use of inserts 376
      15.2.11  Special features and designs 379
   15.3  Off-line cleaning 388
      15.3.1  Manual cleaning 389
      15.3.2  Mechanical cleaning 389
      15.3.3  Steam soaking 396
      15.3.4  Thermal shock 398
      15.3.5  Chemical cleaning 399
   15.4  Concluding remarks 404
   References 405

16. **Fouling Assessment and Mitigation in Some Common
   Industrial Processes** 409
   16.1  Introduction 409
   16.2  Cooling water systems 409
      16.2.1  Cooling system design to reduce fouling potential 412
      16.2.2  Blowdown in recirculating systems 418
      16.2.3  Problems of pollution 420
      16.2.4  Management concepts and water utilisation in recirculating
         systems 421
      16.2.5  The cost penalties of cooling water fouling 428
   16.3  Combustion systems 435
      16.3.1  Coal combustion 437
      16.3.2.1  Slagging assessment 442
      16.3.2.2  Fouling in coal combustion 448
      16.3.3  Mineral oil combustion 450
      16.3.3.1  Assessing fouling potential 454

16.3.4  Gas combustion                                                              455
16.3.5  Waste combustion                                                            455
    16.3.5.1  Fouling assessment                                                    462
16.3.6  Mitigation of fouling and slagging in combustion systems                   463
16.4  Food processing                                                               464
    16.4.1  Food fouling mechanism                                                  466
    16.4.2  Fouling assessment                                                      469
    16.4.3  Cleaning of food fouled surfaces                                        470
References                                                                          473

17.  **Obtaining Data**                                                             479
    17.1  Introduction                                                              479
    17.2  Experience                                                                479
    17.3  Laboratory methods                                                        481
        17.3.1  Experimental techniques                                            482
        17.3.2  Estimating the extent of the fouling                               491
        17.3.3  Comments on laboratory techniques                                  499
    17.4  The acquisition of plant data                                            500
        17.4.1  Assessment of heat exchanger performance                          500
        17.4.2  Sidestream monitoring                                             503
        17.4.3  The use of probes                                                 505
    References                                                                      515

**Index**                                                                          517

# CHAPTER 1

## Introduction

The accumulation of unwanted deposits on the surfaces of heat exchangers is usually referred to as fouling. The presence of these deposits represents a resistance to the transfer of heat and therefore reduces the efficiency of the particular heat exchanger. The foulant may be crystalline, biological material, the products of chemical reactions including corrosion, or particulate matter. The character of the deposit depends on the fluid (liquid or gas) passing through the heat exchanger. It may be the bulk fluid itself that causes the problem of deposit formation, e.g. the decomposition of an organic liquid under the temperature conditions within the heat exchanger. Far more often than not, the fouling problem is produced by some form of contaminant within the fluid, often at very low concentration, e.g. solid particles or micro-organisms.

Fouling can occur as a result of the fluids being handled and their constituents in combination with the operating conditions such as temperature and velocity. Almost any solid or semi solid material can become a heat exchanger foulant, but some materials that are commonly encountered in industrial operations as foulants include:

    Inorganic materials
        Airborne dusts and grit
        Waterborne mud and silts
        Calcium and magnesium salts
        Iron oxide

    Organic materials
        Biological substances, e.g. bacteria, fungi and algae
        Oils, waxes and greases
        Heavy organic deposits, e.g. polymers, tars
        Carbon

Fig. 1.1 is a photograph of the tube plate of a shell and tube heat exchanger fouled with particulate matter deposited from high temperature flue gases passing through the tubes.

The problems associated with heat exchanger fouling have been known since the first heat exchanger was invented. The momentum of the industrial revolution depended on the raising of steam, usually from coal combustion. In the early days serious problems arose in steam raising equipment on account of the accumulation of deposits on the water side of boilers. The presence of these deposits, usually crystalline in character originating from the dissolved salts in the feed water, caused the skin temperature of the boiler tube to reach dangerous levels allowing failure to occur. Development of suitable feed water treatment programmes however has largely eliminated the problem in modern boiler operating technology.

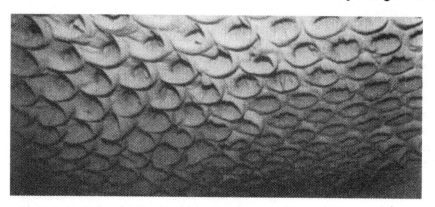

FIGURE 1.1. A fouled tube plate of a shell and tube boiler

In general the ability to transfer heat efficiently remains a central feature of many industrial processes. As a consequence much attention has been paid to improving the understanding of heat transfer mechanisms and the development of suitable correlations and techniques that may be applied to the design of heat exchangers. On the other hand relatively little consideration has been given to the problem of surface fouling in heat exchangers. A review [Somerscales 1988] that traces the history of heat exchanger fouling suggests four epochs in the development of an understanding of the problem of fouling. The chronology follows in general, the development of science and measurement techniques over the same timescale. In the first period up to about 1920, concern was directed towards observing the phenomenon and devising methods of reducing the problem with less emphasis on the scientific understanding of the mechanisms involved. The second period from 1920 - 1935 covered development in the measurement of fouling and representation. The following ten years from 1935 - 1945 saw the extended use of the so-called "fouling factor". The fouling factor may be defined as the adverse thermal effects of the presence of the deposit, expressed in numerical terms. Fouling factors or fouling resistance is discussed in more detail in Chapter 2. From 1945 to the present time a more scientific approach to the problem of fouling has been introduced with detailed investigations into the mechanisms that underline the problem of heat exchanger fouling. Chapters 7 - 12 detail the physical and chemical conditions and interactions that lead to heat exchanger surface fouling.

The review by Somerscales [1988] demonstrates how little has been done towards a better understanding of the problem since the early 1800s except in steam raising technology. It is indeed an anomaly that the accuracy of many sophisticated design techniques is restricted by a lack of understanding of the fouling process likely to be associated with the particular process under consideration.

Energy conservation is often a factor in the economics of a particular process. At the same time in relation to the remainder of the process equipment, the proportion of capital that is required to install the exchangers is relatively low. It is probably for this reason that heat exchanger fouling has been neglected as most fouling problems are unique to a particular process and heat exchanger design.

The problem of heat exchanger fouling therefore represents a challenge [Bott 1992] to designers, technologists and scientists, not only in terms of heat transfer technology but also in the wider aspects of economics and environmental acceptability and the human dimension.

The principal purpose of this book is to provide some insight into the problem of fouling from a scientific and technological standpoint. Improved understanding of the mechanisms that lead to the accumulation of deposits on surfaces will provide opportunities to reduce or even eliminate, the problem in certain situations. Three basic stages may be visualised in relation to deposition on surfaces from a moving fluid. They are:

1. The diffusional transport of the foulant or its precursors across the boundary layers adjacent to the solid surface within the flowing fluid.

2. The adhesion of the deposit to the surface and to itself.

3. The transport of material away from the surface.

The sum of these basic components represents the growth of the deposit on the surface.

In mathematical terms the rate of deposit growth may be regarded as the difference between

$$\phi_D - \phi_R \qquad (1.1)$$

where $\phi_D$ and $\phi_R$ are the rates of deposition and removal respectively.

The extent of the adhesion will influence $\phi_R$.

Fig. 1.2 shows an idealised asymptotic graph of the rate of growth of a deposit on a surface. In region A the process of adhesion is initiated. In some fouling situations the conditioning (or induction) period can take a long time, perhaps of the order of several weeks. In other examples of fouling the initiation period may be only of the order of minutes or even seconds.

Region B represents the steady growth of the deposit on the surface. Under these circumstances there is competition between deposition and removal. The rate of deposition gradually falls while the rate of removal of deposit gradually

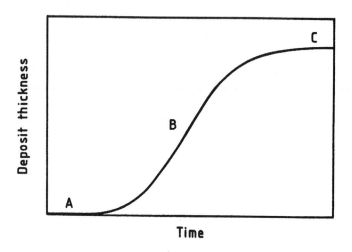

FIGURE 1.2. The change in deposit thickness with time

increases. Finally the rate of removal and the rate of deposition may become equal so that a plateau steady state or asymptote is reached (Region C) when the deposit thickness remains virtually constant.

Many system variables will affect the extent of the various stages and these will be discussed in subsequent chapters.

The world consumption of energy is large, taking into account all sources and methods of utilisation. Fossil fuels are of course, extensively used for the generation of heat used to raise steam for the production of electrical power. Under these circumstances a large fraction of the heat released from the fuel is transferred across various heat exchangers to the cold utility (cooling water). Under the prevailing conditions of operation there is ample opportunity for heat transfer surfaces to become fouled with attendant reductions in the efficiency of energy utilisation. In other processes the primary fuels, coal, oil or natural gas are used for process stream heating. For instance the crude oil vaporisers in petroleum refining are usually heated by means of fuel oil combustion.

Reduced efficiency of the heat exchangers due to fouling, represents an increase in fuel consumption with repercussions not only in cost but also in the conservation of the world's energy resources. The necessary use of additional fossil fuels to make good the shortfall in energy recovered due to the fouling problem, will also have an impact on the environment. The increased carbon dioxide produced during combustion will add to the "global warming" effect.

Although reduced heat transfer efficiency is of prime importance there may also be pressure drop problems. The presence of the foulant will restrict flow that results in increased pressure drop. In severe examples of fouling the exchanger may become inoperable because of the back pressure. Indeed the pressure drop problems may have a more pronounced effect than the loss of thermal efficiency.

In order to help reduce or overcome the problem of fouling, additives may be used. For instance a whole industry has built up around the treatment of water used for cooling purposes. The various chemicals added to the water fall into three categories, i.e. control of biological growth, prevention of scale formation and corrosion inhibition. Careful choice of treatment programmes will do much to reduce the accumulation of deposits on heat exchange surfaces. The addition of chemicals however, brings with it problems for the environment.

It is often the case that the water used for cooling is returned to its source, e.g. a river or lake. Even water taken from some other source such as a bore hole will eventually be returned to the natural environment. The presence of the additives can represent a hazard for the environment since many of the chemicals may be regarded as toxic.

Despite the best efforts of engineers and technologists to reduce or eliminate heat exchanger fouling the growth of deposits will still occur in some instances. Periodic cleaning of the heat exchangers will be necessary to restore the heat exchanger to efficient operation. If the deposits are difficult to remove by mechanical means chemical cleaning may be required. The chemicals used for this purpose will often be aggressive in character and represent an effluent problem after the cleaning operation. Unless this effluent is properly treated it could also represent an environmental problem. Even water used for cleaning can become contaminated and may require suitable treatment before discharge.

Of direct concern to the operator of process equipment are the economic aspects of heat exchanger fouling since this will affect the operating costs that in turn affects the profitability of the operation as a whole. In the first instance the heat exchanger is generally overdesigned to allow for the incidence of fouling. Increasing the size of the exchanger will of course, increase the initial capital cost and hence the annual capital charge.

The restriction to flow imposed by the presence of the deposit means that for a given throughput the velocity will have to increase. The increased velocity represents an increase in pumping energy and hence an increase in costs. Many pumps are electrical and so the increased energy requirement is in terms of the more expensive secondary energy. Other operating costs can accrue from the presence of the deposits such as increased maintenance requirements or reduced output. Emergency shutdown as a direct result of heat exchanger fouling can be particularly expensive. In many examples of severe fouling the frequency of cleaning may not coincide with the planned periodic plant shutdown for maintenance (annual basis) and it might be necessary to install standby heat exchangers for use when the cleaning of heat exchangers becomes necessary. Additional heat transfer capacity provided by standby equipment represents an additional capital charge.

Finally, but by no means least, there is the human dimension to the problem of fouling. Severe fouling can lead to the loss of employee morale [Bott 1992]. The repeated and persistent need to shut down the plant to clean heat exchangers, or difficulties in maintaining the desired output due to the accumulation of deposits will inevitably lead to frustration on the part of those employees whose duty it is to maintain production and product quality. The problem is compounded if financial incentives are linked to quality and volume of production. Repeated subjection to these difficult working conditions can lead to an indifferent attitude that can compound the unsatisfactory character of production brought about by the particular fouling problem.

The fouling of heat exchangers is a wide ranging topic covering many aspects of technology. It represents a challenge not simply in terms of reducing product cost, and hence competitiveness in the market place, but also with the concerns of modern society in respect of conservation of limited resources, for the environment and the natural world, and for the improvement of industrial working conditions. It is the purpose of this book to provide a background and basis that will enable the reader to face up to the challenges presented by the problem of heat exchanger fouling.

## REFERENCES

Bott, T.R., 1992, Heat exchanger fouling. The challenge, in: Bohnet, M., Bott, T.R., Karabelas, A.J., Pilavachi, P.A., Séméria, R. and Vidil, R. eds. Fouling Mechanisms - Theoretical and Practical Aspects. Editions Européennes Thermique et Industrie, Paris, 3 - 10.

Somerscales, E.F.C., 1988, Fouling of heat transfer surfaces: an historical review. 25th Nat. Heat. Trans. Conf. ASME. Houston.

# CHAPTER 2

## Basic Principles

The accumulation of deposits on the surfaces of a heat exchanger increases the overall resistance to heat flow. Fig. 2.1 illustrates how the temperature distribution is affected by the presence of the individual fouling layers.

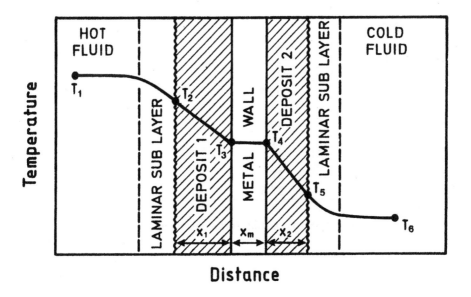

FIGURE 2.1. Temperature distribution across fouled heat exchanger surfaces

$T_1$ and $T_6$ represent the temperatures of the bulk hot and cold fluids respectively. Under turbulent flow conditions these temperatures extend almost to the boundary layer in the respective fluids since there is good mixing and the heat is carried physically rather than by conduction as in solids or slow moving fluids. The boundary layers (the regions between the deposit and the fluid), because of their near stagnant conditions offer a resistance to heat flow. In general the thermal conductivity of foulants is low unlike that of metals which are relatively high. For these reasons, in order to drive the heat through the deposits relatively large temperature differences are required, whereas the temperature difference across the metal wall is comparatively low.

Thermal conductivities of some common foulant-like materials are given in Table 2.1 that also includes data for common construction materials. The effects of even thin layers of foulant may be readily appreciated.

TABLE 2.1

Some thermal conductivities of foulants and metals

| Material | Thermal conductivity $W/mK$ |
|---|---|
| Alumina | 0.42 |
| Biofilm (effectively water) | 0.6 |
| Carbon | 1.6 |
| Calcium sulphate | 0.74 |
| Calcium carbonate | 2.19 |
| Magnesium carbonate | 0.43 |
| Titanium oxide | 8 |
| Wax | 0.24 |
| | |
| Copper | 400 |
| Brass | 114 |
| Monel | 23 |
| Titanium | 21 |
| Mild steel | 27.6 |

The resistance to heat flow across a solid surface is given as

$$\frac{x}{\lambda} \qquad (2.1)$$

where $x$ is the solid thickness

and $\lambda$ is the thermal conductivity of the particular solid.

Referring to the diagram (Fig. 2.1) the resistances of the solids to heat flow are:

For Deposit 1 $\frac{x_1}{\lambda_1}$ where $\lambda_1$ is the thermal conductivity of Deposit 1

For Deposit 2 $\frac{x_2}{\lambda_2}$ where $\lambda_2$ is the thermal conductivity of Deposit 2 and for the metal wall $\dfrac{x_m}{\lambda_m}$ where $\lambda_m$ is the thermal conductivity of the metal

For steady state conditions the heat flux $q$

$$q = \frac{T_2 - T_3}{\left(\dfrac{x_1}{\lambda_1}\right)} = \frac{T_3 - T_4}{\left(\dfrac{x_m}{\lambda_m}\right)} = \frac{T_4 - T_5}{\left(\dfrac{x_2}{\lambda_2}\right)} \tag{2.2}$$

Also $q = \alpha_1(T_1 - T_2) = \alpha_2(T_5 - T_6)$ $\tag{2.3}$

where $\alpha_1$ and $\alpha_2$ are the heat transfer coefficients for the hot and cold fluids respectively.

Equation (2.3) can be rewritten as

$$q = \frac{T_1 - T_2}{\left(\dfrac{1}{\alpha_1}\right)} = \frac{T_5 - T_6}{\left(\dfrac{1}{\alpha_2}\right)} \tag{2.4}$$

$\dfrac{1}{\alpha_1}$ and $\dfrac{1}{\alpha_2}$ represent the resistance to heat flow of the hot and cold fluids respectively.

The total resistance to heat flow will be the sum of the individual resistances, i.e. $R_T$ the total thermal resistance will be given by

$$R_T = \left(\frac{x_1}{\lambda_1}\right) + \left(\frac{x_2}{\lambda_2}\right) + \left(\frac{x_m}{\lambda_m}\right) + \frac{1}{\alpha_1} + \frac{1}{\alpha_2} \tag{2.5}$$

The overall temperature driving force to accomplish the heat transfer between the hot and cold fluids is the sum of the individual temperature differences

i.e. $(T_1 - T_2) + (T_2 - T_3) + (T_3 - T_4) + (T_4 - T_5) + (T_5 - T_6)$

or $T_1 - T_6$

$$\therefore q = \frac{T_1 - T_6}{R_T} \tag{2.6}$$

If the heat exchange area required for the required heat transfer is $A$, then the rate of heat transfer $Q$

$$= \frac{A(T_1 - T_6)}{R_T} \qquad (2.7)$$

In general the design of heat exchangers involves the determination of the required area $A$. The necessary heat transfer, the temperatures and the fluids are generally known from the process specification, the individual heat transfer coefficients of the fluids may be calculated, and values of the fouling resistances on either side of the heat exchanger would have to be estimated. It is the latter that can be difficult and if the resistances are incorrectly estimated difficulties in subsequent operation may be manifest.

At first sight it may be thought possible to calculate the fouling resistance, i.e.

$$x_f / \lambda_f$$

where $x_f$ is the deposit thickness

and    $\lambda_f$ is the foulant thermal conductivity

The difficulty is however, that this involves a knowledge of the likely thickness of the deposit laid down on the heat exchanger surfaces and the corresponding thermal conductivity. In general these data are not available. It is therefore necessary to assign values for the fouling resistance in order that the heat exchanger may be designed.

An alternative way of writing Equation 2.6 for clean conditions when the heat transfer surfaces are clean, is:

$$q = U_C(T_1 - T_6) \qquad (2.8)$$

where $U_C$ represents the overall heat transfer coefficient for clean conditions, i.e.

$$\frac{1}{U_C} = \left(\frac{x_m}{\lambda_m}\right) + \frac{1}{\alpha_1} + \frac{1}{\alpha_2} \qquad (2.9)$$

and allowing for the fouling resistances on either side of the heat transfer surface

$$\frac{1}{U_D} = R_T = \left(\frac{x_1}{\lambda_1}\right) + \left(\frac{x_2}{\lambda_2}\right) + \frac{x_m}{\lambda_m} + \frac{1}{\alpha_1} + \frac{1}{\alpha_2} \qquad (2.10)$$

Rewriting Equation 2.8 for fouled conditions, to give the heat flux

$$q = U_D(T_1 - T_6)$$ (2.11)

Because the temperature driving force across the heat exchanger usually varies along the length of the heat exchanger, it is necessary to employ some mean value of the temperature difference in using Equations 2.8 and 2.11.

If $\Delta T_1$ and $\Delta T_2$ are the temperature difference between the hot and cold fluids at either end of the heat exchanger then the temperature difference may be taken as the arithmetic mean, i.e.

$$\Delta T_m = \frac{\Delta T_1 - \Delta T_2}{2}$$ (2.12)

but more usually the log mean temperature difference is used, i.e.

$$\Delta T_m = \frac{\Delta T_1 - \Delta T_2}{\ln\left(\dfrac{\Delta T_1}{\Delta T_2}\right)}$$ (2.13)

For more background to the use of mean temperature differences in the design of heat exchangers the reader is referred to such texts as Hewitt, Shires and Bott [1994].

The mean temperature difference may be substituted in Equation 2.11 to give the heat flux

$$q = U_D \Delta T_m$$ (2.14)

and if the total available heat transfer area $A$ is taken into account

$$Q = U_D A \, \Delta T_m$$ (2.15)

In the design of heat exchangers $A$ is usually unknown, so rearranging Equation 2.15 provides a means of estimating the required heat transfer area, i.e.

$$A = \frac{Q}{U_D \Delta T_m}$$ (2.16)

The choice of the individual fouling resistances for the calculation of $U_D$ can have a marked influence on the size of the heat exchanger and hence the capital cost.

For a heat exchanger transferring heat from one liquid to another with the individual liquid heat transfer coefficients of 2150 and 2940 $W/m^2K$ and fouling resistances of 0.00015 and 0.0002 $m^2/WK$ on the surfaces of the heat exchanger, the total resistance to heat flow is

$$\frac{1}{2150}+\frac{1}{2940}+0.00015+0.0002 \; m^2/WK$$

$$= 0.00047 + 0.00034 + 0.00015 + 0.0002 \; m^2/WK$$

$$= 0.00116 \; m^2K/W$$

For the given design conditions, i.e. thermal load and temperature difference this fouling resistance represents an increase in the required heat exchanger area over and above the clean area requirements of

$$\frac{0.00035}{0.00081} \; \text{x} \; 100\% = 43.2\%$$

i.e. the cost of the heat exchanger will be increased considerably due to the presence of the fouling on the heat exchanger.

If the same fouling resistances are applied to a heat exchanger transferring heat between two gases where the individual heat transfer coefficients are much lower due to the low thermal conductivity of gases, say 32.1 and 79.2 $W/m^2K$, the situation is quite different.

Under these conditions the total thermal resistance is

$$\frac{1}{32.1}+\frac{1}{79.2}+0.00035 \, m^2K/W$$

$$= 0.0312 + 0.0126 + 0.00035 \; m^2K/W$$

$$= 0.0442 \; m^2K/W$$

In these circumstances the increase in required area in comparison to the clean conditions is

$$\frac{0.00035}{0.0438} \; \text{x} \; 100\% = 0.8\%$$

For the liquid/liquid exchanger the choice of fouling resistances represents a considerable increase in the required surface in comparison within the clean conditions. Using the same fouling resistances for a gas/gas heat exchanger represents negligible additional capital cost.

The traditional method of designing heat exchangers is to consider the potential fouling problem and assign a suitable fouling resistance to correspond. In order to assist with this selection, organisations such as the Tubular Exchanger Manufacturers Association (TEMA) issue tables of fouling resistances for special applications. The first edition appeared in 1941. From time to time these data are reviewed and revised. A review was carried out in 1988 and made available [Chenoweth 1990].

The principal difficulty in this approach to design is the problem of choice. At best the tables of fouling resistances give a range of mean fouling resistances, but in general there is no information on the conditions at which these values apply. For instance there is generally no information of fluid velocity, temperature or nature and concentration of the foulant. As will be seen later these factors amongst others, can have a pronounced effect on the development of fouling resistance. Probably the largest amount of information contained in the tables is concerned with water. Table 2.2 presents the relevant data published by TEMA based on a careful review and the application of sound engineering acumen by a group of knowledgeable engineers, involved in the design and operation of shell and tube heat exchangers.

TABLE 2.2

Fouling resistances in water systems

| Water type | Fouling resistance $10^4 \, m^2K/W$ |
|---|---|
| Sea water (43°C maximum outlet) | 1.75 - 3.5 |
| Brackish water (43°C maximum outlet) | 3.5 - 5.3 |
| Treated cooling tower water (49°C maximum outlet) | 1.75 - 3.5 |
| Artificial spray pond (49°C maximum outlet) | 1.75 - 3.5 |
| Closed loop treated water | 1.75 |
| River water | 3.5 - 5.3 |
| Engine jacket water | 1.75 |
| Distilled water or closed cycle condensate | 0.9 - 1.75 |
| Treated boiler feedwater | 0.9 |
| Boiler blowdown water | 3.5 - 5.3 |

Chenoweth [1990] gives the assumptions underlying the data contained in Table 2.2. For tubeside, the velocity of the stream is at least 1.22 *m/s* (4 *ft/s*) for tubes of non-ferrous alloy and 1.83 *m/s* (6 *ft/s*) for tubes fabricated from carbon steel and other ferrous alloys. For shell-side flow the velocity is at least 0.61 *m/s* (2 *ft/s*). In respect of temperature it is assumed that the temperature of the surface on which deposition is taking place does not exceed 71°C (160°F). It is also assumed that the water is suitably treated so that corrosion fouling and fouling due to biological activity do not contribute significantly to the overall fouling. Chenoweth [1990] comments on the further restrictions of these data. He observes that fouling by treated water is known to be a function of the prevailing velocity, the surface temperature , and the *pH* and is often characterised by reaching an asymptote (see Fig. 1.2). Although asymptotic values could be identified in the tables, the typical values listed for design, reflect a reasonable cleaning cycle and heat exchanger operation without operating upset. The severe limitations imposed by the assumptions will be readily appreciated.

It has to be said however, that without other information, these published data are of value in making an assessment of the potential fouling resistance. At the same time data on fouling resistances have to be treated with caution, they can only be regarded as a guide. A further limitation is that these values only apply to shell and tube heat exchangers. Conditions in plate heat exchangers for instance, could be quite different.

A fundamental flaw in the use of fixed fouling resistances as suggested by the TEMA tables is that they impose a static condition to the dynamic nature of fouling. In fundamental terms the use of Equation 2.5 in conjunction with the tables are not sound unless steady state has been reached. Fig. 1.2 shows that it is only after the lapse of time that a steady fouling resistance is obtained. In other words the heat exchanger does not suddenly become fouled when it is put on stream. For a period of time the heat exchanger will over perform because the overall resistance to heat flow is lower than that used in the design. To allow for this overdesign the heat exchanger operator may adjust conditions that in themselves could exacerbate the fouling problem. For instance, the velocity may be reduced, in turn this could accelerate the rate of deposition. It is possible that the imposed conditions could lead to fouling resistances that are subsequently, greater than those used in the design with attendant operating difficulties. Effects of this kind will be discussed in more detail later.

## REFERENCES

Chenoweth, J., 1990, Final report of the HTRI/TEMA joint committee to review the fouling section of TEMA standards. Heat. Trans. Eng. 11, No. 1, 73.

Hewitt, G.F., Shires, G.L. and Bott, T.R., 1994, Process Heat Transfer. CRC Press, Boca Raton.

# CHAPTER 3

## The Cost of Fouling

### 3.1   INTRODUCTION

In Chapter 1 some of the factors that contribute to the cost of fouling were mentioned. It is the purpose of this chapter to give more detail in respect of these costs.

Attempts have been made to make estimates of the overall costs of fouling in terms of particular processes or in particular countries. In a very extensive study of refinery fouling costs published in 1981 [van Nostrand *et al* 1981] a typical figure was given as being of the order of $10^7$ US per annum for a refinery processing $10^5$ barrels of crude oil per day. Allowing for inflation this figure would be something like $2 - 3 \times 10^7$ in 1993. These authors also report the advantages of using antifoulant chemicals. For instance on the crude unit the use of an additive reduces the annual cost attributable to fouling by almost 50%, even taking into account the cost of the antifoulant.

About the same time it was suggested [Thackery 1979] that the overall cost of fouling to industry in the UK was in the range £3 - 5 $\times 10^8$ per annum. Translating this into costs for 1993 the probable range would be £8 - 14 x $18^8$. An overall cost of fouling for the US published a few years ago [Garrett-Price *et al* 1985] was $8 - 10 \times 10^9$ per annum. The corresponding figures for 1993 would be in the range $15 - 20 \times 10^9$ per annum. A recent study [Chaudagne 1992] for French industry recorded an overall cost of fouling in France to be around $1 \times 10^{10}$ French Francs per annum. Pilavachi and Isdale [1992] conclude over the European Community as a whole the cost of heat exchanger fouling at the time of writing, was of the order of $10 \times 10^9$ ECU and of this total 20 - 30% was due to the cost of additional energy. It is clear from these limited data that fouling costs are substantial and any reduction in these costs would be a welcome contribution to profitability and competitiveness.

### 3.2   INCREASED CAPITAL INVESTMENT

In order to make allowance for potential fouling the area for a given heat transfer is larger than for clean conditions as described in Chapter 2. For the liquid/liquid exchanger discussed in Chapter 2 it was shown that the required area for the given fouling conditions was 1.43 times that for clean conditions. Although the cost of heat exchangers is not strictly pro rata in relation to area it will be appreciated that for a large complex containing several heat exchangers the additional capital cost for all the exchangers will represent a considerable sum of money.

In addition to the actual size of the heat exchanger other increased capital costs are likely. For instance where it is anticipated that a particular heat exchanger is likely to suffer severe or difficult fouling, provision for off-line cleaning will be required. The location of the heat exchanger for easy access for cleaning may require additional pipe work and larger pumps compared with a similar heat exchanger operating with little or no fouling placed at a more convenient location.

Furthermore if the problem of fouling is thought to be excessive it might be necessary to install a standby exchanger, with all the associated pipe work foundations and supports, so that one heat exchanger can be operated while the other is being cleaned and serviced. Under these circumstances the additional capital cost is likely to more than double and with allowances for heavy deposits the final cost could be 4 - 8 times the cost of the corresponding exchanger running in a clean condition.

Additional capital costs for injection equipment will also be involved if it is thought necessary to dose one or both streams with additives to reduce the fouling problem. Consideration of on-line cleaning (see Chapter 15) such as the Taprogge system for cooling water, will also involve additional capital. It has to be said however, that on-line cleaning can be very effective and that the additional capital cost can often be justified in terms of reduced operating costs.

It is important that as the design of a particular heat exchanger evolves to compensate for the problem of fouling, each additional increment in capital cost is examined carefully in order that it may be justified. The indiscriminate use of fouling resistances for instance, can lead to high capital costs, specially where exotic and expensive materials of construction are required. Furthermore the way in which the additional area is accommodated, can affect the rate of fouling. For instance if the additional area results say, in reduced velocities, the fouling rate may be higher than anticipated (see Chapter 13) and the value of the additional area may be largely offset by the effects of heavy deposits.

## 3.3    ADDITIONAL OPERATING COSTS

A number of contributory operating cost factors that result from the accumulation of unwanted deposits on heat exchanger surfaces can be identified.

The function of a heat exchanger as the name implies, is to transfer heat energy between streams. The prime reason for this is to conserve heat which is usually a costly component of any process. Reduced efficiency has to be compensated in some way in the process. If heat is not recovered the shortfall will have to be made up perhaps by the consumption of more primary fuel such as oil, coal or gas. In other operations it is necessary to raise the temperature of a particular stream to facilitate a chemical reaction, for example hydrocracking in refinery operations to produce lower molecular weight products. In power stations the efficiency of the steam condensers at the outlet from the turbines has a direct effect on the cost of the electricity produced (see Chapter 16).    If the cooling of power station condensers is inefficient it will mean that not all the pressure energy available in the steam passing through the turbines may be utilised.

Apart from the problem of reduced energy efficiency other problems may accrue.  For example if the temperature of the feed to a chemical reactor is lower than the optimum called for in the design, the yield from the reactor may be reduced.    The quality of the product may not be acceptable and additional processing may be required to improve the specification of the product.

In the operation of a distillation column where the feed preheater exchanges heat between the bottom product and the feed, inefficient heat exchange will mean additional heat requirements in the reboiler.  In turn this represents a greater "boil up" rate in the column  between the reboiler and the feed inlet that could affect the efficiency of the stripping section of the column due to droplet entrainment and channelling.  Such conditions may affect product quality or throughput may have to be reduced to maintain product specification.  These effects represent a reduced return on investment in terms of the distillation column.  Moreover because the heat removed from the bottom product is reduced additional cooling may be required (at further cost) before the bottom product is pumped to storage. Additional cooling requirements will put an extra load on the cold utility and may adversely affect its operating cost.

The presence of deposits on the surface of heat exchangers restricts the flow area.  As a consequence for a given throughput the velocity of flow increases.  In approximate terms.

$$\Delta p \ \alpha \ u^2 \tag{3.1}$$

where  $\Delta p$ is the loss of pressure through the exchanger

and    $u$ is the fluid velocity

so that even small changes in velocity can represent substantial increases in $\Delta p$. Fouling deposits are usually rough in comparison with standard heat exchanger surfaces.  The roughness increases the friction experienced by the fluid flowing across the surface so that for a given velocity $\Delta p$ is greater than in the clean condition.  The larger the $\Delta p$ the higher the pumping energy required and hence a greater pumping cost.  A more extensive discussion of pressure drop is given in Chapter 5.

The presence of fouling on the surface of heat exchangers may be the cause of additional maintenance costs.  The more obvious result of course, is the need to clean the heat exchanger to return it to efficient operation.  Not only will this involve labour costs but it may require large quantities of cleaning chemicals and there may be effluent problems to be overcome that add to the cost. If the cleaning agents are hazardous or toxic, elaborate safety precautions with attendant costs, may be required. Cleaning of heat exchangers is discussed in more detail in Chapter 15.

Additional maintenance costs may derive from the higher pressure drop across the exchanger due to the presence of the deposit. The higher inlet pressure may cause failure of joints and place a heavier load on the associated pump. It is possible that the presence of the deposit will accelerate corrosion of the heat exchanger. In turn this may lead to earlier replacement of the whole exchanger or at least, heat exchanger components. Failure of heat exchanger joints may lead to hazardous conditions due say to leaking flammable or toxic substances.

The presence of a deposit on the "cold side" of high temperature heat exchangers such as might be found in steam raising plant, may give rise to high metal temperatures that can increase corrosion or even loss of integrity of the metal with costly consequences.

The frequent need to dismantle and clean a heat exchanger can affect the continued integrity of the equipment, i.e. components in shell and tube exchangers such as baffles and tubes may be damaged or the gaskets and plates in plate heat exchangers may become faulty. The damage may also aggravate the fouling problem by causing restrictions to flow and upsetting the required temperature distribution.

## 3.4   LOSS OF PRODUCTION

The effects of fouling on the throughput of heat exchangers due to restrictions to flow and inefficient heat transfer, have already been mentioned. The need to restore flow and heat exchanger efficiency will necessitate cleaning. On a planned basis the interruptions to production may be minimised but even so if the remainder of the plant is operating correctly then this will constitute a loss of output that, if the remainder of the equipment is running to capacity still represents a loss of profit and a reduced contribution to the overall costs of the particular site. The consequences of enforced shutdown due to the effects of fouling are of course much more expensive in terms of output. Much depends on a recognition of the potential fouling at the design stage so that a proper allowance is made to accommodate a satisfactory cleaning cycle. When the seriousness of a fouling problem goes unrecognised during design then unscheduled or even emergency shutdown, may be necessary. For example, in the particular fouling situation illustrated by Fig. 1.1, three heat exchangers were designed and installed. It was anticipated that two would be operating while the other was being cleaned on a six month cycle. Under this arrangement production would have been maintained at a satisfactory and continuous level. In the event the heat exchangers required cleaning every 10 - 14 days! The problem became so difficult that at certain times all three exchangers were out of operation with severe penalties in terms of cost and loss of production.

Production time lost through the need to clean a heat exchanger can never be recovered and it could in certain situations, mean the difference between profit and loss.

## 3.5    THE COST OF REMEDIAL ACTION

The use of additives to eliminate or reduce the effects of fouling has already been mentioned.  An example of the effectiveness of an antifoulant on the preheat stream of a crude oil distillation unit has been described [van Nostrand *et al* 1981]. These data show that considerable mitigation of the fouling can be achieved by this method.   Fig.  3.1 demonstrates the fall off in heat duty with and without antifouling additives.  At the time of publication (1981) the annual cost of these chemicals was $1.55 \times 10^5$ for a crude unit handling 100,000 barrels per day.

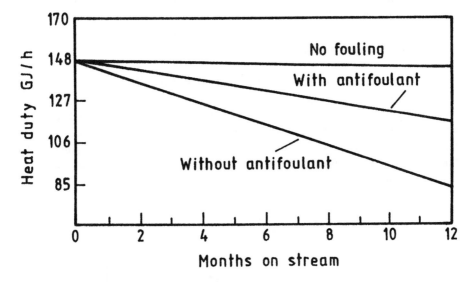

FIGURE 3.1.  The reduction of heat duty with and without antifoulant

Treatment of cooling water to combat corrosion, scale formation and biofouling can be achieved by a suitable programme.  The cost may be high and for a modest cooling water system the cost may run into tens of thousands of pounds.

If the fouling problem cannot be relieved by the use of additives it may be necessary to make modifications to the plant.  Modification to allow on-line cleaning of a heat exchanger can represent a considerable capital investment. Before capital can be committed in this way, some assessment of the effectiveness of the modification must be made.  In some examples of severe fouling problems the decision is straightforward, and a pay back time of less than a year could be anticipated.  In other examples the decision is more complex and the financial risks involved in making the modification will have to be addressed.

## 3.6    FINANCIAL INCENTIVES

The opportunities and financial incentive to tackle the problem of fouling were illustrated by a comprehensive investigation into the cost of condenser fouling for a hypothetical 600 *MW* coal fired power station [Curlett and Impagliazzo 1981]. The study not only considered cooling water velocity and temperature, but also the design of the condenser and materials of construction and the load on the respective turbine.   Turbine design also had an effect.   Changes in wet bulb temperature as they affect the performance of the cooling water system were also taken into account.

Among the interesting conclusions drawn from the study, the authors found that the magnitude of the effect of condenser tube fouling on unit output is sensitive to:
1.  Wet bulb temperature.   The effect at the summer peak load was twice the yearly average for every degree of fouling investigated, for the particular power station location (Dallas, Texas).

2.  The characteristics of the turbine.

Simplified annual cost data [1987 prices] taken from Curlett and Impagliazzo [1981] are presented on Figs. 3.2 and 3.3 for a lightly and heavily loaded turbine respectively at different condenser fouling resistances.   The data are calculated for summer conditions and a water velocity through the tubes of 1.83 *m/s*.   The base fouling resistance is 3.5 x $10^{-5}$ $m^2K/W$ for which the condenser was designed.   Any fouling above this value will represent a cost penalty either in terms of output (capability) or additional fuel costs.

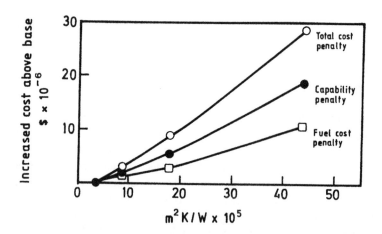

FIGURE 3.2.   Change in costs with fouling for a 600 *MW* power station with a lightly loaded turbine

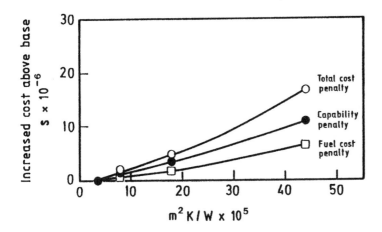

FIGURE 3.3. Change in costs with fouling for 600 *MW* power station with a heavily loaded turbine

These data show that for a lightly loaded turbine where the extent of the fouling resistance is 5 times higher than the design figure, the annual total cost penalty is $8.4 m and $4.7 m for lightly and heavily loaded turbines respectively. The magnitude of these figures are indicative of the degree of effort justified to reduce tube fouling on an existing condenser.

## 3.7    CONCLUDING COMMENTS ON THE COST OF FOULING

A number of contributions to the cost of fouling have been identified, however some of the costs will remain hidden. Although the cost of cleaning and loss of production  may be recognised and properly assessed, some of the associated costs may not be attributed directly to the fouling problem. For instance the cost of additional maintenance of ancillary equipment such as pumps and pipework, will usually be lost in the overall maintenance charges. The additional energy used to accommodate the increased pressure drop or the shortfall in heat recovery that requires an additional energy input, are unlikely to be recognised. Furthermore because the fouling process is dynamic, i.e. the fouling effects generally increase with time, the effect on the associated services, e.g. hot and cold utilities may not be apparent for a considerable time.

## REFERENCES

Chaudagne, D., 1992, Fouling costs in the field of heat exchange equipment in the French Market, in: Bohnet, M., Bott, T.R., Karabelas, A.J., Pilavachi, P.A., Séméria, R. and Vidil, R. eds. Fouling Mechanisms - Theoretical and Practical Aspects. Editions Européennes Thermique et Industrie, Paris.

Curlett, P.L. and Impagliazzo, A.M., 1981, The impact of condenser tube fouling on power plant design and enconomics, in: Chenoweth, J.M. and Impagliazzo, A.M. eds. Fouling in Heat Exchange Equipment. HTD, Vol. 17, ASME.

Garrett-Price, B.A. *et al*, 1985, Fouling of Heat Exchangers, Characteristics, Costs, Prevention, Control, Removal. Noyes Publications, New Jersey.

van Nostrand, W.L., Leach, S.H. and Haluska, J.L., 1981, in: Somerscales, E.F.C. and Knudsen, J.G. eds. Fouling of Heat Transfer Equipment. Hemisphere Publishing Corp. Washington.

Thackery, P.A., 1979, The cost of fouling in heat exchanger plant, in: Fouling - Science or Art? Inst. Corrosion Science and Technology and Inst. Chem. Engineers, Guildford.

# CHAPTER 4

## General Models of Fouling

### 4.1    INTRODUCTION

In Chapter 1 it was stated that the rate of build up of deposit on a surface could be defined by the simple concept of the difference between the rates of deposition and removal.  In more precise mathematical terms

$$\frac{dm}{dt} = \phi_D - \phi_R \qquad (4.1)$$

where  $m$ is the mass of deposit say per $m^2$

and     $\phi_D$ and $\phi_R$ are the deposit and removal mass flow rates per unit area of surface respectively

The equation is a statement of the mass balance across the fluid/solid interface, i.e.

Accumulation = Input - Output          (4.2)

In addition to the increased heat transfer resistance of the foulant layer its presence can have two further effects.  If the deposit thickness is appreciable, then the area for fluid flow, the cross-sectional area of a tube for instance if the deposition takes place within a tube, is reduced (see Chapter 3)  Under certain conditions, this reduction may be considerable.  For the same volume flow, therefore, the fluid velocity will increase and for identical conditions the Reynolds number will increase.

If the clean tube diameter is $d_1$ and the volumetric flow rate is $V$ then the original velocity $u_1$ is given by

$$u_1 = V \frac{4}{\pi d_1^{\,2}} \qquad (4.3)$$

If due to the fouling process $d_1$ is reduced to $d_1/2$ the new velocity $u_2$ for the same mass flow rate is given by

$$u_2 = \frac{V 4 \times 4}{\pi d_1^{\,2}} \qquad (4.4)$$

$$= \frac{16V}{\pi d_1^{\ 2}} \tag{4.5}$$

i.e. a fourfold increase in velocity.

The corresponding Reynolds number will be

$$\frac{d_1}{2} \frac{16V}{\pi d_1^{\ 2}} \cdot \frac{\rho}{\eta} \text{ i.e. } \frac{8V\rho}{\pi d_1 \eta}$$

where $\eta$ and $\rho$ are the fluid viscosity and density respectively

Compared with the original Reynolds number

$$d_1 \frac{4V}{\pi d_1^{\ 2}} \cdot \frac{\rho}{\eta} \text{ i.e. } \frac{4V\rho}{\pi d_1 \eta}$$

i.e. the Reynolds number has been doubled due to the presence of the deposit.

In addition the roughness of the deposit surface will be different from the clean heat exchanger surface roughness (usually greater) which will result in a change in the level of turbulence particularly near the surface. Greater roughness will produce greater turbulence with its enhancement of heat transfer or a smoother surface may reduce the level of turbulence. An alternative statement describing the effects of fouling may be made on this basis [Bott and Walker 1971].

| Change in | = | Change due to | + | Change due | + | Change due | |
|---|---|---|---|---|---|---|---|
| heat transfer | | thermal | | to roughness | | to change in | |
| coefficient | | resistance of | | of foulant | | *Re* caused by | (4.6) |
| | | foulant | | | | the presence | |
| | | | | | | of the foulant | |

The purpose of any fouling model is to assist the designer or indeed the operator of heat exchangers, to make an assessment of the impact of fouling on heat exchanger performance given certain operating conditions. Ideally a mathematical interpretation of Equation 4.6 would provide the basis for such an assessment but the inclusion of an extensive set of conditions into one mathematical model would be at best, difficult and even impossible.

Fig. 1.2 provided an idealised picture of the development of a deposit with time. Other possibilities, still ideal, are possible and these are shown on Fig. 4.1. Curve C represents the asymptotic curve of Fig. 1.2. Curve A represents a straight line relationship of deposit thickness with time, i.e. the rate of development

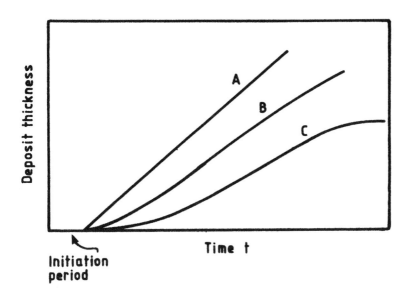

FIGURE 4.1. Idealised deposition curves

of the fouling layer is constant once the initiation of the process has taken place. Curve B on the other hand, represents a falling rate of deposition once initiation has occurred. It is possible that in effect, Curve B is essentially part of a similar curve to C and if the process of deposition were allowed to progress sufficiently an asymptote would be produced.

General models of the fouling process are essentially the fitting of equations to the curves illustrated in Fig. 4.1. The curves A, B and C on Fig. 4.1 are shown to have an initiation or induction period, but in some examples of fouling, e.g. the deposition of wax from waxy hydrocarbons during a cooling process, the initiation period may be so short as to be negligible. It is often extremely difficult or impossible to predict the initiation period even with the benefit of experience, so that most mathematical models that have been developed ignore it, i.e. fouling begins as soon as fluid flows through the heat exchanger.

The inaccuracy in ignoring the initiation period is not likely to be great. For severe fouling problems the initiation of fouling is usually rapid. Where the establishment of the fouling takes longer it is usually accompanied by a modest rate of fouling. Under these circumstances where long periods between heat exchanger cleans are possible, the induction period represents a relatively small percentage of the cycle. Errors in ignoring it are therefore small particularly in the light of the other uncertainties associated with the fouling process. Typical initiation periods may be in the range 50 - 400 hours.

## 4.2     A SIMPLE GENERAL MODEL

The simplest model is that of Curve A in Fig. 4.1 but ignoring the induction period and would have the form

$$x_f = \frac{dx}{dt} \cdot t \tag{4.7}$$

where $x_f$ is the thickness of deposit at time $t$.

If the induction time (or initiation period) is $t_i$ then the Equation 4.7 becomes

$$x_f = \frac{dx_f}{dt}(t - t_i) \tag{4.8}$$

The difficulty of course in using this model is that without experimental work $dx/dt$ is unknown and the use of $x_f$ to determine the fouling resistance to heat transfer is also a problem since the thermal conductivity of the foulant is not usually known (see Chapter 2). In terms of fouling resistance Equation 4.8 would take the form

$$R_{ft} = \frac{dR_f}{dt}(t - t_1) \tag{4.9}$$

where $R_{ft}$ is the fouling thermal resistance at time $t$ and

$$R_{ft} = \frac{x_{ft}}{\lambda_f} \tag{4.10}$$

where $x_{ft}$ is the thickness and time $t$

Even in this form the model is difficult to use unless $dR/dt$ is known from experimental determinations the conditions of which can also be applied to the fouling problem in hand.

## 4.3     ASYMPTOTIC FOULING

One of the simplest models to explain the fouling process was put forward by Kern and Seaton [1959].

$$R_{ft} = R_{f\infty}(1 - e^{\beta t}) \tag{4.11}$$

where $R_{ft}$ is the fouling thermal resistance at time $t$

$R_{f\infty}$ is the fouling resistance at infinite time - the asymptotic value.

$\beta$ is a constant dependent on the system properties

The model is essentially a mathematical interpretation of the asymptotic fouling curve, Fig. 1.2 (or Curve C on Fig. 4.1, but ignoring the initiation time). It is an idealised model and does little for the designer of a heat exchanger unless specific values for $R_{f\infty}$ and $\beta$ are to hand. The actual values of these constants will depend upon the type of fouling and the operating conditions. In general there will be no way of predicting these values unless some detailed experimental work has been completed. Such research is often time consuming and therefore, expensive. The Kern and Seaton model does, however, provide a mathematical explanation of the simple fouling concept. A compromise solution was proposed [Bott and Walker 1973] which employs limited data gathered over a much shorter time span, but the results of such an approach would need to be treated with caution.

Kern and Seaton [1959] proposed a mathematical restatement of Equation 4.1 with tubular flow in mind of the form

$$\frac{dx_f}{dt} = K_1 c' M - K_2 \tau x_{ft} \tag{4.12}$$

where $K_1 c' M$ is the rate of deposition term similar to a first order reaction

$K_2 x_{ft}$ is the rate of removal (or erosion) term

and $K_1$ and $K_2$ are constants

$c'$ is the foulant concentration

$M$ is the mass flow rate

$x_t$ is the foulant layer thickness at time $t$

$\tau$ is the shearing stress $= f \rho u^2$

where $f$ is a so-called friction factor (see Chapter 5 for more detail)

By assuming that $c'$ and $M$ are constant which is reasonable for a steady state flow heat exchanger, and $x_f$ the thickness, is very much less than the tube diameter for deposition in a tube, it is possible to integrate Equation 4.12.

$$x_f = \frac{K_1 c' M}{K_2 \tau}\left(1 - e^{-K_2 n}\right)$$ \hfill (4.13)

The equation is similar to equation 4.11 in form with $\dfrac{K_1 c' M}{K_2 \tau}$ a constant for a given set of operating equations and is equivalent to $R_{f\infty}$ in Equation 4.11. $K_2 \tau$ is also a constant and equivalent to $\beta$.

The initial rate of deposition and the asymptotic fouling resistance can be obtained by putting $x = 0$ and $\dfrac{dx_f}{dt} = 0$ in Equation 4.12.

then $\left(\dfrac{dx_f}{dt}\right)_{t=0} = K_1 c' M$ \hfill (4.14)

$K_1 c' M$ is a constant for a constant set of operating conditions

The asymptotic thickness $x_{f\infty} = \dfrac{K_1 c' M}{K_2 \tau}$ \hfill (4.15)

and is also constant for given conditions

Kern and Seaton [1959] developed the theory further using the Blasius relationship, to make allowance for the change in flow area caused by the deposition process.

i.e.    where $f = \dfrac{\tau}{\rho u^2} = K_f Re^{-0.25}$ \hfill (4.16)

where $K_f$ is the Blasius constant

and    $\Delta p = 4\dfrac{l}{d_i \rho 2g}$ \hfill (4.17)

where $d_i$ is the inside diameter of the tube

$l$ is the length of tube in the direction of flow

Under these conditions for turbulent flow

$$x_{f\infty} = \frac{2K_1c'}{K_2} \left[ \frac{\pi^2 g \rho l^4 M^3}{K_f (\Delta p_\infty)^4} \right] \tag{4.18}$$

$\Delta p_\infty$ is the pressure drop at the asymptotic value of the foulant thickness

$\dfrac{K_1c'}{K_2}$ characterises the fouling qualities of the fluid and generally will remain constant. Should practical data be available for one set of conditions, the thickness of the asymptotic value of the fouling thickness at a different set of conditions may be obtained from the ratio:

$$\frac{x_{f\infty 1}}{x_{f\infty 2}} = \frac{\left[ \dfrac{l^4 M^3}{\Delta p_\infty^{\,4}} \right]^{\!1/5}_{2}}{\left[ \dfrac{l^4 M^3}{\Delta p_\infty^{\,4}} \right]^{\!1/5}_{1}} = \left[ \frac{l_2}{l_1} \right]^{0.8} \left[ \frac{\Delta p_{\infty 1}}{\Delta p_{\infty 2}} \right]^{0.8} \left[ \frac{M_2}{M_1} \right]^{0.6} \tag{4.19}$$

The subscripts 1 and 2 refer to the two sets of conditions.

The model which Kern and Seaton proposed is an attempt to provide a generalised equation for fouling; that is to say with no reference to the mechanism of deposition. In general, it can be assumed that the mechanism of removal will be similar in most situations since it will depend upon the conditions at the fluid/foulant interface, although the cohesive strength of the foulant layer will be different in different examples.

A generalised equation for asymptotic fouling for any mechanism based on the "driving force" for deposit development has been proposed [Konak 1973]; the "driving force" is suggested as the difference between the asymptotic fouling resistance and the fouling resistance at time *t.*, i.e. the driving force = $(R_{f\infty} - R_{ft})$.

Assuming a power law function

$$\frac{dR_{ft}}{dt} = K \left( R_{f\infty} - R_{ft} \right)^n \tag{4.20}$$

where $K$ is a constant

$n$ is an exponent

The final equation becomes

$$\left[1-\frac{R_{ft}}{R_{f\infty}}\right]^{1-n} -1 = K R_{f\infty}{}^{n-1}t \tag{4.21}$$

for $n \neq 1$

when $n = 1$

$$-\ln\left[1-\frac{R_{ft}}{R_{f\infty}}\right] = Kt \tag{4.22}$$

Both Equations 4.21 and 4.22 satisfy the boundary condition $x_f \rightarrow 1$ and $t \rightarrow \infty$

Equation 4.22 is a form of Equation 4.11 proposed by Kern and Seaton [1959]

## 4.4  FALLING RATE FOULING

Epstein [1988] presents a mathematical analysis of falling rate fouling as exemplified by Curve B on Fig. 4.1. He assumes that

$$\frac{dR_f}{dt} \text{ is proportion to (some driving force)}^n \tag{4.23}$$

i.e. proportional to $q^n$              (4.24)

where $n$ is an exponent

   $q$ is the heat flux

For constant surface coefficient of heat transfer $\alpha$, the heat flux is given by

$$q = U_D \Delta T = \frac{\Delta T}{R_C + R_f} \tag{4.25}$$

where $U_D$ is the overall heat transfer coefficient for fouled conditions

   $R_C$ is the resistance to heat transfer for clean conditions

and  $R_C$ is $\dfrac{1}{U_C} = \dfrac{1}{\alpha}$ when no fouling has occurred

where $U_C$ is the overall heat transfer coefficient for unfouled conditions

Assuming that the overall temperature difference remains constant with time a combination of Equation 4.24 and 4.25 yields

$$\frac{dR_f}{dt} = \frac{K}{(R_c + R_f)^n} \qquad (4.26)$$

where $K$ is a constant

$$\text{Integrating} \int_{t=o}^{t=t} \frac{dR_f}{dt} = \int_{R_o}^{R_f} \frac{K}{(R_c + R_f)^n} \qquad (4.27)$$

i.e. $\left(R_c + R_f\right)^{n+1} - R_c^{n+1} = K(n-1)t \qquad (4.28)$

which yields a non-asymptotic falling rate curve $R_f$ vs $t$

An alternative way of writing Equation 4.28 is

$$\frac{1}{U_D^{n+1}} - \frac{1}{U_C^{n+1}} = Kt \qquad (4.29)$$

The values of $K$ and $n$ that will be necessary to allow an assessment of fouling to be made, will depend on the mechanism responsible for the fouling, e.g. whether or not the fouling is caused by chemical reaction or mass transfer of particles. Such data are not in general, readily available.

## 4.5    CONCLUDING REMARKS

Attempts have been made to develop the generalised models that were devised several decades ago. For instance Taborek *et al* [1972] took the general equation

$$\frac{dm}{dt} = \phi_D - \phi_R \qquad (4.30)$$

where $m$ is the mass deposit

and attempted to write equations that could be the basis of a determination of $\phi_D$ and $\phi_R$. These authors recognised that a specific fouling mechanism must modify the general equation, and proceeded to outline the form of the equations that might be used to take account of this fact. Thus they introduced expressions that took account of chemical reaction, mass transfer and settling. The removal term was written in terms of fluid shear and the bond resistance in the deposit that affected removal. Despite these refinements however, these models still left a great deal to be assumed about the particular fouling problem under consideration.

The use of general models for fouling analysis has many attractions but with the present state of knowledge and the severe limitations on the generation of suitable data, their application to specific problems is unlikely to be significant at least in the immediate future. The fact that the references to general models are roughly in the period 1960 - 1975, with little published since that time is not without significance. The recent initial work of Anjorin and Feidt [1992] on the analysis of fouling using entropy concepts however, shows promise. The next two chapters illustrate the complexities that are "hidden" in the terms $\phi_D$ and $\phi_R$. As the work on fouling develops, in the longer term, generalised relationships may assume more importance.

The present development of general theories to the problem of fouling, however, has the very definite advantage that it has drawn attention to the underlying phenomena and seeks to make a logical analysis of the problem. The undoubted worth of this approach is to emphasise the factors which need to be considered in any development of a theory and model of any particular system.

Specific models that have been developed for particular mechanisms will be discussed in the appropriate chapters.

## REFERENCES

Anjorin, M. and Feidt, M., 1992, Entropy analysis applied to fouling - a new criteria, in: Bohnet, M., Bott, T.R., Karabelas, A.J., Pilavachi, P.A., Séméria, R. and Vidil, R., eds. Fouling Mechanisms Theoretical and Practical Aspects. Editions Européennes Thermique et Industrie, Paris, 69 - 77.

Bott, T.R. and Walker, R.A., 1971, Fouling in heat transfer equipment. Chem. Engr. No. 251, 391 - 395.

Bott, T.R. and Walker, R.A., 1973, An approach to the prediction of fouling in heat exchanger tubes from existing data. Trans. Inst. Chem. Engrs. 51, No. 2, 165.

Kern, D.O. and Seaton, R.E., 1959, A theoretical analysis of thermal surface fouling. Brit. Chem. Eng. 14, No. 5, 258.

Konak, A.R., 1973. Prediction of fouling curves in heat transfer equipment. Trans. Inst. Chem. Engrs. 51, 377.

Epstein, N., 1981, in: Somerscales, E.F.C. and Knudsen, J.G. eds. Fouling of Heat Transfer Equipment. Hemisphere Publishing Corp. Washington.

Taborek, J., Aoki, T., Ritter, R.B., Palen, J.W. and Knudsen, J.G., 1972, Predictive methods for fouling behaviour. Chem. Eng. Prog. 68, No. 7, 69 - 78.

# CHAPTER 5

## Fluid Flow and Mass Transfer

### 5.1 INTRODUCTION

Heat exchangers are designed to handle fluids, either gases or liquids. It is to be expected therefore, that the flow of the fluids through the exchanger will influence to a lesser or greater degree, the laying down of deposits on the surfaces of heat exchangers. In particular the behaviour of fluids in respect of the transport of material, whether it be particles, ions, microbes or other contaminating component, will affect the extent and the rate of deposit accumulation.

A short résumé of the basic concepts of fluid flow and mass transfer will be given here as a basis for further discussion in depth, of the different fouling mechanisms.

### 5.2 THE FLOW OF FLUIDS

Two properties of fluids influence the way fluids behave. They are density and viscosity. Most gases have a relatively low density and low viscosities. On the other hand liquids can display a range of densities and viscosities, for instance the density and viscosity of light organic liquids are relatively low, but other liquids such as mercury have a high density and liquids with a high viscosity include fuel oils and treacle.

Viscosity is not apparent till the fluid is in motion. For a fluid in motion a force is required to maintain flow. In order to spread a viscous paint on a solid surface the necessary force is applied by the paint brush. In simple terms the bottom of the paint adheres to the surface while the layers of paint remote from the surface adhere to the brush. The force applied through the brush - a shear force - maintains the layers of paint between the brush and the surface in motion. The brush may be moved at constant speed, but the layers of paint move at different speeds. It may be visualised that the molecules of paint adjacent to the surface are stationary while there is a gradual increase in velocity through the individual layers till the layer in contact with the brush moves at the speed (or velocity) of the brush. In other words a velocity gradient is established throughout the layers.

The velocity gradient is proportional to the shear force per unit

i.e. $\tau \alpha \dfrac{du}{dx}$ 　　　　　　　　　　　　　　　　　　　(5.1)

where $\tau$ is the shear force

u is the velocity

x is the distance perpendicular to the surface.

In order to make relationship 5.1 into an equation a constant of proportionality is required, i.e.

$$\tau = \eta \frac{du}{dx} \qquad (5.2)$$

when $\eta$ is termed the coefficient of viscosity (or more usually "the viscosity" of the fluid).

The viscosity is defined as the shear force per unit area necessary to achieve a velocity gradient of unity. Equation 5.2 applies to the majority of fluids, and they are generally known as Newtonian fluids, or fluids that display Newtonian behaviour. There are exceptions, and some fluids (usually liquids) do not conform to Equation 5.2, and these are generally classified as non-Newtonian fluids although within this grouping there is a sub classification with distinctly different "viscosity" behaviour for the fluids within the different groups.

Osborn Reynolds in 1883 in a classical experiment, observed two kinds of fluid flow within a pipe namely laminar or streamline flow (sometimes called viscous flow) and turbulent flow. In the former a thin filament of dye in the centre of the pipe remained coherent, whereas for turbulent flow the filament of dye was broken up by the action of the turbulence or turbulent eddies.

The criterion for whether the fluid is flowing under laminar or turbulent conditions is the so-called Reynolds number (*Re*) for pipe flow defined as

$$Re = \frac{du\rho}{\eta} \qquad (5.3)$$

where d is the inside pipe diameter

and $\rho$ is the fluid density

The physical significance of the Reynolds number is essentially that it represents the ratio

$$\frac{\text{momentum forces}}{\text{viscous forces}}$$

Below about $Re = 2000$ the flow is streamline above $Re$ 3000 the flow is turbulent. The region between $Re = 2000$ and 3000 is more uncertain and is usually called the transition region.

It will be apparent how the fluid properties influence the magnitude of the Reynolds number. In general terms for instance, a liquid with a high viscosity will have a low Reynolds number and hence it is likely to flow under laminar conditions, and a liquid with a high density is likely to be turbulent under flowing conditions.

The velocity distribution is different under these two regimes. Under laminar conditions the velocity profile is a parabola (see Fig. 5.1) and the mean velocity of flow is half the velocity at the centre of the tube (the maximum velocity). Under turbulent conditions the velocity profile is no longer parabolic but is as shown on Fig. 5.2. For turbulent flow the mean velocity of the fluid in the pipe is 0.82 x the velocity at the centre.

Even under turbulent conditions there remains near the fluid/solid interface a slow moving layer (usually referred to as the viscous sub-layer) resulting from the "drag" between the fluid and the solid surface.

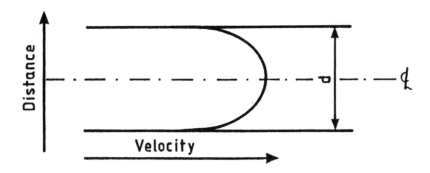

FIGURE 5.1. Velocity profile in a tube with laminar flow

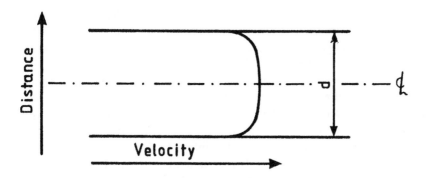

FIGURE 5.2. Velocity profile in a tube with turbulent flow

In addition to heat transfer, of concern to designers and operators of heat exchangers is the pressure drop experienced in the fluid as it passes through the exchanger. Often it is the increased pressure drop brought about by the presence of deposits, rather than the reduced heat transfer efficiency that forces the shut down of a heat exchanger for maintenance and cleaning (see Chapter 3).

If the shear stress at the wall of a pipe is $\tau$ the frictional force at the wall is given by

$$F = \tau \pi d l \qquad\qquad (5.4)$$

where $l$ is the length of the pipe

In order for fluid to flow this must be balanced by the force driving the fluid through the pipe, i.e.

$$F = \Delta p \frac{\pi d^2}{4} \qquad\qquad (5.5)$$

$$\therefore \Delta p \frac{\pi d^2}{4} = \tau \pi d l \qquad\qquad (5.6)$$

$$\text{or } \Delta p = \frac{4 \tau l}{d} \qquad\qquad (5.7)$$

The dimensionless group $\dfrac{\tau}{\rho u^2}$ is dependent on Reynolds number and for turbulent conditions also on the roughness of the surface.

## 5.3    MASS TRANSFER

The transport of material towards a surface that is being fouled depends upon the principles of mass transfer. When a concentration gradient of a particular component within a fluid exists there is a tendency for that component to move so as to reduce the concentration gradient. The process is known as mass transfer. In a stationary fluid or a fluid flowing under streamline conditions, with a concentration gradient of a component at right angles to the direction of flow, the mass transfer occurs as a result of the random motion of the molecules within the fluid system. The motion is often referred to as "Brownian motion". In a turbulently flowing fluid the situation is quite different. Under these conditions "eddy diffusion" is superimposed on the Brownian motion. Eddy diffusion results from the random physical movement of particles of fluid brought about by the turbulent conditions. The "parcels" of fluid physically transfer molecules of the diffusing component down the concentration gradient.

If we consider a mixture in which component A is diffusing through a fluid B towards a surface. Diffusion is at right angles to the general flow of fluid B, and by Fick's Law [Fick 1855].

$$N_A = -D_{AB}\frac{dc_A}{dx} \tag{5.8}$$

where $N_A$ is the rate of diffusion of molecules of $A$

$c_A$ is the concentration of $A$ at a distance $x$ from the surface

$D_{AB}$ is the "diffusivity" of $A$ through $B$

The diffusion of $A$ through $B$ will depend not only on the physical properties of $A$ and $B$ but also the prevailing fluid flow conditions.

If the flow is turbulent then the diffusivity of $A$ in $B$ is augmented by eddy diffusion. Under these conditions Equation 5.8. becomes

$$N_A = -(D_{AB} + E_D)\frac{dc_A}{dx} \tag{5.9}$$

where $E_D$ is the eddy diffusion

In general for turbulent flow $E_D \gg D_{AB}$ and the latter can often be neglected in the assessment of mass transfer.

Not only will the transfer of a foulant to a heat exchanger surface depend upon the physical properties of the constituents of the system, it will also depend upon the concentration gradient between the bulk fluid and the fluid/surface interface.

It has to be remembered that even under turbulent flow conditions there is a laminar sub-layer of slow moving fluid adjacent to the solid surface (see Section 5.2). The transport of material across this "boundary layer" will in general, only be possible by Brownian or molecular motion. The viscous sub-layer represents a resistance not only to heat transfer, but also the transfer of mass.

The reverse is also true. The removal of material from a surface by mass transfer, e.g. the waste materials from biofilm activity, will also depend upon the flow conditions, the physical properties of the constituents in the system and importantly the flow conditions (i.e. laminar or turbulent)

## 5.4    REMOVAL OF DEPOSITS

In Chapter 1 it was suggested that the net accumulation of deposits on surfaces could be considered to be the result of two competing phenomena, namely the rate of deposition and removal. In Section 5.3 the mass transfer of material to the surface was discussed and although the same principles will apply to the removal of material from a surface, the situation is far more complex than simple mass transfer theory would suggest.

In general the mass transfer to the surface may be regarded as the transport of relatively simple "particles", i.e. microbes, ions, soluble organic molecules or particulate matter itself. Once on the surface with adhesion between components of the depositing species and with the surface (see Chapter 6), the deposit can no longer be likened to the material that originally approached the surface. There are many examples of such transformations, but the following Table 5.1 illustrates some possibilities.

TABLE 5.1

Deposit transformations

| Fouling mechanism | Approaching species | Deposit transformation |
|---|---|---|
| Particulate deposition | Small particles | Agglomeration and bonding at the surface. |
| Crystallisation | Ions or crystallites | Crystallisation and orientation of crystals into a coherent structure. |
| Freezing | Molecules either in solution or liquid form | Continuous structure of solid material. |
| Corrosion | Aggressive ions or molecules | Chemical reaction with the surface producing "new" chemical compounds that may form a continuous structure. |
| Chemical reaction | Ions, free radicals or molecules | Larger molecules or polymers. |
| Biofouling | Micro-organisms and nutrients | Matrix of cells and extracellular polymers. |
| Mixed systems | Any mixture of the above | Complex matrix of particles and chemicals held together in extracellular products. |

Except in all but the simplest of examples, e.g. the removal of small particles form a loosely packed accumulation, the deposits will have a resistance to removal. With the passage of time the fouling layers may age that could increase the resistance to removal or facilitate re-entrainment. For instance over a period of time a crystalline structure may develop through re-orientation, into a more robust configuration. On the other hand the layers of a biofilm nearest the solid surface may be starved of nutrients causing death and possibly a reduction in the adhesion qualities of the biofilm. In other situations planes of weakness may develop within the deposit structure, or large foreign bodies may be included that weaken the fouling layers. There are many examples of changes in the morphology of a deposit with time.

The changes may be influenced by temperature changes brought about by the presence of the deposit itself. For example the surface of a boiler tube heated by gases from combustion, will initially be at a relatively low temperature, i.e. near the temperature of the boiling water in the tube. As the accumulation of deposit increases, the outer surfaces of the deposit will gradually increase in temperature, i.e. approaching the flue gas temperature in the boiler due to the increasing thermal resistance of the deposit. Chemical transformations depend on temperature, so that the chemical composition of the deposit is likely to change as the fouling develops. Furthermore depending on the physical character of the deposit, it is possible that as the outer surface temperature rises the outer layers become molten.

The prevailing velocity in the system may also be a factor. For instance a biofilm may be more compact and dense under high velocities. In the presence of low velocity the structure may be more open and "fluffy".

It is clear from these few examples and brief discussion, that during accumulation of deposits on surfaces considerable changes are likely to occur in the character of the fouling precursors that arrive at the surface. Such changes may be brought about by system variables such as temperature and velocity, in combination with the character of the precursors and deposits.

Equations 5.1 and 5.4 relate the physical properties of a fluid with the velocity in the system. In general terms as the bulk velocity is increased the velocity gradient increases since the velocity near the solid surface is extremely low or zero. Increased velocity gradient in turn increases the shear force near the deposit. It is this shear force that is often regarded as the force for the removal of solid material from the surface. Indeed if the velocity across a deposit of loosely bound particles is increased, the deposit thickness decreases [Hussain 1982]. Biofilm thickness decreases with velocities above about 1 $ms^{-1}$ [Miller 1979] and deposits in combustion systems are dependent on velocity [Tsados 1986].

Considering the interface region between the flowing fluid and the deposit in more detail and the foregoing discussion on the quality of the deposit, the effect of fluid shear brought about by the conditions quantified in Equation 5.1 may by itself be insufficient to cause substantial rates of removal. It has to be remembered also that for steady state conditions, the deposit was formed while these same shear forces considered to be responsible for removal were present. The situation is illustrated in Fig. 5.3.

At the same time it is likely that the surface of the deposit in contact with the flowing fluid will be rough in character so that the idealised conditions described by Equation 5.1 will be modified. In general for a given Reynolds number, the rougher the surface the higher the shearing force involved. Comments on the effects of roughness on heat transfer are made in Chapter 4 and included in the statement of Equation 4.6.

With these complications in mind attempts have been made to try to improve the understanding of the mechanism of removal of deposits from surfaces in flowing systems. Attention to the conceptual difficulty of deposition and removal occurring at the same time, was drawn by Epstein [1981]. Certainly the existence of deposition and removal occurring at the same time with magnetite particles suspended in water flowing through stainless steel tubes was demonstrated by Hussain *et al* [1982]. Nevertheless there would appear to be an inherent fundamental contradiction in the understanding of the fouling process.

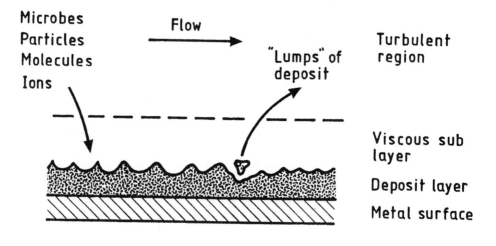

FIGURE 5.3. Deposition and removal occurring at the same time

Although re-entrainment does occur at shear stresses above a critical value [McKee 1991], it is not yet established as to how a shear stress (or drag force) acting parallel to a solid surface could provide a lifting force acting perpendicular to the surface sufficient to transfer material to the bulk fluid. A simplified diagram of the forces acting on a particle residing on a surface in contact with a flowing fluid is shown on Fig. 5.4.

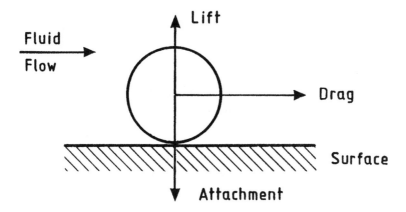

FIGURE 5.4. Forces acting on a particle on a surface

It was against this background that Cleaver and Yates [1973] suggested that the removal might be effected by instabilities in the laminar or sub-layer of a fluid flowing across a surface. Amongst the complex instabilities that can occur in the regimes close to the wall are the so-called "turbulent bursts". Fig. 5.5 is an idealised sketch of the phenomenon. In essence a "tornado" of fluid is ejected outwards towards the faster moving bulk fluid, to be replaced by "downsweeps" of fluid towards the surface. The former could conceivably provide sufficient lift force for particle removal, and the latter could play a part in transporting material towards the surface being fouled.

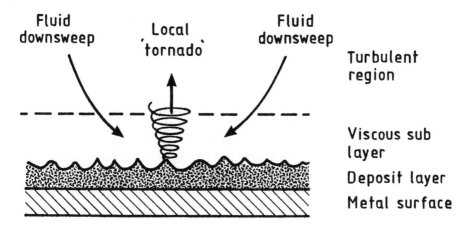

FIGURE 5.5. Flow disturbances near a solid surface

Because by their very nature instabilities are random, the burst process will be random and localised. This means that at any moment of time a proportion of the surface is experiencing burst activity while the remainder will be subject to the more normal laminar sub-layer conditions as discussed in Section 5.2. Over a period of time however, the whole of the surface will have experienced burst activity in addition to the more stable sub-layer conditions.

It has not been confirmed however, that the so-called turbulent bursts can penetrate sufficiently deep into the laminar sub-layer to cause re-entrainment to occur. In addition "the strength" of individual bursts will vary depending on the flow velocity and the geometry which again, will add to the complexity of the phenomenon.

Work by Yung [1986] was inconclusive, but work by McKee [1991] has demonstrated that turbulent bursts can influence the retention of deposits on surfaces. It is clear that much more detailed work needs to be carried out before the exact rôle of instabilities in the laminar sub-layer and associated turbulent bursts on deposit removal can be established.

In this section it has generally been assumed that in addition to deposition processes, there are removal mechanisms occurring at the same time. Such a situation may not apply to all fouling processes. In some examples the mass transfer of fouling precursors towards the surface may be the only transport mechanism; removal processes may be non-existent. The general shape of the curve of fouling thickness with time (Fig. 1.1) may arise, not because of the interaction of the two mass transfer processes, but because the deposition process is suppressed, i.e. an auto retardation of the deposition process till it becomes virtually zero. Such effects might arise due to electro-chemical conditions (see Chapter 6), or solution saturation conditions near the deposit/fluid interface.

Some scale forming systems (e.g. in evaporators) show a straight line correlation between the amount of deposit and time that might suggest the mass transfer to the surface is the dominating or controlling mechanism, if not the sole transport process affecting fouling in the particular system.

## 5.5    CONCLUDING REMARKS

The transport mechanisms that interact and influence the deposition and accumulation of fouling substances on the surfaces of heat exchangers are extremely complex. The complexities make it difficult to come to a full understanding of the precise mechanisms that prevail in the fouling process. It is perhaps in potential removal mechanisms that the principal difficulties arise. A search for a general explanation is frustrated by the possibilities that in some fouling processes removal does not occur whereas in others, it represents a substantial proportion of the transport processes involved.

Although there are these difficulties of interpretation of the fouling process, the discussion has drawn attention to some of the salient system features of the complex fouling process. Each fouling mechanism, i.e. biofouling, crystallisation, particle deposition and so on, is likely to respond differently to the imposed operating conditions. Furthermore the detail of particular fouling processes within a generalised mechanism, could be quite different. As the various mechanisms are considered in subsequent chapters, reference to the underlying transport principles will be made. In the end when there are no data or other information available, it is necessary to revert to empiricism, i.e. to consider the global situation rather than the precise details of the interactions at the fluid/solid interface.

REFERENCES

Epstein, N., 1981, Fouling in heat exchangers. 4th Int. Conf. Fouling of heat transfer equipment. New York, 235 - 253.

Cleaver, J.W. and Yates, B., 1973, Mechanism of detachment of colloidal particles from flat substrate in a turbulent flow. J. Colloid and Interface Science, 44, 3, 464 - 473.

Fick, A., 1855, Ueber Diffusion. An. Phys. 94, 59.

Hussain, 1982, Particulate magnetite fouling from flowing suspensions in simulated heat exchanger tubes. PhD Thesis, University of Birmingham.

McKee, C.J., 1991, The interaction between turbulent bursts and fouled surfaces. PhD Thesis, University of Birmingham.

Miller, P.C., 1979, Biological fouling film formation and destruction. PhD Thesis, University of Birmingham.

Tsados, A., 1986, Gas side fouling studies. PhD Thesis, University of Birmingham.

Yung, B.P.K., 1986, Particle deposition and re-entrainment in relation to the hydrodynamics at a surface. PhD Thesis, University of Birmingham.

# CHAPTER 6

## Adhesion

### 6.1    INTRODUCTION

In general terms the phenomenon of fouling represents the interaction between solids.   In the initiation of the fouling process the interaction is between the surfaces of the heat exchanger and "particles" of foulant.   Subsequently as the deposit thickens, the interaction is between the foulant material clinging to the surface and the fresh "particles" of foulant arriving at the fluid/solid interface.   The term "particle" is general and may refer to particulate matter, bacteria, corrosion products and so on.   The term "particle" also refers to particles that may be generated at the surface such as the products of chemical reaction or crystallisation.   Adhesion therefore is clearly a complex process as far as heat exchanger fouling is concerned, particularly where mixed fouling mechanisms are present, i.e. cooling water fouling where crystallisation, microbial attachment corrosion and particle deposition may occur simultaneously.   It is useful however, to consider the underlying theory and the factors that influence the phenomenon of adhesion.   To this end adhesion must normally be considered in terms of a solid sphere residing on a flat plate [Krupp 1967].

Total forces of adhesion of small particles on surfaces can range from $10^{-5}$ to $10^2$ *dynes*.  This total force can correspond to a very large force per unit area perhaps > $10^9$ *dynes/cm²* for micron sized particles [Bowling 1988].

### 6.2    CLASSIFICATION OF INTERACTIONS

The interactions between solid particles and surfaces can fall into a number of different categories of forces:

1.  Long range attractive forces serve to bring the particle to the surface and to provide the basis for the contact.   The forces involved may include van der Waals forces, magnetic attraction, and electrostatic forces.

2.  Bridging effects.   A definite interaction is involved in this process that may include mutual diffusion or "alloying" between the substance of the particle and the surface.   Liquid/solid bridging may be involved at the interface that invokes capillary forces.

3.  Short range forces can only be effective once there is intimate contact between the "particle" and the surface, i.e. after the long range forces and to a lesser extent the bridging effects, have established a physical contact.   The short range forces may include chemical bonding and hydrogen bonds [Krupp 1967].

As far as heat exchanger fouling is concerned it is the long range forces that are of prime importance since they are responsible for bringing the "particles" and surface into initial contact. In this respect a general discussion of adhesion is acceptable whereas the bridging effects and short term forces will be very specific to a particular condition.

The long range forces have to be considered in terms of particle size. Large particles, i.e. > 50 $\mu m$ are generally influenced by electrostatic forces and van der Waals forces generally affect smaller particles in a lower $\mu m$ range.

Bowling [1988], Visser [1988] and Oliveira [1992], have reviewed particle surface interaction and their work is used extensively in the development of the discussion in this chapter.

## 6.3    VAN DER WAALS FORCES

Bowling [1988] describes van der Waals forces in the following way. At absolute zero temperature solids can exhibit local electric fields and above this temperature additional contributions come from the excitation of the atoms and molecules making up the solid material. Van der Waals forces include forces between molecules possessing dipoles and quadrapoles produced by the polarisation of the atoms and molecules in the material. These dipoles and quadrapoles may be present naturally or by induced polarity. Non-polar attractive forces may also be present. The non-polar van der Waals forces may also be referred to as London-van der Waals dispersion forces because London associated these forces with the cause of optical dispersion, i.e. spontaneous polarisation [Corn 1966]. Such dispersion forces will make the major contribution to the intermolecular forces, except where the opportunity to polarise is small and the dipole moment is large.

Hamaker [1937] extended the theory of London to the interaction between solid bodies. The theory is based on the interaction between atoms and molecules and calculates the attraction between larger bodies as an integration involving geometry and the physical and chemical nature of the interacting bodies. The method employs the so-called Hamaker constant $A_H$ that "takes care" of the properties of the bodies. Visser [1972] has reviewed Hamaker constants.

For the case of a sphere resting on a flat plate (Fig. 6.1)

$$F_w = \frac{A_H r}{6x^2}$$  (6.1)

where $F_w$ is the van der Waals force

  $r$ is the radius of the sphere

  $x$ is the separation distance

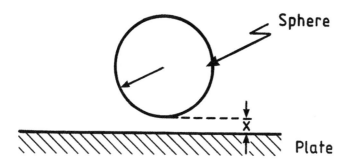

FIGURE 6.1. Sphere resting on a plate

For two flat plates (see Fig. 6.2) the interaction force per unit area is given by

$$p = \frac{A_H}{6\pi x^3} \tag{6.2}$$

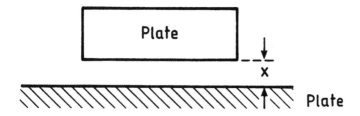

FIGURE 6.2. Two plates in contact

The simple additive approach of Hamaker has been criticised as being inaccurate [Kitchener 1973] and Oliveira [1992] states that it is not valid for condensed media interactions. Visser [1988] has explained the reason why the original Hamaker approach was inaccurate being due to the false assumption that the molecular forces are additive. It cannot be so, since the interactions between molecules with larger separations than near or direct contact, are screened by any molecules at a closer distance. As a consequence, layers of adsorbed materials for instance can influence the interaction between the two interacting macroscopic bodies. Langbein [1969] has demonstrated that when the thicknesses of the surface layers on microscopic bodies are larger than the separation distance, their interaction is determined by the properties of the layer rather than the underlying material. As Visser [1988] remarks this has implications for fouling and its prevention.

The objections to the Hamaker theory were overcome by Lifshitz and his co-workers [Lifshitz 1956, Dzyaloshinskii 1961] using the bulk optical properties of the interacting bodies. The approach employed by Lifshitz uses the so-called Lifshitz-van der Waals constant $h$ that depends only on the materials involved provided the separation distance is relatively small. Under some conditions the constant $h$ can be related to the Hamaker constant by

$$h = \frac{4\pi A_H}{3} \tag{6.3}$$

Rumpf [1977] states that the value of $h$ generally ranges between about 0.6 and 9.0 $eV$, depending on the combination of materials, but Visser [1972] presents data that clearly show that values may be well outside this stated range. Table 6.1 gives some values for the Lifshitz-van der Waals constant under vacuum conditions and in water. The presence of the water has a pronounced effect on the constant - i.e. a reduction in the van der Waals forces. The reduction is higher, the weaker the van der Waals interaction in vacuum [Visser 1988].

TABLE 6.1

Lifshitz-van der Waals constants for various materials in contact with similar material in vacuum and in water [Visser 1972]

| Material | $h\ eV$ in vacuum | $h\ eV$ in water |
|---|---|---|
| Gold | 14.3 | 9.85 |
| Silver | 11.7 | 7.76 |
| Copper | 8.03 | 4.68 |
| Diamond | 8.60 | 3.95 |
| Magnesium oxide | 3.03 | 0.47 |
| Polystyrene | 1.91 | 0.11 |

Negative values of the Lifshitz-van der Waals constant are possible. For instance Visser [1988] gives the constant for $Si$ in contact with $Al_2O_3$ in water as -0.78 $eV$. According to Oliveira [1992] negative values of the constant will produce spontaneous separation.

For the combination of two different materials 1 and 2 in vacuum $h_{12}$ can be obtained in good approximation by

$$h_{12} = \sqrt{h_{11} \cdot h_{22}} \tag{6.4}$$

## 6.4    ELECTROSTATIC DOUBLE LAYER FORCES

Solid materials immersed in an aqueous medium generally attain a surface charge. The charge arises due to the adsorption of specific ions present in the medium or the dissociation of molecules or groups on the surface of the solid.

The charge may be demonstrated by imposing an electric field on a suspended colloid when the anode and cathode attract negatively and positively charged particles respectively. It is possible to measure the so-called Zeta-potential of such a system by balancing the surface charge of the particle of investigation with a counter charge of ions of opposite sign in order to make the system electrostatically neutral.

The redistribution of ions or charged particles, brought about by the immersion of a charged surface in an aqueous environment will involve particles of similar charge (co-ions) being repelled from the surface while those of opposite charge (counter ions) will be attracted to the surface. The overall effect, i.e. a combination of the distribution of charged particles and Brownian motion, is the so-called Poisson-Boltzmann distribution of the ions (or particles) throughout the water phase. A diffuse layer is produced which together with the charged solid surface, is called the electrical double layer. Fig. 6.3 shows a simplified sketch of the situation that applies to a spherical particle immersed in water.

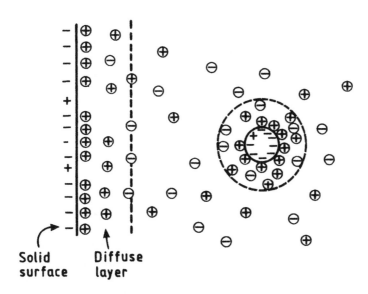

FIGURE 6.3. Simplified sketch of an electrostatic double layer and a solid spherical particle

Independently Deryagin and Landau [1941] and Verwey and Overbeek [1948] developed a theory, usually referred to as the DVLO theory, to explain the complex interactions when electrostatic double layer systems are in contact. On the one hand there is the force of repulsion $F_R$ due to the double layer effect, but on the other there is the attractive van der Waals force ($F_w$). It is usual to describe this balance between $F_w$ and $F_R$ in the form of an energy-distance diagram as shown on Fig. 6.4. The potential energy of interaction

$$V_T = V_w + V_R \qquad\qquad\qquad\qquad (6.5)$$

where $V_w$ and $V_R$ are the energy associated with the van der Waals and double layer interactions respectively.

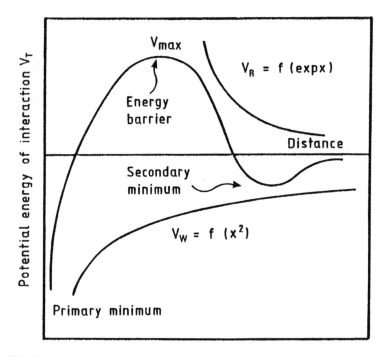

FIGURE 6.4. Energy-distance profiles

As a result of the charges in the relationship between $V$ and $x$ the summed curve shows a maximum at $V_{max}$ at a short separation distance. $V_{max}$ is an energy barrier. At large distances the energy is a minimum ($V_{min}$). If the thermal energy of the particles is smaller than $V_{max}$, no interaction will take place and fouling will not occur. The problem is more complex than the simple approach taking only the van der Waals forces and the double layer forces into account, since other interactions are also likely to be present.

Reversible and irreversible adhesion may be explained in terms of the primary minimum and $V_{min}$ [Oliveira 1992]. Reversible adhesion will occur at $V_{min}$ and irreversible adhesion at short distances and high negative values of $V$. The implications for the fouling process and deposit removal may be clearly identified.

The DVLO theory assumes that the surfaces of two interacting solid bodies have the same charge and sign. The result is a double layer that is repulsive. There are situations where surfaces are oppositely charged and electrostatic charge effects will reinforce the attractive van der Waals forces. Visser [1988] cites clay particles as an example. Clay particles, e.g. kaolin, have a platelet structure and it is possible for the edges to acquire a positive charge and the surface to be negative.

Another example is given by Visser [1988] where two negatively charged surfaces are linked by positively charged ions [see Fig. 6.5].

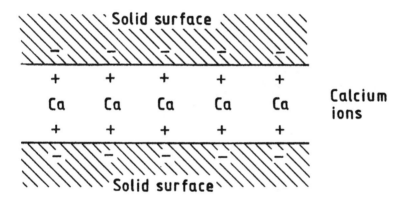

FIGURE 6.5. Calcium bridging

Other similar situations where ions are involved may be visualised.

Visser [1988] also draws attention to the effects of surface hydration and steric hindrance that also prevent close approach of particles and surfaces thereby reducing the opportunity for adhesion. In both examples interpenetration of the adsorbed layers, i.e. a physical barrier of where say, a polymer is adsorbed, prevents collision and hence adhesion.

In non-immersed systems such as dust particles in flowing air streams, van der Waals forces will operate leading to adhesion and since there are no barriers, prevention of fouling is extremely difficult. In gas streams where there are reacting species and possibly changes in phase such as in combustion flue gases, the situation is extremely complex.

## 6.5    PRACTICAL SYSTEMS

In the theoretical analysis ideal, smooth, surfaces were considered, i.e. flat plates or spheres on flat plates.  In reality the surfaces of industrial equipment will be far from smooth, and the adhesion forces will be enhanced due to the availability of larger contact areas.  Fig. 6.6 illustrates a spherical particle resting in a "hole" on a plate; even this is still a very idealised picture.

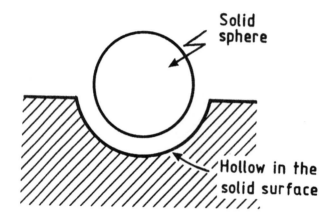

FIGURE 6.6.  A sphere resting in a hollow on a plate

For real particles on real surfaces it is conceivable that particles will move over the surface, under the influence of the fluid flow and viscous sub-layer instabilities, till it falls into a hole to suit its own particular shape.   Particle and surface morphology are therefore crucial in  fouling processes.

## 6.6    CONCLUDING REMARKS

In summary the principal factors responsible for adhesion include:
1.  van der Waals forces of attraction

2.  Electrostatic forces in systems of oppositely charged surfaces

3.  The larger the contact area the greater the total attractive force

In many respects the theories are limited to simple and ideal situations.  In industrial equipment where conditions are far from ideal and the geometry complex, it will be difficult to apply the theories outlined in this chapter. Furthermore Mozes [1994] who has considerable experience in the application of adhesion theory to biofilms, has made some valuable observations on the attachment of micro-organisms. She states that the following features are sometimes ignored: A microbial cell

1. is not a homogeneous, rigid, smooth spherical ideal colloid particle

2. is not an inert system and never in equilibrium with its environment

3. may be motile

4. may sense contact with a foreign surface and respond physiologically to this stimulus

She goes on to say that a true understanding of the adhesion/ flocculation/flotation phenomena must combine and integrate consideration of biological characteristics of cells as living entities with the simple physico-chemical approach regarding cells as colloidal particles. Similar specific comments may be made about all fouling phenomena. Theory has to make simplifying assumptions that may be unjustified in the context of the phenomena concerned. The purpose however, is to provide a background to the discussions on the various mechanisms by which surfaces become fouled.

REFERENCES

Bowling, R.A., 1988, A theoretical review of particle adhesion, in: Mittal, ed. Particles on Surfaces 1. Detection, Adhesion and Removal. Plenum Press, New York.

Corn, M., 1966, in: Davies, C.N. ed. Aerosol Science Academic Press, New York.

Derjaguin, B.V. and Landau, L.O., 1941, Theory of the stability of strongly charged lyophobic solutions and of adhesion of strongly charged particles in solutions of electrolytes. Acta Physicochim. USSR, 14, 633.

Dzyaloshinskii, I.E., Lifshitz. E.M., Pitaevski, L.P., 1961, Adv. Phys. 10, 165.

Hamaker, H.C., 1937, The London-van der Waals attraction between spherical particles. Physica (Utrecht) 4, 1058.

Kitchener, J.A., 1973, J. Soc. Cosmet. Chem. 24, 709.

Krupp, H., 1967, Ad. Colloid Interface Science. 1 (2), 111.

Langbein, D., 1969, van der Waals Attraction between macroscopic bodies. J. Adhes. 1, 237.

Lifshitz, E.M., 1956, Sov. Phys. JETP. 2, 73.

Mozes, N., 1994, Living cells and interfacial phenomena. 3ème Réunion Européennes Adhesion des micro-organismes aux surfaces. Universitée Paris-Sud, Faculté de Pharmacie Laboratoire de Microbiologie Industrielle. Châtenay-Malabry, France C.I.

Oliveira, D.R., 1992, Physico-chemical aspects of adhesion, in: Melo, L.F., Bott, T.R., Fletcher, M. and Capdeville, B. eds. Biofilms - Science and Technology, Kluwer Academic Publishers, Dordrecht.

Rumpf, H., 1977, in: Sastry, K.V.S. ed. Proc. 2nd Int. Symp. on Agglomeration. Am. Inst. Mining, Metallurgy and Petroleum Engineers Inc. New York.

Vervey, E.J.W., Overbeek, J.T.G., 1948, Theory of the Stability of Lyophobic Colloids. Elsevier, Amsterdam.

Visser, J., 1972, On Hamaker constants. A comparison between Hamaker constant and Lifshitz constants. Adv. Coll. Int. Sci. 3, 331.

Visser, J., 1988, Adhesion and removal of particles I and II, in: Melo, L.F., Bott, T.R. and Bernardo, C.A. eds. Fouling Science and Technology. Kluwer Academic Publishers, Dordrecht.

# CHAPTER 7

## Particulate Deposition

### 7.1   INTRODUCTION

Particulate deposition can occur in both liquid and gas systems.  In liquids the particulate matter can be corrosion products brought forward from other points in the process plant.  In other liquid systems the particles may arise from the fluids being processed, i.e. so-called oligomers in polymer manufacture.  Some processes involve slurries that can give rise to deposition problems in heat exchangers. Cooling water taken from a river or lake is likely to contain particulate matter in the form of silt or decomposing organic matter, that may eventually become deposited on heat transfer surfaces.  Cooling water problems are discussed in Chapter 16.

In air blown coolers, usually employing extended surfaces, dust particles carried forward in the air are potential foulants for the heat exchangers.  In combustion systems, particularly fuels such as coal containing incombustible mineral matter, the combustion gases will contain particles of this material.   The problem is particularly acute when pulverised firing is employed.  The fine particles of coal are intermixed with the finely ground mineral matter originally in the coal, and the whole burning mass is carried forward through the equipment by the combustion gases with the risk that deposition on the walls of the combustion chamber and heat transfer surfaces will occur.  The high temperatures involved coupled with the complex chemical composition of the fuel components and the reactive flue gases, makes for a multitude of chemical reactions and particle sizes that will affect the deposition process.   Even in fluidised bed boilers where the combustion temperatures are much lower, there is still a substantial carry-over of fine particles of non-combustible material that can cause fouling problems.  A more detailed discussion of combustion fouling are presented in Chapter 16.

In a series of discussions on the mechanisms by which surfaces become fouled it is useful to begin with particulate deposition since in most mechanisms some aspect of particulate transport is involved.  For instance in crystallisation, it could well be that the deposition although primarily crystal formation on the surface, is dependant on the transport of solid crystallites across the boundary layers.  A similar situation is likely to be present in freezing fouling where small crystals of the process liquid move to the surface to be incorporated in the solid layer on the surface.  In biofouling the micro-organisms responsible for slime formation on surfaces represent particles (of small size) that must encounter the surface before colonisation can take place.  In chemical reaction fouling some reactions may take place remote from the surface that give rise to particles that eventually deposit on the surface. Some corrosion fouling is initiated by the presence of particles on the surface subject to corrosion.

The theory associated with the transport of particles towards and onto surfaces is extensive and complex. Sufficient theory however will be presented in this chapter to provide a basic understanding of the principles as they affect heat exchanger fouling. Practical aspects will be considered towards the end of the chapter.

## 7.2    FUNDAMENTAL CONCEPTS

The arrival of a particle at a surface can be by two mechanisms, i.e. gravitational settling and particle transport within a fluid as it moves across the surface onto which the particles deposit. Both mechanisms have relevance to the fouling of heat exchangers but, because in general the fluid in a heat exchanger is moving the transport of particles to the surface assumes significant importance in any consideration of particulate fouling. Settling is usually associated with near stationary fluids. The discussion developed in this chapter will largely be concerned with the transport of particles in relation to a flowing fluid. In some instances, in high temperature systems particularly, the particles may begin life as a vapour only achieving a solid state as the vapour approaches a "cool" surface.

Two things must occur before a particle in suspension in a fluid deposits on a surface to become part of the foulant layer. First the particle has to be transported to the surface by one or a combination of mechanisms including Brownian motion, turbulent diffusion, as described in Chapter 5, or by virtue of the momentum possessed by the particle. The size of the particle will have a large influence on the dominant mechanism. For instance very small particles would be expected to be subject to Brownian diffusion and eddy diffusion whereas the larger particles because of their mass would move under momentum forces. Having arrived at the surface the particle must stick if it is to be regarded as part of the foulant layer residing on the surface.

Other factors that must be borne in mind in a consideration of particle deposition, although difficult to incorporate in any mathematical model, include agglomeration of particles in the bulk or at or near the surface, and the complex interactions of the forces near the wall discussed in Chapter 6.

In Chapter 5 the need for a concentration gradient to drive the mass transfer process was discussed. Fig. 7.1 shows the particle concentration profile within a fluid flowing through a pipe under turbulent conditions. In terms of the rate of deposition $\phi_D$ the transport of particles to a surface can be defined by

$$\phi_D = K_t(c_b - c_s) \tag{7.1}$$

where $K_t$ is a transport coefficient

$c_b$ and $c_s$ are the particle concentrations in the bulk and at the surface respectively

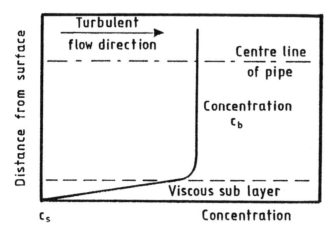

FIGURE 7.1. Particle concentration profile in turbulent flow in a pipe

If the concentration at the surface can be considered zero, i.e. the particles arriving at the surface stick and are therefore taken out of the fluid stream then

$$\phi_D = K_d c_b \tag{7.2}$$

where $K_d$ is the deposition coefficient. Sometimes $K_d$ is referred to as the deposition velocity

$K_d$ is only identical to $K_t$ when all the particles that arrive at the surface remain on the surface.

Where some particles hit the surface and rebound or are taken back into the flow stream by a removal mechanism $K_d$ will be $< K_t$.

Under these circumstances when some of the particles that arrive at a surface fail to stick and return to the bulk flow or at least to the laminar sub-layer, the term "sticking probability" is sometimes applied. As the name suggests it attempts to quantify the retention of particles on the surface.

i.e. $K_d = PK_t \tag{7.3}$

where $P$ is the sticking probability

The sticking probability will have a value $<$ unity and its magnitude will depend on the conditions associated with the flowing fluid, the size and nature of the particles and the topography and the physical character of the surface. For these reasons it is extremely difficult or impossible to predict the sticking probability for a given situation, although Beal [1970] did attempt, not too successfully, to include such a term into his deposition model. In certain combustion situations (see Chapter 16) particles including molten ash, arrive at a sticky surface so that under these conditions the sticking probability may be regarded as unity. The difficulty here though, is that the sticking condition is difficult to predict and is likely to change with time as the deposit develops. For instance initially the temperature of the thin layers of deposit on a heat transfer surface, in contact with hot combustion gases are really that of the surface itself, i.e. very close to the fluid temperature on the other side of the surface. As the deposit thickens the temperature of its outer surface approaches the combustion gas temperature and this could give rise to molten layers on the outside of the deposit. Thus as the fouling progresses the sticking probability will change and is likely to approach unity.

Particle deposition on surfaces has been reviewed by various authors including Friedlander and Johnstone [1957], Kneen and Strauss [1969], Beal [1970], Browne [1974], Hidy and Heisler [1979], Gudmundsson [1981], Wood [1981] and Epstein [1988].

## 7.3   PARTICLE TRANSPORT - ISOTHERMAL CONDITIONS

Much of the theory developed for the transport of particles has been for isothermal conditions. Fig. 7.2 shows a particle approaching a surface. It is assumed that the particle is approaching the surface at right angles from a distance equal to the stopping distance.

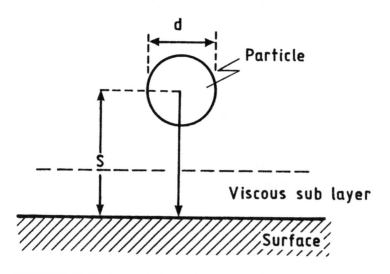

FIGURE 7.2.  Idealised approach of a particle towards a surface

The distance a particle will travel through a stationary fluid, when it has an initial velocity $u_o$ is given by

$$s = \tau u_o \qquad (7.4)$$

where $s$ is the stopping distance

and $\tau$ is the so-called particle relaxation time

$\tau$ for particles subject to aerodynamic resistance in the Stokes drag regime is given by

$$\tau = \frac{\rho_p d_p^{\;2}}{18\eta} \qquad (7.5)$$

where $\rho_p$ and $d_p$ are the particle density and diameter respectively

The dimensionless particle relaxation time $\tau^*$ is given by

$$\tau^* = \frac{\tau(u^*)^2}{\nu} \qquad (7.6)$$

where $u^*$ is the friction velocity

$\nu$ is the kinematic viscosity

$$u^* = u\sqrt{\frac{f}{2}} \qquad (7.7)$$

Epstein [1988] develops the concept of a particle in free flight with an initial velocity of $v_o$ subject to Stokes drag forces as it travels through a fluid.

Making a force balance on the particle

i.e. force = rate of change of momentum

$$3\pi\eta u_p d_p = -m_p \frac{du_p}{dt} \qquad (7.8)$$

where $m_p$ is the mass of particle diameter $d_p$

and $u_p$ is the particle velocity

But $\quad m_p = \dfrac{\pi d_p^{\,3} \rho_p}{6}$ $\hspace{6cm}$ (7.9)

where $\rho_p$ is the particle density

and $\quad dt = \dfrac{ds}{u_p}$

where $s$ is the stopping distance

Substituting in Equation 7.8 gives

$$3\pi\eta u_p d_p = -\frac{\pi d_p^{\,3} \rho_p}{6} \cdot dv_p \frac{u_p}{d_s} \hspace{4cm} (7.10)$$

Simplifying and rearranging yields

$$ds = -\frac{d_p^{\,2} \rho_p}{18\eta} \cdot du_p \hspace{5cm} (7.11)$$

Integrating

$$\int_{s=0}^{s=s} ds = -\int_{u=u_o}^{u=o} \frac{d_p^{\,2} \rho_p du_p}{18\eta} \hspace{4cm} (7.12)$$

or $s = \dfrac{d_p^{\,2} \rho_p u_o}{18\eta}$ $\hspace{5cm}$ (7.13)

$s$ the particle stopping distance is sometimes referred to as the Stokes particle stopping distance

The particle relaxation time $\tau$ is given by

$$\frac{s}{u_o} = \frac{d_p^{\,2} \rho_p}{18\eta} \hspace{5.5cm} (7.14)$$

It becomes the dimensionless relaxation time if

$\dfrac{s}{u_o}$ is divided by $\dfrac{v}{(u^*)^2}$ (see Equation 7.6) i.e. $\tau^* = \dfrac{d_p^2 \rho_p}{18\eta} \cdot \dfrac{(u^*)^2}{v}$ (7.15)

or $\tau^* = \dfrac{\rho_p d_p^2 (u^*)^2}{18\rho v^2}$ (7.16)

Similarly a dimensionless transport coefficient may be defined if $K_t$ is divided by $u^*$

i.e. $K_t^* = \dfrac{K_t}{u^*}$ (7.17)

There is a strong correlation between $K_t^*$ and $\tau^*$ [Montgomery and Corn 1970]. Gudmundsson and Bott [1977] have demonstrated that the magnitude of $\tau^*$ is an indication of the regime under which the particle approaches the surface (see Table 7.1).

TABLE 7.1

Values of $\tau^*$ and associated regime

| $\tau^*$ | Regime |
|---|---|
| < 0.1 | Diffusion |
| 0.1 - 10 | Inertia |
| > 10 | Impaction |

It is convenient to discuss isothermal particle deposition under these different regimes.

The three regimes may also be demonstrated by plotting $K_t^*$ against $\tau^*$ (see Equations 7.16 and 7.17). Epstein [1988] presents the curve reproduced as Fig. 7.3.

### 7.3.1 Diffusion regime

Particles suspended in the fluid are carried by the fluid as it flows across the surface. If the fluid is flowing under laminar conditions the transport of the particles across the fluid layers to the surface will be by Brownian motion. Under turbulent conditions particles will be brought to the laminar sub-layer by eddy diffusion, but the remainder of the journey to the surface, across the laminar sub-layer is generally ascribed to Brownian motion. Under these conditions for the small particles involved, they may be treated as molecules. In other words mass

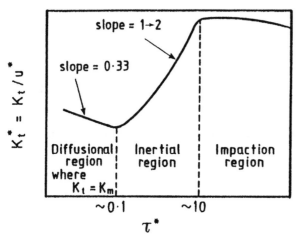

FIGURE 7.3. Particle transport regimes

transfer theory developed for miscible or soluble substances can be applied. Under these circumstances $K_t$ in Equation 7.1 becomes the mass transfer coefficient $K_m$.

The Brownian diffusivity $D_B$ may be calculated by the so-called Stokes-Einstein Equation [Einstein 1956], i.e.

$$D_B = \frac{K_B T}{3\pi\eta d_p} \qquad\qquad (7.18)$$

where  $T$ is the fluid temperature $K$

$\qquad K_B$ is the Boltzmann constant 1.38 x $10^{-23}$ J/K

The mass transfer coefficient may be estimated by equations of the form

$$Sh = K(Re)^n (Sc)^m \qquad\qquad (7.19)$$

where  $Sh =$ is the Sherwood number $\dfrac{K_m d}{D_B}$

$\qquad$ and $d$ is the pipe diameter or flow duct equivalent diameter

$\qquad Re$ is the Reynolds number

$\qquad Sc$ is the Schmidt number $\dfrac{\eta}{\rho D_B}$

$\qquad K$ is a constant

$\qquad m$ and $n$ are numerical indices

Gudmundsson [1981] has quoted a correlation for the dimensionless mass transfer coefficient $K_t^*$ in terms of the particle Schmidt number for particles in the diffusion regime in turbulent pipe flow.

$$K_t^* = \frac{0.112}{Sc^{0.75}} \qquad (7.20)$$

where $Sc$ is $\dfrac{\eta}{\rho D_B} \gg 1$.

The form of Equation 7.19 used by Hussain *et al* [1986] for an assessment of the deposition of magnetite onto the inner wall of an aluminium tube, was that proposed by Berger and Hau [1977], i.e.

$$Sh = 0.0165\, Re^{0.86}\, Sc^{0.33} \qquad (7.21)$$

Epstein [1988] suggests that for diffusion controlled deposition

$$K_t = K_m \qquad (7.22)$$

i.e. $K_t$ is proportional to $u^* dp^{-0.67}$ $\qquad (7.23)$

The discussion has largely been confined to smooth surfaces and the theory is complicated by the presence of roughness elements on the surface. Browne [1974] states that particle deposition rates are extremely sensitive to roughness even when this is too small to produce aerodynamically rough flow. This is particularly so for particles which give minimum deposition rates on smooth tubes. A reasonable approach may be made by estimating the value of $u^*$ for the surface involved. The magnitude of $u^*$ increases with roughness and the stopping distance is reduced.

### 7.3.2 Particles with inertia

Small particles tend to follow the flow lines in a fluid moving across the surface and travel across the boundary layers by Brownian motion. Friedlander and Johnstone [1957] noted that the deposition of aerosol particles onto a surface was greater than predicted by diffusion theory and mass transfer correlations. Larger particles will acquire sufficient energy from the fluid to enable them to move through the viscous sub-layer. Particles with insufficient energy will be slowed and finally arrested by the drag forces in the sub-layer to become subject to Brownian motion. Others will have sufficient energy so that they can reach the surface.

Liu and Agarwal [1974] presented their data for the inertia regimes in terms of the dimensionless mass transfer coefficient $K_t^*$ and the dimensionless relaxation time $\tau^*$. The equation has the form:

$$K_t^* = 6 \times 10^{-4} (\tau^*)^2 \qquad\qquad (7.24)$$

A similar equation was given by Papavergos and Hedley (1984) for the range $\tau^* = 0.2 - 20$

i.e. $K_t^* \approx 3.5 \times 10^{-4} (\tau^*)^2 \qquad\qquad (7.25)$

Epstein [1988] points out that this expression shows that

$K_t$ is proportional to $d_p^4 (u^*)^5 \qquad\qquad (7.26)$

The important effects of particle diameter and the resistance to particle movement on mass transport towards the surface, are made clear by this relationship.

### 7.3.3  Impaction of particles

In the impaction regime, the particle velocity towards the wall attains a similar magnitude to the friction velocity $u^*$ [Gudmundsson 1981, Epstein 1988]. In the range $\tau^* = 10 - 100$ the component of velocity of a particle along a pipe radius is reasonably constant and the stopping distance becomes of the same order as the pipe diameter [Epstein 1988]. The effect of turbulent fluctuations on relatively large particles is limited so that the value of $K_t^*$ remains virtually constant. For larger particles the response to changes in turbulence is even less so that the values of $K_t^*$ are reduced.

Papavergos and Hedley [1984] suggest that for this regime

$$K_t^* \approx 0.18 \qquad\qquad (7.27)$$

Epstein [1988] comments that this suggests

$K_t$ is proportional to $(d_p)^0 u^* \qquad\qquad (7.28)$

i.e. very different to the similar expression for the inertia regime, and a simple dependence on the friction velocity $u^*$.

### 7.3.4  The important variables

Beal [1970] attempted mathematically to analyse particle transport. Although the basis of the mathematics is ideal in concept involving assumptions, it does demonstrate the importance of fluid velocity and particle diameter on particulate deposition. Figs. 7.4 and 7.5 are based on the mathematical approach of Beal [1970]. The effect of sufficiently high velocities (see Fig. 7.4) where the momentum becomes dominant (the steep part of the curves), even for particles of small diameter is apparent. The sharp discontinuities in slope result from the different correlations used by Beal [1970] to determine the eddy diffusivity for integration across the boundary layer. Fig. 7.4 also demonstrates that neglecting particle momentum could lead to significant error at high velocities, while neglect of Brownian motion would lead to error at low velocities.

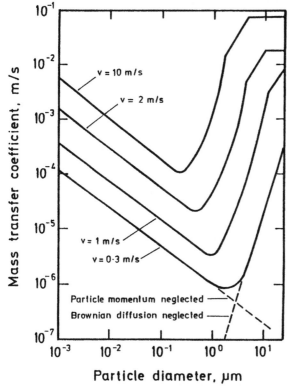

FIGURE 7.4. Effect of particle size on deposition mass transfer

For small particles (colloidal) the transport is entirely diffusion controlled, and the rate of diffusion decreases as the particle diameter increases. As the particle size increases so does the momentum, so that the momentum effect gradually becomes more important. Beal [1970] shows that the deposition rate reaches a minimum below which diffusion is the dominant transport mechanism and above the minimum momentum assumes the greater importance. At very large particle diameter where the stopping distance is greater than the assumed boundary layer thickness, particle size no longer has any effect, and the transport rate remains constant for a given velocity.

An interesting aspect of Fig. 7.4 is that it illustrates there is a particle size range that is likely to remain entrained, i.e. minimising the fouling under the prevailing conditions. From Fig. 7.4 the particle size for minimum fouling is in the approximate range 0.2 - 2 $\mu m$ for velocities in the range 0.3 - 10 $m/s$.

Fig. 7.5 is based on the data of Fig. 7.4 and plots the mass transfer coefficient for deposition at different velocities for 10 *μm* particles; these data neglect Brownian motion. The conditions represented on Fig. 7.5 are idealised conditions that could apply to an air blown heat exchanger. The figure shows that there is a rapid increase in mass transfer coefficient for deposition as velocity increases up to around 3 *m/s*. Above this velocity there is a reduced but steady, rate of increase of mass transfer coefficient with velocity.

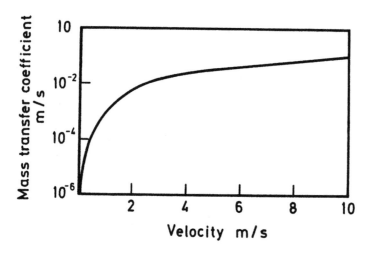

FIGURE 7.5. Effect of velocity on deposition mass transfer

## 7.4    NON-ISOTHERMAL PARTICLE DEPOSITION

When suspended particles are subject to a temperature field there is a tendency for small particles to travel down the temperature gradient. The phenomenon is usually called thermophoresis, but it may also be referred to as thermal diffusion. The process has been used as a separation process for high value or heat sensitive substances [Bott and Khoo 1967] although in general the separation process is not of particles but of molecules.

Thermophoresis arises as a result of the "bombardment" on the "hot" side of a particle in comparison to a lower bombardment effect on the cooler side of the particle, that in combination this tends to "push" the particle towards the cooler region of the temperature field. The magnitude of the thermal force depends on the properties of both the fluid and the particles, as well as the magnitude of the temperature gradient acting on the particle. Although the thermophoretic effect is possible in both liquids and gases, it is usually considered to be significant in respect of fouling in gas systems. The reason for this is that the relatively large temperature differences necessary are not common in liquid systems and the viscosity and density of liquids is not conducive to particulate transport under these conditions. On the other hand in gas systems, particularly in boiler plant burning conventional fuels where large temperature gradients can exist, i.e. in the extreme from flame temperatures in excess of 1500°C to boiling water temperatures, thermophoresis can make a significant contribution to particle transport. Moreover the physical properties of the gas stream do not generally create the same level of resistance to particle transport as in a liquid system, but since the physical properties of a gas depend on the pressure the effects of pressure on thermophoretic transport are often quoted rather than viscosity and density.

Using data obtained from an advanced gas cooled nuclear reactor Owen *et al* [1987] developed a model for particle transport to rough surfaces. The model included the effects of thermophoresis and eddy diffusion. In the context of this discussion the effect of thermophoresis on the transport of particles to a surface (i.e. fouling) is paramount but the movement of particles away from a surface due to thermophoresis may also be an important consideration that may not be neglected. Fig. 7.6 [from Owen *et al* 1987] shows that the ratio of particle deposition velocity with thermophoresis $(u_T)$ to the velocity with no thermophoresis present $(u_o)$ is a function of the temperature difference between the gas temperature $(T_g)$ and the surface temperature $(T_w)$ for a given roughness. As may be seen from the diagram thermophoresis enhances particle transport towards a cold surface and away from a heated surface. Owen *et al* [1987] discovered that the thermophoretic effects are reduced as the surface roughness increases.

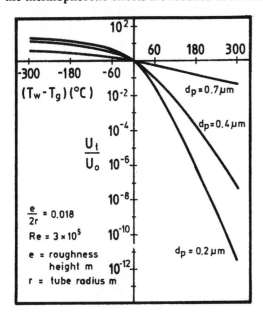

FIGURE 7.6. The dependence of velocity due to thermophoresis on temperature difference

The work of Owen *et al* [1987] was based on particles < 1 $\mu m$ and the enhancing effect of small particle diameter is clear from Fig. 7.6 and in general thermophoresis is not significant as a transport mechanism in gas systems for particles greater than about 10 $\mu m$ but it is significant for particles below about 2 or 3 $\mu m$ diameter [Singh and Bryers 1972]. Gökoglu and Rosner [1982] state that thermophoresis can increase the deposition rate of 1 $\mu m$ diameter particles by more than 1000-fold (compared to diffusion only) on a solid object under certain conditions. Owen [1960] devised an equation from which it is possible to calculate the thermophoretic velocity $u_t$ with which the particle travels down a temperature gradient.

$$\frac{u_t}{u^*} = \frac{0.09}{\pi} \cdot \frac{u^*}{u} \left( \frac{T_1 - T_2}{T_2} \right) \qquad (7.29)$$

where $T_1$ and $T_2$ are the extremes of absolute temperature with $T_1 > T_2$ creating the temperature gradient.

The importance of the thermophoresis effect may be considered in the ability to propel particles across the viscous sub-layer adjacent to the target surface, where other transport mechanisms, e.g. diffusion, inertia and impaction, are limited in so far as they reduce the ability of the particle to reach the surface (i.e. the magnitude of the stopping distance). For diffusion the transport depends on Brownian motion and concentration difference. For inertia and impaction the distance a particle can travel depends on the energy it possess as it enters the viscous sub-layer. Where high temperature gradients occur across the laminar sub-layer as might well be the situation in turbulent combustion gases, the final journey of the particle would be under the influence of the temperature gradient; with the arrival at the viscous sub-layer boundary region achieved by eddy diffusion.

El-Shobokshy and Ismail [1981] noted that there was a critical particle size in deposition in relation to thermophoresis and inertial impaction. Deposition was a minimum at the critical particle size $d_c$. For particle diameters < $d_c$ the mechanism of deposition was largely due to thermophoresis above $d_c$ the deposition was associated with inertia.

Singh and Bryers [1972] made the point that in flow along the inside of for instance, a vertical tube, a selection process is likely to occur, i.e. the smaller particles experiencing a higher thermal force are likely to deposit near the entrance; larger particles travelling further downstream before depositing. There could be implications here for heat exchanger fouling where the quality of the deposit changes not only with time but with location or geometry of the equipment.

Where fine particles are likely to be present in a fluid passing through heat exchange equipment subject to a high temperature difference, it is necessary to be aware that the diffusional and inertial deposition of particles is likely to be augmented by thermophoresis effects.

## 7.5    ELECTROMAGNETIC EFFECTS

Lister [1981] in a discussion of fouling in power generating systems has drawn attention to the possibility that electric fields can influence the deposition of particles onto surfaces. From Stokes law the terminal velocity of a particle $u_T$ is given by

$$u_T = \frac{eE}{3\pi\eta d_p} \qquad\qquad (7.30)$$

where  $e$ is the electrostatic charge

$E$ is the field strength through which the particle moves

$d_p$ is the particle diameter

$\eta$ is the fluid viscosity

Based on work by Samms and Watt [1966], Lister [1981] concludes that electromagnetic forces are unlikely to affect deposition from combustion gases, but they are pertinent to particle/liquid systems associated with power generation.

According to Lister [1981] deposition may be induced by thermoelectric potentials. The passage of heat normal to a surface (heat flux) can induce a "thermoelectric" effect if the liquid stream flowing across the surface contains a sufficient concentration of charged particles. The induced EMF will generate a particle flux $\phi_f$ according to the equation

$$\phi_f = \frac{\varepsilon u_e c q}{\lambda_f} \qquad\qquad (7.31)$$

where  $\varepsilon$ is the thermally induced EMF

$u_e$ is the electrophoretic mobility of the charged particles

$c$ is the particulate concentration

$\lambda_f$ is the thermal conductivity of the fluid

Lister [1981] states that Equation 7.31 has formed the basis of some reasonable predictions of corrosion product deposits in various high temperature water systems [Brusakov 1971].

## 7.6    PARTICLE DEPOSITION MODELS

Despite the complexity of the phenomenon attempts have been made to devise models that accurately describe the mechanism of particulate deposition.   In general, although not exclusively, the models build on the original Kern and Seaton [1959] approach (see Chapter 4).  Furthermore the proposed models are often put forward in connection with a discussion of experimental results, so that there is often an element of empiricism in the final equations to accommodate the experimental data.

It is not proposed to review in some detail the models that have been developed since although they can provide an insight into the underlying mechanisms and the influence of specific variables, they are not generally universally applicable. Reviews of particulate fouling with reference to heat exchangers have been provided by Gudmundsson [1981], Lister [1981], Beal [1983] and Epstein [1988].

Watkinson and Epstein [1971] considered that the deposition term in the Kern and Seaton [1959] relationship would be dependent upon the mass transfer to the surface coupled to the sticking probability.   Watkinson and Epstein [1971] assumed that the sticking probability is a function of the adhesive forces present and the drag forces acting on the particles resident on the surface.   The incorporation of a sticking probability into the assessment means that the particulate concentration at the wall could not be regarded as zero and consequently that the driving force could not be equated to the bulk concentration.

Under these conditions the mass flux to the wall is given by the expression

$$\frac{dm}{dt} = K_t(c_b - c_w) \tag{7.32}$$

where $c_b$ and $c_w$ are the concentrations of the particles in the bulk fluid and at the wall respectively

Equation 1.1 gives, in effect, the rate of change of deposit mass with time as

$$\frac{dm}{dt} = \phi_D - \phi_R \tag{7.33}$$

If the sticking probability is $P$ then the concentration of particles at the wall will be given by

$$c_w = c_b(1 - P) \tag{7.34}$$

The mass transfer to the wall, i.e. allowing for particles that actually remain on the wall is given by

$$\phi_D = K_t(c_b - c_b(1 - P)) \tag{7.35}$$

i.e. $\phi_D = K_t c_b P$ (7.36)

It would be possible therefore to combine $K_t P$ into a revised mass transfer coefficient say $K_d$, that allows for the sticking probability but also relies on the bulk concentration, i.e.

$$\phi_D = K_d c_b \tag{7.37}$$

Equation 7.37 gets around the problem of knowing the concentration of particles at the interface but it requires a knowledge of the sticking probability. There is no way in which the sticking probability may be calculated and it is in this respect that many mathematical models have to resort to empiricism. Furthermore the model makes no distinction between the sticking probability of particles to the virgin surface and particles arriving at a surface already covered with particles. It is likely that as the deposit layer thickens the sticking probability will change, if for no other reason than the adhesion forces change (see Chapter 6).

Combining Equations 7.36 and 7.37

$$K_d c_b = K_t c_b P \tag{7.38}$$

that leads to the rather obvious conclusion

$$P = \frac{K_d}{K_t} \tag{7.39}$$

Reference has already been made to the pioneering work of Cleaver and Yates [1973, 1975, 1976] in Chapter 5 in connection with removal mechanisms, associated with disturbances in the laminar sub-layer or "turbulent bursts" as they are sometimes called. The Cleaver and Yates theory suggests that the bursts, occurring at a known average rate, each completely clean a limited area of the surface. The deposit asymptote is reached when the wall has received sufficient particles that the rate of deposition balances the rate of removal by turbulent bursts.

The discussion of the development of the model by Watkinson and Epstein [1971] revealed the need to consider a variable sticking probability to account for the differences between a clean surface and particulate layers in the retention of further particles. Thomas [1973] and Thomas and Grigull [1974] considered that the sticking probability and the adhesion forces acting on the particles within a deposit of particles, would be lower than that of particles on an unfouled surface. Under this mechanism an asymptote would be reached when the shear forces at the surface are equal to the adhesive forces acting on the particles, as they become reduced due to the presence of successive layers of particles on the surface.

Thomas and Grigull [1974] also suggested that the presence of particles on the surface could reduce the surface roughness making the surface more hydrodynamically smooth.    Such an assumption runs counter to the usually observed rough topography of most deposits. Under these smooth conditions the laminar sub-layer would be thicker for a given flow rate, and would increase the resistance to the mass transfer of particles towards the surface with an attendant reduction in the particulate deposition rate.

Autoretardation of the fouling process was also employed by Rodliff and Means [1979] to explain the reduction in deposition of magnetite particles approaching a surface.   The chemiphoresis mechanism effectively generates an autoretardation effect by inhibiting corrosion of the metal wall as the deposit thickness increases, and so reduces the number of ions (charged) of corrosion products diffusing through the deposit and into the fluid (water) stream.   The theory assumes that these particles tend to neutralise the electrical double layer (see Chapter 6) on the magnetite particles in suspension, and so remove the repulsive forces which would normally inhibit the close packing of like charged particles making up the deposit. As the deposit grows less neutralising ions are able to diffuse through the deposit into the laminar sub-layer and the sticking potential of the magnetite particles approaching the surface is reduced.   The model only considers deposition and does not take into account any removal mechanism.

Bowen [1978] and Bowen and Epstein [1979] have also considered autoretardation in the deposition process.   In their approach the surface charge of particles deposited on the surface remain unchanged, so there is repulsion of further particles as they approach the surface.   It is considered that the particles will only stick to "bare" areas of the target surface.   The theory therefore suggests that a deposit will remain as a mono-particulate layer.   It is likely to apply to highly charged particles that retain their charge after deposition.

Glen *et al* [1988, 1992] have described a model developed essentially for particulate deposition in gas streams, in particular for high temperature systems. Under these conditions material carried by the gas phase may exist as vapour, liquid or solid.   The transport and deposition will depend on the nature of the contaminating material.   The mechanisms already identified as being responsible for particle deposition include:

> inertial impaction
> diffusion (Brownian and eddy)
> gravitation
> thermophoresis

to which Glen *et al* [1988] add:
>       interception
and      diffusiophoresis

All particles carried in a fluid stream are subject to inertia forces. Particles that follow the streamlines round an object may still collide with the object by interception because of the finite size of the particle. Diffusiophoresis is the net force on the particles caused by the flux of a condensing vapour acting towards a cooler surface.

In a practical system it would be expected that a range of particle size would be encountered. The particle size, as has already been described, affects the mechanism responsible for deposition and a model that is designed to describe the deposition process must take particle size into account.

Glen *et al* [1988, 1992] suggest that it is convenient to describe each mechanism in terms of a dimensionless deposition parameter $N_i$ that incorporates the variables affecting the deposition process. Some of the parameters are given in Table 7.2. Fig. 7.7 shows the effect of particle size on these parameters.

TABLE 7.2

Dimensionless deposition parameters

| Mechanism | Parameter |
|---|---|
| Inertial impaction $N_I$ | $\dfrac{C\rho_p d_p u_g}{18\eta D}$ |
| Interception $N_C$ | $\dfrac{d_p}{D_c}$ |
| Diffusion $N_D$ | $\dfrac{D_p}{u_g D}$ |
| Thermophoresis $N_T$ | $\dfrac{u_T}{u_g}$ |

where  $C$ is the Cunningham coefficient
$D_c$ is the collector diameter
$u_g$ and $u_T$ are the gas velocity and the particle
velocity attributable to thermophoresis
respectively
$D_p$ is the diffusion coefficient for particles

The previous discussion has concluded that all particles that arrive at a surface do not necessarily remain at the surface. Glen *et al* [1988, 1992] allow for this phenomenon by the inclusion of a particle collection efficiency $\eta_i$ defined as the fraction of the free stream particle mass flow (through the projected area of the target body or object, onto which the particles impinge) that actually strikes the body.

For each mechanism

$$\eta_i = f(N_i) \tag{7.40}$$

but since $N_i$ depends on particle size

$$\eta_i = \sum f\{N_i(d_p)\} \tag{7.41}$$

where the summation covers the whole particle range over which the model applies.

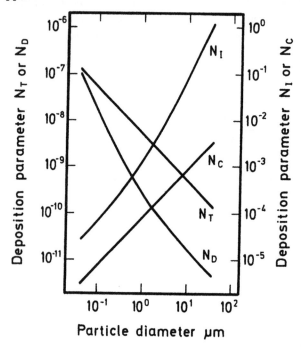

FIGURE 7.7. The effect of particle size on dimensionless deposition parameters

The particle size distribution must either be known or assumed in order that further development of the model may be attempted.

The overall particle collection efficiency $\eta_{tot}$ can then be written in the following form:

$$\eta_{tot} = P_s\left(1 - (1 - a_I \eta_I)(1 - a_c \eta_c)(1 - a_D \eta_D ...)\right) \tag{7.42}$$

$a_I, a_c, a_D$ etc. are vector switches, taking values between 0 and 1.

$P_s$ is the overall sticking probability

The need for the so-called vector switch is to approximate the influence of particle direction on sticking probability, since the $N_i$ values should incorporate velocity, but in practice only speed is used.

Glen *et al* [1988] make the important point that the model is only valid in this form if the sticking probability is the same for all the particles in the given size range for each active mechanism. Another major assumption is that the different mechanisms do not interfere with each other and that they are additive in the way described. Both are major assumptions and considerable errors may be anticipated in some specific applications. For instance Equation 7.42 may be modified to take account of chemical reactions taking place in the particulate systems such as might occur in gases from the combustion of fuel. For bulk incineration of domestic refuse Glen *et al* [1988] re-wrote Equation 7.42 as:

$$\eta_{tot} = 1 - (1 - P_I a_I \eta_I)(1 - P_o a_o \eta_o) \qquad (7.43)$$

where $\eta_o$ is a combined collection efficiency for all the non-impacting mechanisms of the form:

$P_I$ is the sticking probability of impacting particles

$$\eta_o = 1 - \sum (1 - a_i \eta_i) \qquad (7.44)$$

Where the summation extends over all the mechanisms other than impaction (i.e. interception, diffusion and thermophoresis).

Rosner and Tassopoulos [1989] also recognise the importance of particle size and its distribution on the deposition from particulate suspensions. They also point out that in most engineering systems the particle size distribution is a function of the mechanism by which the particles were formed, e.g. combustion systems, crystal formation and so on. Rosner and Tassopoulos [1989] therefore conclude that in order to calculate the total mass deposition rates, it is necessary to sum (or integrate) over both a function covering capture efficiency and a function describing mass distribution. Based on this background these authors have suggested a method for estimating particle mass deposition rates. The technique described depends on the mean particle volume $\bar{V}$ defined by

$$\bar{V} = \frac{\phi_p}{N_p} \qquad (7.45)$$

where $\phi_p$ is the particle volume described in terms of a particular number density function

and $N_p$ is the particle number density

First an expected (or reference) mass deposition flux $-\left(\dfrac{dm}{dt}\right)_{ref}$ is calculated as
if all the particles in the mainstream had the same volume $\bar{V}$. The reference deposition rate is corrected by a factor, which the authors state to be near unity, and is expected to depend on the character of the functions describing the capture efficiency and the particle size distribution. Correlations are presented that may be used to estimate deposition rates for convective diffusion, eddy impaction and inertial deposition. The authors [Rosner and Tassopoulos 1989] suggest that the technique could be developed for "computationally-efficient-finite analytic methods" for computing total mass deposition rates from aerosol population distributions of arbitrary shape, captured in accord with an efficiency function of arbitrary shape.

In connection with their work the authors stress that the basic assumption, i.e. that each particle size class does not influence the deposition rate of other size classes. They cite work by Rosner and co-workers to emphasise the limitations of the assumption and it is worthwhile stating these here since they are of importance in their own right in any consideration of particle deposition. The following observations are made:

1. The diffusional capture of small particles influences the inertial capture of larger particles [Rosner and Nagarajan 1987].

2. Appreciable particle-particle coagulation occurs in the immediate vicinity of the collector [Park and Rosner 1989a, Biswas 1988].

3. A portion of the pre-existing aerosol population coupled with the host fluid flow to modify the transfer coefficient of the smaller particles [Park and Rosner 1989b].

4. The depositing particles scavenge an appreciable mass of co-existing vapour within the thermal boundary layer adjacent to the collector (Castillo and Rosner 1988).

5. Appreciable particle production (from a supersaturated vapour) occurs within the thermal boundary layer adjacent to the collector [Castillo and Rosner 1989, Liang *et al* 1989].

Despite these reservations the simple analysis presented by Rosner and Tassopoulos [1989] is still useful and instructive in terms of the effects of the different variables.

The quality of the deposit, e.g. in terms of structure and porosity has been considered by Tassopoulos *et al* [1989]. Using a theoretical discrete stochastic model they found that the polydispersity of the depositing particles can significantly increase the openness of the deposit structure. On the other hand a large particle mean free path, i.e. larger than the actual particle size, results in a more compact less porous deposit. These authors have also determined pore size distributions for the resulting deposits in both two and three dimensions, and determined that they are independent of deposit size, if they are properly normalised and conform to a power law with an exponent depending on the associated Peclet number *Pe*.

Tassopoulos *et al* [1989] admit that these concepts are perhaps a first step towards a clearer understanding of the effect of system variables on the character of the deposit.

Important to the development of the ideas are:

1. An understanding of how deposition mechanisms evolve with time.

2. Data on the characteristics of deposits in terms of effective permeability and thermal conductivity.

3. The potential effect of deposit age on the qualities of the deposit, where the character of the deposit may change with time or thickness. For instance the changing temperature on the outside of deposits in a high temperature system (combustion) is more than likely to result in differences of porosity for example, within the deposit.

A numerical method to predict the deposition of particles from turbulent non-isothermal water flow has been developed by Adomeit and Renz [1992]. The interaction forces between particles and the surface being fouled are considered in the light of the DVLO theory and applicable transport equations. These authors conclude that the approach can reasonably predict deposition rates provided the repulsive interactions are small, i.e. for high adhesion probabilities. For combinations of *pH* value and temperature that give strong repulsion, the predictions can only be regarded as qualitative. The numerical modelling technique has merit but improvements could be introduced by taking into account surface heterogeneticies and extending the procedure to polydisperse systems.

Epstein [1988] has drawn attention to two hydrodynamic effects that he terms "neglected" and are not generally taken into account in models. They include the Magnus effect and the induced fluid drainage force.

The Magnus effect is the force imposed on a rotating object in a shear flow. When a fluid and a solid particle are flowing downwards across a surface, due to the density difference the particle will be travelling at a higher velocity than the fluid in the vicinity of the particle, the result is a force directed towards the wall. For the corresponding upflow the particle would be directed away from the wall.

The fluid drainage force arises when a solid particle approaches close to a solid surface. As a particle moves to the vicinity of the surface it experiences increased viscous resistance to motion because the drag between the fluid and the two surfaces as they approach each other, gradually increases the force necessary to force the fluid from the front of the particle. The general effect of this force is to reduce deposition.

Epstein [1988] concludes that for gases the Magnus effect and the fluid drainage force are likely to be small, but for liquids they could assume importance. The empirical sticking probability is a way of taking these unrecognised influences into account.

The few models discussed here serve to demonstrate the complexity of the problem of attempting to describe the mechanism of particulate deposition in mathematical terms. Much of the difficulty stems from the inability to recognise the importance of specific variables and a lack of understanding of the interaction between the variables. Furthermore the lack of knowledge of the true composition of flow streams and flow conditions in industrial heat exchangers, makes the application of models for design and operation analysis, almost impossible except in all but the simplest examples. In the following section the effect of some of the contributory factors will be demonstrated in relation to some experimental data.

## 7.7    SOME EXPERIMENTAL DATA ON PARTICULATE DEPOSITION

In this section some limited experimental data are presented in the light of the theories and models discussed earlier in this chapter. Two groups of studies are considered, namely liquid and gas systems.

### 7.7.1  Liquid systems

Considerable effort has been made in the examination of the deposition of oxides of iron onto surfaces, since this has considerable relevance to the operation of boiler plant. Williamson [1990] has reviewed some experimental magnetite ($Fe_3O_4$) deposition data. He draws attention to the wide discrepancies in the results even for similar systems as may be seen from examination of Table 7.3. He attributes the discrepancies to inadequate control of one or more of the less obvious variables in the system such as the water chemistry (even deionised water is likely to "pick up" $CO_2$ from the atmosphere with an effect on the *pH*). A fixed particle size distribution is extremely difficult to maintain in an experimental system due to potential agglomeration and is likely to be time dependent. Even particle concentration may not be uniform due to settlement in the parts of the equipment where velocities are low. Experimentally the consistency of these variables are difficult to determine.

Newson *et al* [1981] reported studies of magnetite particulate deposition from water flowing on the inside of vertically mounted aluminium tubes using an X-ray technique. The range of the experimental data are presented in Table 7.4.

TABLE 7.3

Data on deposition from magnetite/water systems [Williamson 1989]

| Author | Tube material | Particle concentration $mg/kg$ | Temperature °C | System pressure $bar$ | Water velocity $m/s$ | Reynolds number | Derived relationships | |
|---|---|---|---|---|---|---|---|---|
| | | | | | | | Initial Deposition rate $\phi_D$ dependent on | Asymptotic Deposit mass $M^*$ dependent on |
| Thomas [1973] Thomas and Grigull [1974] | Stainless | 2.2 | 291 | 246 | 1.05 - 1.77 | 45000 - 77000 | $u_m$ | $u_m^{-0.5}$ |
| Burill [1977, 1978, 1979] | Zircaloy 4 | 300 | 27 | 1 - 2 | 1.00 - 100 | 20000 - 400000 | $u_m^2$ | - |
| Gudmundsson [1977] | Aluminium | 280 | 40 | 1.1 | 1.20 - 200 | 35000 - 53000 | $u_m^{-1}$ | - |
| Newson [1979] | Aluminium | ~100 | 40 | 1.1 | 0.40 - 1.30 | 13000 - 38000 | $u_m^{-2.3}$ | $u_m^{-2.5}$ |
| Hussain [1982] | Aluminium | 200 - 600 | 12 - 75 | 1.1 | 0.30 - 2.00 | 7800 - 89000 | $u_m^{0.7}$ | $u_m^{-0.7}$ |

TABLE 7.4

Magnetite deposition from flowing aqueous suspensions (Newson *et al* 1981]

| Temperature °C | Reynolds number | Particle concentration mg/kg | Particle size μm |
|---|---|---|---|
| 12 - 75 | 7800 - 88500 | 200 - 600 | ~2 |

Graphs of some typical experimental data are given on Fig. 7.8.

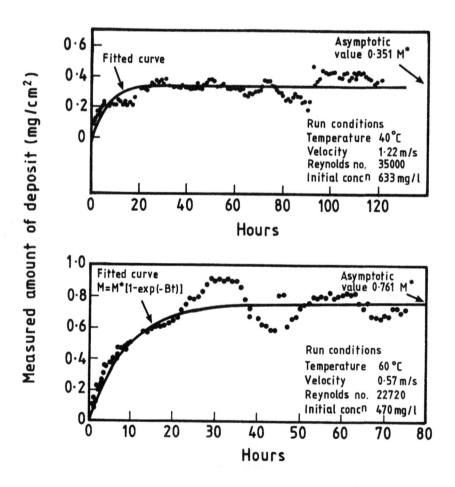

FIGURE 7.8. Particulate magnetite deposition on the inside of aluminium tubes

The data suggest that both the initial deposition rate and the asymptotic deposit mass are both dependent upon the bulk velocity $u$ raised to the power 0.6 - 0.7. The results were also compared with the mass transfer rates of Cleaver and Yates [1975] and Metzner and Friend [1958]. Although the dimensionless particle relaxation times (see Section 7.3) were below 0.1, the inertial deposition rates calculated from the theory of Cleaver and Yates were of an order of magnitude higher than the diffusional rates calculated and indeed measured. The measured power on velocity of 0.7 compared to a theoretical value of 0.875 for diffusion and 2 for inertial particle transfer, suggest a diffusion controlled mechanism.

Melo and Pinheiro [1988] have also studied the deposition of magnetite and magnetite/kaolin from flowing aqueous streams. The data were analysed in terms of the Kern and Seaton [1959] models, i.e. Equation 4.11. The experimental conditions and the data obtained are given in Table 7.5.

It would appear that the presence of the magnetite reduces the deposition flux of the kaolin, but the value of $\frac{1}{\beta}$ is increased when both species of particle are present. The fact that the particle relaxation time $\tau^* > 0.1$ for the magnetite particles suggests that the particle transport will probably be determined by an inertial mechanism [Wood 1981], resulting in higher values of the transport flux.

Melo and Pinheiro [1988] conclude that in these experiments adhesion is the controlling process, and that the adhesion process is "more difficult" (or "slower") in the formation of pure magnetite deposits than in the formation of pure kaolin deposits. In addition in spite of the higher total concentration of particles in the kaolin/magnetite experiments, the fouling rate is lower than when only kaolin particles are present in suspension. The experimental values of the fouling resistance after 10 days confirm the relative slowness of the fouling process when magnetite particles are in suspension.

TABLE 7.5

Results with suspensions containing magnetite [Melo and Pinheiro 1988]

| Re | pH | Suspended particles | Deposit thickness $\mu m$ | $\phi_D \times 10^9$ $m^2 K / J$ | $\left(\frac{1}{\beta}\right) \times 10^{-6} s$ | $R_{f\infty} \times 10^4$ $m^2 K / W$ | $R_f \times 10^4$ $m^2 K / W$ after 10 days |
|---|---|---|---|---|---|---|---|
| | | | | Eq. 4.1 | Eq. 4.11 | Eq. 4.11 | |
| 6900 | 7.5 | Kaolin 2.2 $Kg/m^3$ | 81 | 1.19 | 0.52 | 5.0 | 4.4 |
| | | Magnetite 0.3 $kg/m^3$ | 76 | 0.79 | 0.42 | 3.9 | 3.2 |
| | | Magnetite 0.3 $kg/m^3$ + Kaolin 2.2 $kg/m^3$ | 69 | 0.75 | 0.66 | 5.0 | 3.6 |

$\phi_D$ is defined here as the deposition flux in thermal units

Newson *et al* [1988] have studied the deposition of small diameter particles (~0.2 *μm*) of haematite (*Fe₂O₃*) from a water system onto aluminium and 316 stainless steel tubes. Particle concentrations up to 210 *mg/kg* and Reynolds numbers in the range 11,000 - 140,000 were employed. These authors found a significant effect of *pH* on the extent of the deposition and also that it would appear that agglomeration of particles in suspension could also be significant.

Fig. 7.9 shows the effect of *pH* on the measured haematite deposit mass per unit area with time. In these experiments the particulate concentration was 100 *mg/kg* and the Reynolds number 11,000. The high levels of deposit mass per unit area at a *pH* of around 6 may also be compared to the data on Fig. 7.10. The *pH* controls the magnitude and sign of charges on particles and substrate, as shown in the work of Matijevic [1982]. It may be possible with systems having these characteristics that particle deposition could be limited by suitable *pH* control. It is also interesting that the work of Newson *et al* [1988] with turbulent flow conditions gave similar results to Kuo and Matijevic [1980] for laminar flow through a packed bed.

The initial deposition rate $\phi_D$ was determined for a series of experiments and divided by the particulate concentration, i.e. representing the driving force (see Equation 7.2) to give a deposition $K_D$. The mass transfer coefficient is a maximum at a *pH* of about 6.2. It would also appear that there is little difference between "fresh" and "used" haematite particles that may have experienced agglomeration.

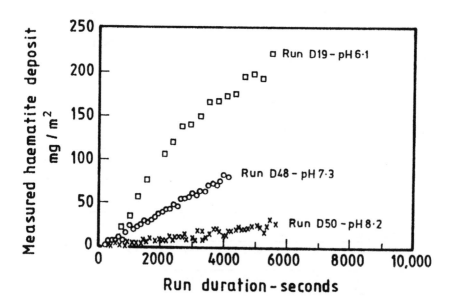

FIGURE 7.9. The effect of *pH* on haematite deposition

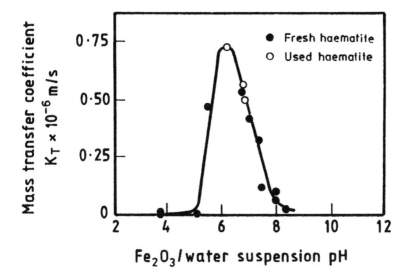

FIGURE 7.10. The effect of *pH* on experimentally determined mass transfer coefficient

At a *pH* of 6.2 $\phi_D$ = 0.073 *mg/m²s* with a corresponding value of the mass transfer coefficient $K_D$ of 0.74 x 10⁻⁶ *m/s*. The experimental value of the mass transfer coefficient was based on the deposition rate $\phi_D$ and in effect, assumed that the sticking probability *P* is unity. Using the correlation of Berger and Hau [1977] for the Sherwood number in terms of the Reynolds and Schmidt numbers, Newson *et al* [1988] calculated the mass transfer coefficient as 0.74 x 10⁻⁶ *m/s* which is precisely the derived experimental value and suggests that at *pH* = 6.2 the sticking probability *P* was indeed unity at the commencement of the experiments.

In their comprehensive studies of the haematite particle deposition Newson *et al* [1988] demonstrated using a radioactive technique, that the asymptotic fouling deposit was not governed by an equation of the form of Equation 4.1 where $\phi_D = \phi_R$, i.e. the rate of deposition = the rate of removal. For the conditions of these experiments a falling sticking probability model such as suggested by Thomas and Grigull [1974] would be applicable. An equation similar to Equation 7.35 would apply in this situation except that *P* will be time dependent and it would be necessary to determine the exact relationship before a suitable model could be devised.

The work reported earlier [Melo and Pinheiro 1988] was extended by Oliveira *et al* [1988] to examine the effects of *pH* and the chemicals used to control the *pH* on the deposition of kaolin. Fig. 7.11 presents some data. As a general conclusion increasing the *pH* above 7 results in a decrease of deposit thickness. Increasing the *pH* above 9 using *NaOH*, results in a dramatic fall in the deposit thickness.

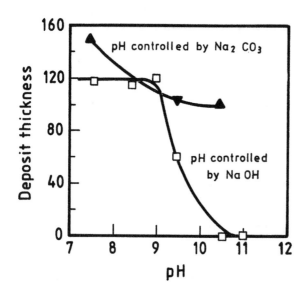

FIGURE 7.11.  The effect of *pH* on kaolin deposition

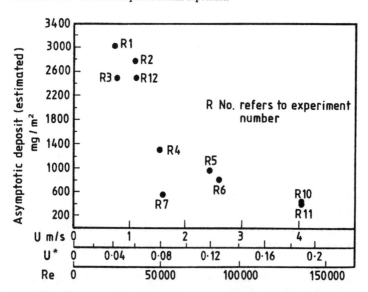

FIGURE 7.12.  The effect of velocity, friction velocity and Reynolds number on the asymptotic fouling resistance

For the magnetite/kaolin mixtures the magnetite appears to increase the mechanical strength of the deposit.

Williamson [1990] has presented data that show the dependence of the haematite deposition (i.e. 0.2 $\mu m$ particles) on velocity. Fig. 7.12 illustrates the rapid fall in asymptotic deposit mass ($mg/m^2$) as velocity is increased. Fig. 7.12 also records the corresponding values of friction velocity $u^*$ and Reynolds number. The data suggest that employing velocities > 2 $m/s$ would minimise the fouling due to particle deposition.

Oliveira *et al* [1988] also observe a reduction in deposit thickness with increased Reynolds number in an aqueous suspension of mixed magnetite and kaolin particles.

Work on ferric oxide particulate deposition onto heated stainless steel tubes was reported by Hopkins and Epstein [1974]. The research demonstrated that other variables in addition to those already discussed, influence the extent of deposition and include particulate concentration, heat flux, temperature and the presence of small residual deposits on surfaces generally considered to be clean.

Thermal fouling could not generally be detected for ferric oxide concentrations up to 750 $mg/l$ even after 7 days operation. Above this concentration falling rate deposition curves were obtained with time, but these were not reproducible till a ferric oxide suspension of 1750 $mg/l$ was exceeded, when higher concentrations produced consistently higher fouling thermal resistances.

Hopkins and Epstein [1974] also noted that deposition of ferric oxide decreased as the heat flux was raised. At extremely high heat fluxes (up to 292 $kW/m^2$) little deposit was observed on the stainless steel surface. The researchers suggested that the effects of thermophoresis could have held the particles off the surface. At high heat flux the wall temperature is correspondingly raised for given flow conditions and as wall temperature increases, the local fouling resistance falls.

### 7.7.2 Gas systems

The problem of gas side deposition is present in a wide range of industries where particulate matter is carried either intentionally or unintentionally in a gas stream. Difficulties due to fouling can be manifest in dryers, gas lines, reactors and combustion equipment as well as heat exchangers involved in gas processing. The examples cited above indicate that the temperature range over which deposition can be encountered is extremely wide, covering temperatures from ambient to well in excess of 1000°C.

The particulate matter carried forward in gas streams can also cover a wide range of materials including inorganic mineral matter, organic particles and "natural" materials like wood particles. Inorganic substances possibly with carbon particles, are common foulants in combustion systems where fossil fuels like oil and coal or waste materials like domestic refuse are burnt. Even with "clean" fuels such as natural gas, problems of fouling may arise as a result of poor combustion conditions. The deposition of organic particulate matter may arise in food processing. Problems in combustion and food processing are more fully discussed in Chapter 16.

Some early sources of data on turbulent gas side fouling together with an indication of the theoretical considerations, are to be found in Table 7.6.

An extremely valuable and detailed review of many aspects of gas side fouling has been provided by Marner [1990]. The review discusses theory, experimental data, and methods of obtaining data with well over 200 references.

TABLE 7.6

Some early sources of data on turbulent gas side fouling adapted from Bott [1988]

| Geometry | Particle size $\mu m$ | Theory | Reference |
|---|---|---|---|
| Verticle tubes | 0.8 - 3 monolayer | Stokes stopping distance | Friedlander and Johnstone [1957] |
| Smooth surfaces | 1 | Eddy diffusivity and inertial coasting | Owen [1960] |
| Pipes | Up to 50 | Eddy diffusivity and inertial coasting | Davies [1965] |
| Heat exchanger | various | Thermophoresis | Hawes and Garton [1967] |
| Smooth pipes | 0.5 - 50 | Stopping distance | Kneen and Strauss [1969] |
| Tubes | 0.3 - 1.3 | Thermophoresis | Byers and Calvert [1969] |
| Vertical tubes | Up to 30 | Effective particle eddy diffusivity and re-entrainment | Sehmel [1970] |
| Tubes | 1 - 40 | Transverse lift force | Rouhiainen and Stachiewicz [1970] |
| Channels or pipes | 0.1 - 100 | Brownian diffusion and inertia | Beal [1972] |
| Tubes | 0.8 - 125 | Eddy and particle diffusivity | Hutchinson *et al* [1971] |
| Vertical annuli | 0.65 - 5 | Brownian motion and impaction | Kitamoto and Takashima [1974] |
| Surface | 0.01 - 1000 | Combined | Gardner [1975] |

It is useful to discuss separately deposition at low temperature and high temperature since the mechanisms of deposition are different.

*Deposition at near ambient temperatures*

Because of the low heat transfer coefficient usually associated with gases, the gas side surface of heat exchangers are often extended; air blown coolers generally employ finned tubes for example. Fig. 7.13 shows how the deposit accumulates on a bank of finned tubes [Bemrose 1984].

It will be appreciated that the presence of the deposit will greatly affect the flow pattern of the air as it passes across a bank of finned tubes. The flow distribution is already complex due to the presence of the finned tubes at right angles to the flow direction. The flow patterns will of course be dependent upon the free area available for flow. Bemrose and Bott [1983] analysed the available free flow area for three different configurations. Fig. 7.14 gives the free flow area for different distances through an element as defined in Fig. 7.15. Clearly to accommodate the flow area changes a given air flow has to accelerate or decelerate as it passes through the finned tube bundle. The figures show the marked differences that can occur due to simple changes in tube layout.

Bott and Bemrose [1983] reported work on the particulate fouling of an experimental finned tube heat exchanger using calcium carbonate dust with a particle size range of 3 - 30 $\mu m$. The spirally wound finned tubes had the following geometry:

| | |
|---|---|
| Tube outside diameter | 2.54 *mm* |
| Fin outside diameter | 5.7 *mm* |
| Fin thickness | 0.41 *mm* |
| Fin density | 410 *fins/m* |
| Tube material | steel |
| Fin material | aluminium |
| Fin bonding | mechanically peened into a spiral groove in the tube surface |

Normalised data for a particular test are given in Fig. 7.16 and the friction factor *f* shows a steady increase with time although it would appear that the curve is asymptotic in character. The shape is consistent with the asymptotic fouling curve discussed in Chapter 1, i.e. the friction factor change mirrors the change in fouling resistance. On Fig. 7.16 also the change in heat transfer efficiency with time shows a steady decline in terms of the group $St\ Pr^{2/3}$.

where $St$ is the Stanton number defined as $\dfrac{\alpha}{\rho u c_p}$

and      $\alpha$ is the heat transfer coefficient

$\rho$ is the air density

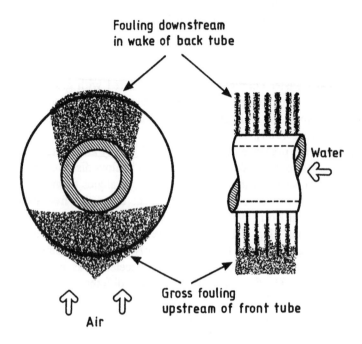

FIGURE 7.13. Deposit accumulation on finned tubes

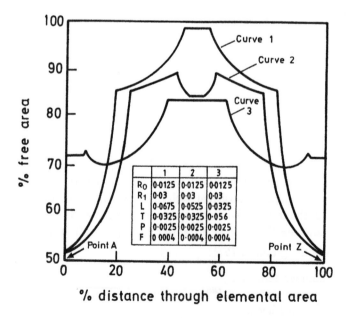

FIGURE 7.14. Free flow area as a function of distance through an element

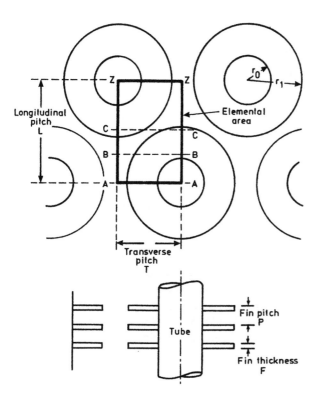

FIGURE 7.15. Distance definitions

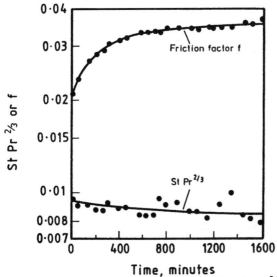

FIGURE 7.16. The change in friction factor and the group $St\,Pr^{2/3}$ with time due to particulate fouling

*u* is the air velocity

$c_p$ is the specific heat of the air

*Pr* is the Prandtl number defined as $\dfrac{c_p \eta}{\lambda}$

$\eta$ is the viscosity of the air

$\lambda$ is the thermal conductivity of air

The observation of many thousands of points led to the conclusion that the group $St\ Pr^{2/3}$ tended to fall by 10 - 20% during the test as shown on Fig. 7.16.

Bott and Bemrose [1983] translated these data into process plant terms, and demonstrated that a degree of fouling experienced in the tests would lead to a considerable increase in the fan power necessary to maintain the airflow through the exchanger. The impact is clearly seen on Fig. 7.17 where the data have been applied to a 3.1 x 3.1 *m* heat exchanger. With an air velocity of 3.7 *m/s* the fan power increases from 6 *kW* to 12 *kW* before reaching the asymptote. For such a heat exchanger in continuous service this represents at 3p per *kWh*, an annual increase in operating cost of approximately £1575. In addition there would be the cost of reduced heat transfer efficiency and periodic cleaning.

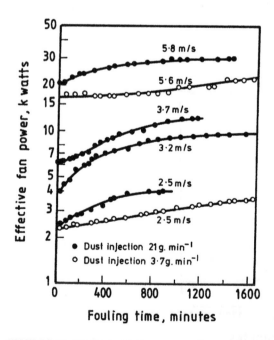

FIGURE 7.17. The increase in fan power requirement due to particulate fouling

*Deposition at high temperature*

Deposition of particles onto surfaces from high temperature gas streams, particularly from combustion gases, has an extremely complex mechanism. In Section 7.4 thermophoresis was discussed as being a contributory factor but it is difficult to separate this phenomenon from other mechanisms of deposition that may also be occurring during combustion operations. Vapour condensation, liquid droplet formation and coalescence, and chemical reactions will interact in the accumulation of deposits in combustion equipment. Under these high temperature conditions the deposition process becomes extremely complex. Rosner and Sheshadri [1981] identified the three regimes given in Table 7.1 but also suggested the likely mechanisms that will apply giving special consideration to high temperature deposition and in relation to particle size. Table 7.7 lists their observations.

Because of the complex character of deposition from flue gases the subject is addressed separately in Chapter 16.

TABLE 7.7

Characteristics of deposition and particle size (Rosner and Seshadri [1981])

| Particle size range (*nm*) | Mass transport | Deposition species | Transport mechanism |
|---|---|---|---|
| 0.1 - 1 | Vapour diffusion | Atoms and molecules | Fick, Soret, and eddy diffusion |
| $1 - 10^3$ | Vapour diffusion transition | Heavy molecules (condensate aerosols, cluster, and particles $< \mu m$) | Brownian and eddy diffusion and thermophoresis |
| $10^3 - 10^5$ | Inertial | Macroscopic particles | Inertial and eddy impaction |

REFERENCES

Adomeit, P.L. and Renz, U., 1992, Numerical simulation of the influence of temperature on particle deposition, in: Bohnet, M., Bott, T.R., Karabelas, A.J., Pilavachi, P.A., Séméria, R. and Vidil, R. eds. Fouling Mechanisms Theoretical and Practical Aspects. Editions Européennes Thermique et Industrie, Paris.

Beal, S.K., 1970, Deposition of particles in turbulent flow on channel or pipe walls. Nuclear Sci. and Engng. 40, 1 - 11.

Beal, S.K., 1972, Turbulent agglomeration of suspensions. J. Aerosol Sci. 3, 113 - 125.

Beal, S.K., 1983, Particulate fouling of heat exchangers, in: Bryers, R.W. and Cole, S.S. eds. Fouling of Heat Exchanger Surfaces. Proc. Eng. Found. Conf., White Haven.

Bemrose, C.R. and Bott, T.R., 1983, Theory and practice in gas side particulate fouling of heat exchangers, in: Bryers, R.W. and Cole, S.S. eds. Fouling of Heat Exchanger Surfaces. Proc. Eng. Found. Conf., White Haven, 257 - 275.

Berger, F.P. and Hau, K.F., 1977, Mass transfer in turbulent pipe flow measured by the electrochemical method. Int. J. Heat and Mass Trans. 20, 11, 1185 - 1194.

Biswas, P., 1988, Differential impaction of aerosols. J. Aerosol Sci. 19, 603.

Bott, T.R., 1988, Gas side fouling, in: Melo, L.F., Bott, T.R. and Bernardo, C.A. eds. Fouling Science and Technology, Kluwer Academic Publishers, Dordrecht, 191 - 203.

Bott, T.R. and Khoo, Y.K., 1967, Some separations in a liquid thermal diffusion column. Trans. Inst. Chem. Engrs. 45, T115 - 118.

Bott, T.R. and Bemrose, C.R., 1983, Particulate fouling on the gas side of finned tube heat exchangers. Trans. ASME, 105, 178 - 183.

Bowen, B.D., 1978, Fine Particle deposition in smooth channels. PhD Thesis, University of British Columbia.

Bowen, B.D., and Epstein, N, (1979) Fine particle deposition in smooth parallel plate channels. J. Colloid Interface Sci. 72, 81 - 87.

Browne, E.W. B., 1974, Deposition of particles on rough surfaces during turbulent gas flow in a pipe. Atmos. Environment 8, 8, 801.

Brusakov, V.P., 1971, Laws for the deposition of materials on heat transmitting surfaces under the action of thermoelectric effects Atom. Energiya 30, 1, 10.

Burrill, K.A., 1977, Corrosion product transport Pt. 1. Can. J. Chem. Eng. 55, 54 - 61.

Burrill, K.A., 1978, Corrosion product transport Pt. 2. Can. J. Chem. Eng. 56, 79 - 86.

Burrill, K.A., 1979, Corrosion product transport Pt. 3. Can. J. Chem. Eng. 57, 211 - 224.

Byers, R.L. and Calvert, S., 1969, Particle deposition from turbulent streams by means of thermal force. Ind. Eng. Chem. Fundamentals 8, 4, 646 - 655.

Castillo, J.L. and Rosner, D.E., 1988, Non-equilibrium theory of surface deposition from particle laden dilute condensable vapour containing steam, allowing for particle thermophoresis and vapour scavenging within the laminar boundary layer. Int. J. Multiphase Flow 14, 1, 99.

Castillo, J.L. and Rosner, D.E., 1989, Theory of surface deposition from a binary dilute vapour containing stream allowing for equilibrium condensation within the laminar boundary layer. Int. J. Multiphase Flow 15, 1, 97.

Cleaver, J.W. and Yates, B., 1973, Mechanism of detachment of colloidal particles from flat substrate in a turbulent flow. J. Colloid. Interface Sci. 44, 3, 464 - 474.

Cleaver, J.W. and Yates, B., 1975, A sub-layer model for the deposition of particles from a turbulent flow. Chem. Eng. Sci. 30, 983 - 992.

Cleaver, J.W. and Yates, B., 1976, The effect of re-entrainment on particle deposition. Chem. Eng. Sci. 31, 147 - 151.

Davies, C.N., 1965, Deposition of aerosols from turbulent flow through pipes. Proc. Roy. Soc. Series A, No. 1417, Vol. 289, 235 - 246.

Einstein, A., 1956, The theory of the Brownian movement. Dover, New York.

El-Shobokshy, M.S., Ismail, I.A., 1980, Deposition of aerosol particles from turbuelnt flow onto a rough pipe wall. Atmos. Environ. 14, 3, 297 - 304.

Epstein, N., 1988, Particulate fouling of heat transfer surfaces: Mechanisms and models, in: Melo, L.F., Bott, T.R. and Bernardo, C.A. eds. Fouling Science and Technology. Kluwer Academic Publishers, Dordrecht, 143 - 164.

Friedlander, S.K. and Johnstone, H.F., 1957, Deposition of suspended particles from turbulent gas streams. Ind. Eng. Chem. 49, 7, 1151 - 1156.

Gardner, G.C., Deposition of particles from a gas flowing parallel to a surface. Int. J. Multiphase Flow, 2, 213 - 218.

Glen, N.F. and Howarth, J.H., 1988, Model refuse incineration fouling. Second UK Nat. Heat Trans. Conf. Vol. 2, 401 - 420.

Glen, N.F., S. Flynn, Grillot, J.M. and Mercier, P., 1992, Measurement and modelling of fouling on finned tubes, in: Bohnet, M., Bott, T.R., Karabelas, A.J., Pilavachi, P.A., Séméria, R. and Vidil, R. eds.. Fouling mechanisms theoretical and practical aspects. Editions Européennes Thermique et Industrie.

Gökoglu, S. and Rosner, D.E., 1982, Correlations for predicting fouling rates with thermophoresis, viscous dissipation and transpiration cooling. Int. J. Heat and Mass Trans., 639 - 646.

Gudmundsson, J.S., 1977, Fouling of surfaces. PhD Thesis, University of Birmingham.

Gudmundsson, J.S., 1981, Particulate fouling, in: Somerscales, E.F.C. and Knudsen, J.G. eds. Fouling in Heat Transfer Equipment, 357 - 387. Hemisphere Publ. Corp., Washington.

Gudmundsson, J.S., and Bott, T.R., 1977, Particle eddy diffusivity in turbulent pipe flow. Aerosol Sci. 8, 317 - 319.

Hawes, R.I. and Garton, D.A., 1967, Heat exchanger fouling with dust suspensions, 48, 8, 143 - 145.

Hidy, G.M. and Heisler, S.L., 1979, in: Shaw, D.T. ed. Recent developments in aerosol science. Wiley Interscience, London.

Hopkins, R.M. and Epstein, N., 1974, Fouling of stainless steel tubes by a flowing suspension of ferric oxide in water. Proc. 5th Intl. Heat Trans. Conf. Tokyo, 5, 180 - 184.

Hussain, C.I., 1982, Particulate magnetite fouling from flowing suspension on simulated heat exchanger surfaces. PhD Thesis, University of Birmingham.

Hussain, C.I., Newson, I.H. and Bott, T.R., 1986, Diffusion controlled deposition of particulate matter from flowing slurries. Proc. 7th Int. Heat Trans. Conf. 5, 2 573 - 2579. Hemisphere Pub. Corp., Washington.

Kern, D.Q. and Seaton, R.E., 1959, A theoretical analysis of thermal surface fouling. Brit. Chem. Eng. 4, 5, 258 - 262.

Kitamoto, A. and Takashima, Y., 1974, Deposition rate of aerosol particles from turbulent flow through vertical annuli. Bul. Tokyo Inst. Tech. No. 121, 67 - 82.

Kneen, T. and Strauss, W., 1969, Deposition of dust from turbulent gas streams. Atmospheric Environment 3, 55 - 67.

Liang, B., Gomez, A., Castillo, J.L., Rosner, D.E., 1989, Experimental studies of nucleation phenomena within thermal boundary layers - influence of chemical vapour deposition rate processes. Chem. Eng. Comm.

Lister, D.H., 1981, Corrosion products in power generating systems, in: Somerscales, E.F.C. and Knudsen, J.G. eds. Fouling in Heat Exchanger Equipment. 135 - 200. Hemisphere Pub. Corp., Washington.

Liu, B.Y.H. and Agarwal, J.K., 1974, Experimental observations of aerosol deposition in turbulent flow. J. Aerosol Sci. 5, 145 - 155.

Marner, W.J., 1990, Progress in gas side fouling of heat transfer surfaces. Appl. Mech. Rev. 43, No. 1, 35 - 66.

Melo, L.F. and Pinheiro, J.D., 1988, Fouling by aqueous suspensions of kaolin and magnetite: hydrodynamic and surface phenomena, in: Melo, L.F., Bott, T.R. and Bernardo, C.A. eds. Fouling Science and Technology, 173 - 189, Kluwer Academic Publishers, Dordrecht.

Metzner, A.B. and Friend, W.L., 1958, theoretical analogies between heat, mass and momentum transfer. Can. J. Chem. Eng. 36, 6, 236.

Montgomery, T.L. and Corn, M., 1970, Aerosol Deposition in a Pipe with Turbulent Airflow. Aerosol Sci. 1, 185 - 213.

Newson, I.H., 1979, Studies of particulate deposition from a flowing suspension. Proc. Conf. Fouling - science or Art? Inst. Corr. Sci. and Tech. and I. Chem. E. University of Guildford.

Newson, I.H., Miller, G.A., Haynes, J.W., Bott, T.R. and Williamson, R.D., 1988, Particulate fouling: studies of deposition, removal and sticking mechanism in a haematite/water system. 2nd Nat. Heat Trans. Conf. Glasgow, I.Mech.E. 137 - 160.

Newson. I.H., Bott, T.R. and Hussain, C.I., 1981, Studies in magnetite deposition from a flowing suspension. 20th Nat. Heat. Trans. Conf. Milwaukee, Fouling in Heat Exchange Equipment, ASM HTD, 17, 73 - 81.

Oliveira, D.R., Melo, L.F. and Pinheiro, J.D., 1988, Fouling by aqueous suspensions of kaolin and magnetite. 2nd Nat. Heat Trans. Conf. Glasgow, I.Mech.E. 137 - 160.

Owen, I., El-Kady, A.A. and Cleaver, J.W., 1987, Fine particle fouling of roughened heat transfer surfaces. Proc. ASME/JSME Thermal Eng. Joint Conf., 3, 95 - 101.

Owen, P.R., 1960, Dust deposition from a turbulent air stream. Int. J. Air Pollution 3, 1 - 3, 8 - 25.

Papavergos, P.G. and Hedley, A.B., 1984, Particle deposition behaviour from turbulent flows. Chem. Eng. Res. Des. 62, 275 - 295.

Park, H.M. and Rosner, D.E., 1989a, Boundary layer coagulation effects on the size distribution of thermophoretically deposited particles. Chem. Eng. Sci. 44, 2225 - 2231.

Park, H.M. and Rosner, D.E., 1989b, Combined inertial and thermophoretic effects on particle deposition rates in highly loaded dusty gas streams. Chem. Eng. Sci. 44, 2233 - 2244.

Rodliffe, R.S. and Means, F.A., 1979, Factors Governing Particulate Corrosion Product Adhesion to Surfaces in Water Reactor Coolant Circuits. Report CEGB RD/B/N4525.

Rosner, D.E. and Tassopoulos, M., 1989, Deposition rates from polydispersed particle population of arbitrary spread. AIChE J. 35, 9, 1497 - 1508.

Rosner, D.E. and Nagorajan, R., 1987, Towards a mechanistic theory of net deposit growth from ash laden flowing combustion gases: self regulated sticking of impacting particles and deposit erosion in the presence of vapour glue. AIChE Symp. Ser. 83, 289.

Rouhiainan, P.O. and Stachiewicz, J.W., 1970, On the deposition of small particles from turbulent streams. J. Heat Trans. 92, 169 - 177.

Samms, J.A.C. and Watt, J.D., (1966), Physical mechanisms in the formation of boiler deposits. Monthly Bulletin of BCURA, 30, 7, 225.

Sehmel, G.A., 1970, Particle deposition from turbulent air flow. J. Geophys. Res. 75, 9, 1766 - 1781.

Singh, B. and Bryers, R.L., 1972, Particle deposition due to thermal force in the transition and near-continuous regimes. Ind. Eng. Chem. 11, 1, 127.

Tassopoulos, M., O'Brien, J.A. and Rosner, D.E., 1989, Simulation of microstructure/mechanism relationships in particle deposition. AIChE J, 35, 6, 967 - 980.

Thomas, D., 1973, Experimentelle untersuchung über die Ablagerung von suspendientem Magnetit bei Rohrströmungen in Dampfevzeugern und über den Einfluss der Mangetiablagerungen auf den Wörmeübergang. Dr. Ing. Dissertation Technischen Universität München.

Thomas, D. and Grigull, U., 1974, Experimental investigation of the deposition of suspended magnetite from the fluid flow in steam generating tubes. Brennst-Wärme-Kraff 26, 3, 109 - 115.

Watkinson, A.P. and Epstein, 1971, Particulate fouling of sensible heat exchangers. Proc. 4th Int. Heat Trans. Conf. Vol. 1. Paper HE 1.6, Elsevier Science Publishers BV, Amsterdam.

Williamson, R.D., 1989, The deposition of iron oxide particles on surfaces from turbulent aqueous suspension. PhD Thesis, University of Birmingham.

Wood, N.B., 1981, A simple method for the calculation of turbulent deposition to smooth and rough surfaces. J. Aerosol Sci. 12, 3, 275 - 290.

# CHAPTER 8

## Crystallisation and Scale Formation

### 8.1    INTRODUCTION

The formation of scale on heat transfer surfaces is a common phenomenon where aqueous solutions are involved, e.g. the use of "natural" waters for cooling purposes or evaporative desalination.   It involves the deposition of salts from solution, either by crystallisation on the heat transfer surface or in the bulk liquid phase.  The result can be a tenacious scale that may be difficult to remove generally referred to as "hard scale", or a softer deposit that is sometimes called a "sludge". In either example the result is a loss of heat transfer efficiency.  Unless suitable preventative measures are taken the problem can give rise to serious consequences. In combustion steam generators (boilers) for instance the presence of scale on the water side (usually tubes) can give rise to high metal temperatures that may result in metal failure.  When scale causes a rapid rise in tube wall temperature a violent rupture occurs as plastic flow conditions are reached at temperatures in the region of 975 - 1100 $K$ [Betz 1976].  The prevention of scale formation and other forms of deposit in boilers is essential for safe operation.

Hanlon *et al* [1979] commented that the potential for scale precipitation in industrial equipment is very high.    As an example he observed that for a 1 million gallon/day desalination plant under normal concentration conditions a maximum of about 1400 $kg$ of $Ca\ CO_3$ or 800 $kg\ Mg\ (OH)_2$ could be precipitated each day.  In terms of scale thickness this could represent a build up of 0.1 $mm$ per day on the total heat exchanger surface within a typical plant.  Although this may be regarded as something of an extreme example it does illustrate the magnitude of scaling or crystallisation fouling problems.

The origin of the water will have a considerable influence on the species of dissolved salts in the water.  In general, although not exclusively, the origin of water used for industrial purposes is water evaporated from land masses and oceans and subsequently precipitated from the atmosphere as rain, snow or hail due to the prevailing meteorological processes and climatic conditions.

Initial contamination of the precipitating water comes from the atmosphere.  As the droplets of water, or particles of ice fall they pass through air that contains solid particles (dust) derived from industrial operations (e.g. smoke and fumes) or natural processes (e.g. volcanic action).  Furthermore the atmosphere contains gases (e.g. $CO_2$) that are soluble.  The water reaching the surface of the earth therefore will contain dissolved substances giving it a *pH* somewhere in the range

*Fouling of Heat Exchangers*

3.5 - 7.1. It will be seen that this range tends to be acidic (low *pH*) which has given rise to the term "acid rain", so much the concern of environmentalists.

As the water flows across the land surface and into the soil and rock structures some soluble salts will be dissolved. The *pH* of the water will influence the solubilisation process. The final composition of the resulting solution will depend to a large extent on the geology of the area through which the water flows. Water that tends to remain on the land surface is generally referred to as "surface water" whereas the water that penetrates deeper into the soil and rock surfaces is generally termed "ground water".

The analysis of the water from some UK rivers is given in Table 8.1 [Kemmer 1988]. The change in composition of the River Dee from its source to a point downstream is significant and demonstrates the "pick up" by the water as it flows across rocks and soil.

TABLE 8.1

Analysis of water in some UK rivers [based on Kemmer 1988]

| River | Tees | Tyne | Dee* | Dee** | Great Ouse | Severn |
|---|---|---|---|---|---|---|
| Constituent | | | | | | |
| Calcium | 74 | 77 | 8 | 78 | 357 | 136 |
| Magnesium | 30 | 33 | 1 | 21 | 25 | 45 |
| Sodium | 15 | 20 | 10 | 39 | 58 | 30 |
| Total Electrolyte | 119 | 130 | 19 | 138 | 440 | 211 |
| Bicarbonate | 77 | 80 | 3 | 62 | 242 | 71 |
| Carbonate | 0 | 0 | 0 | 0 | 0 | 0 |
| Hydroxyl | 0 | 0 | 0 | 0 | 0 | 0 |
| Sulphate | 24 | 32 | 2 | 38 | 144 | 70 |
| Chloride | 17 | 17 | 13 | 37 | 51 | 68 |
| Nitrate | 1 | 1 | 1 | 1 | 3 | 2 |
| Alkalinity† | 77 | 80 | 3 | 62 | 242 | 71 |
| Carbon Dioxide | 3 | 3 | 5 | 3 | Nil | 3 |
| *pH* | 7.6 | 7.5 | 6.3 | 7.7 | 8.1 | 7.7 |
| Silica | 4 | - | 1.5 | 8.0 | - | 5 |
| Iron | 0.5 | 0.6 | 0.6 | 0.5 | 0.1 | Nil |

Concentrations in *mg/l*

\*      Head waters in Wales

\*\*     Downstream

†      End point based on methyl orange indicator

The composition of geothermal water is very different from river water as Table 8.2 indicates. These data were taken from two geothermal locations in Iceland [Gudmundsson and Bott 1977]. The high values of the concentrations are due to three major factors namely the relatively high temperatures and pressures encountered in the ground from which the geothermal water originated, and the long contact times. Some geothermal water is considered to be "fossil water", i.e. having been isolated in the depths of the earth for possibly millions of years. The relatively high silicon content is noteworthy.

TABLE 8.2

The analysis of geothermal water a two locations in Iceland

| Measurement | Svartsengi | Hveragerdi |
|---|---|---|
| $pH/°C$ | 7.65/20 | 9.46/23 |
| $SiO_2$ | 569.5 | 305 |
| $Na^+$ | 8,100 | 155.4 |
| $K^+$ | 1,630 | 12.9 |
| $Ca^{++}$ | 1,182 | 3.21 |
| $Mg^{++}$ | 1.3 | 0.25 |
| $Cl^-$ | 15,920 | 126.2 |
| $F^-$ | 0.1 | 2.3 |
| $SO_4^-$ | 33.3 | 54.9 |
| $CO_2^*$ | 22.7 | 119.2 |
| $H_2S^{**}$ | 0.1 | 4.74 |
| Dissolved solids | 27,856 | 777.0 |

Concentrations in *mg/l*

\*     $H_2CO_3 + HCO_3^- + CO_3^{--}$

\*\*   $H_2S + HS^- + S^{--}$

Although the emphasis of this introduction has been towards the use of "natural" water for industrial cooling purposes and the source (after purification) for boiler feed water, crystallisation fouling is not confined to these uses. Whenever a solution of whatever substance, e.g. organic or inorganic in an aqueous or non-aqueous phase, is subject to conditions that cause precipitation then there is a potential for deposition onto heat transfer surfaces.

Practical problems specifically associated with cooling water are discussed in Chapter 16.

## 8.2    THEORETICAL BACKGROUND TO CRYSTALLISATION FOULING

The discussion will concentrate on the common perception of the crystal state, i.e. the solid crystal, but it is necessary, for the sake of completion to note that it is possible for there to exist so-called "liquid crystals". The liquid crystalline state exhibits the flow properties of a liquid but at the same time, possesses some of the properties of the crystalline state. Mullin [1972] considers that the name "liquid crystal" is inappropriate because the very word "crystal" suggests some sort of lattice structure. He points out that a lattice structure is not possible in the liquid state, although some form of molecular orientation can occur with certain molecules under certain conditions. Mullin [1972] suggests that the name "anisotropic liquid" is to be preferred to "liquid crystal". From the point of view of heat exchanger technology however, this distinction is largely academic as the presence of anisotropic liquids as foulants on heat transfer surfaces in only likely to occur in exceptional circumstances. It is solid crystal formation that constitutes a potential for heat exchanger fouling.

### 8.2.1   Crystalline solids

A crystal represents a coherent rigid lattice of molecules, or arrangements of atoms or ions the juxtaposition of which are characteristic of the substance. For example crystals of sodium chloride form a simple cubic structure involving the ions $Na^+$ and $Cl^-$ (see Fig. 8.1).

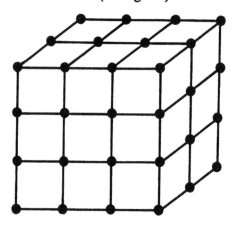

FIGURE 8.1.  The cubic structure of sodium chloride

The regular repeating "construction" of the crystal results in the crystal having a characteristic shape. As the crystal grows, smooth surfaces often referred to as "faces", appear. In general seven crystal systems exist and these are listed in Table 8.3 together with some common examples.

TABLE 8.3

The seven crystal systems [based on Mullin 1972]

| System | Other names | Examples |
|---|---|---|
| Regular | Cubic | Sodium chloride |
|  | Octahedral | Potassium chloride |
|  | Isometric | Alums |
|  | Tesseral | Diamond |
| Tetragonal | Pyramidal | Rutile |
|  | Quadratic | Zircon |
|  |  | Nickel sulphate |
| Orthorhombic | Rhombic | Potassium permanganate |
|  | Prismatic | Silver nitrate |
|  | Isoclinic | Iodine |
|  | Trimetric | $\alpha$-Sulphur |
| Monoclinic | Monosymmetric | Potassium chlorate |
|  | Clinorhombic | Sucrose |
|  | Oblique | Oxalic acid |
|  |  | $\beta$-Sulphur |
| Triclinic | Anorthic | Potassium dichromate |
|  | Asymmetric | Copper sulphate |
| Trigonal | Rhombohedral | Sodium nitrate |
|  |  | Ruby |
|  |  | Sapphire |
| Hexagonal | None | Silver Iodide |
|  |  | Graphite |
|  |  | Water (ice) |
|  |  | Potassium nitrate |

Mullin [1972] stresses that crystals of the same substance rarely look alike and that any two crystals of a given substance will in general, vary in shape and size. The reason for these differences is largely due to the restrictions imposed on the crystals during formation, giving rise to limited growth in one direction and exaggerated growth in another. These variations are called "modification of the crystal habit". Geometric limitations that pertain to heat exchangers will cause such distortions to the "free" growth of crystals on surfaces.

Fig. 8.2 shows three different habits of crystals belonging to the hexagonal system. It will be seen that although the general characteristics are similar, the final shape of the crystals is markedly different depending on the restrictions imposed during the crystal growth. At the one extreme flat plate-like crystals appear whereas at the other extreme needle-like crystals occur.

Rapid crystallisation from super-cooled melts and supersaturated solutions often results in a "fluffy" tree-like crystal formation called dendrites. The main crystal stem grows quite rapidly followed at a later stage by a slower process of primary branching. Secondary branches may grow out of the primary branches. A good example of this structure is the beautiful pattern of the snowflake.

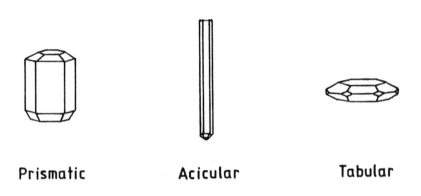

Prismatic            Acicular            Tabular

FIGURE 8.2. Possible crystal habits of a hexagonal crystal

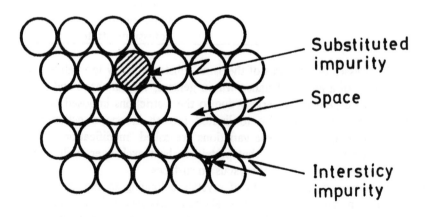

FIGURE 8.3. Common defects in a crystal

Crystals for one reason or another, are generally not perfect but contain various imperfections. A common defect is a "missing" unit from a lattice that leaves a "hole" in the structure. The vacant site may be occupied by a substitutional impurity as illustrated on Fig. 8.3. The spaces between the basic crystal units may also accommodate "foreign" bodies quite different from the crystalline material. It is possible under certain circumstances for individual ions to rearrange themselves within the lattice. It is also possible for an ion not consistent with the ions making up the general crystal structure, to be incorporated in the structure. All these modifications to the pure crystal habit are likely to lead to distortions that can have an influence on the stability of the crystalline structure.

Other defects that affect crystal structure include line and surface defects. Line imperfections result from the inability of the regular crystal structure to be maintained throughout the whole solid. The condition is illustrated on Fig. 8.4.

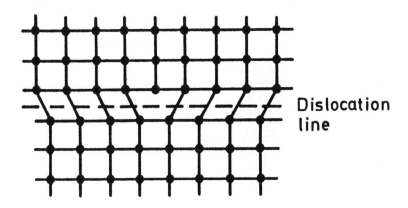

FIGURE 8.4. Line dislocation within a crystal structure

The dislocations are responsible for shearing within crystals under the influence of a suitable force.

Surface imperfections that include mismatch of crystal boundaries from the irregular formation of crystals at different points on the growing boundaries, prevents the individual crystals coming together to form a coherent whole, and give rise to structural weaknesses.

8.2.2   Solubility

A solution that is in equilibrium with the soluble solid phase is said to be saturated in respect of the solid. It is possible however to have a concentration in solution that is greater than that referred to as saturated. The term supersaturation is applied to this enhanced solubility of the solid. In the laboratory considerable supersaturation can be obtained (for a normal solubility salt) by carefully cooling a saturated solution in an impurity-free environment (i.e. pure salt in distilled water in a dust-free atmosphere). Supersaturation is a prerequisite for any crystallisation process and is therefore necessary before scaling can occur on heat transfer surfaces.

Two forms of solubility are possible. Some salts have greater solubilities as the temperature is raised; these salts are called normal solubility salts and examples include $NaCl$ and $NaNO_3$. Other salts usually termed inverse solubility salts have lower solubilities as the temperature is raised. Examples of inverse solubility salts include $CaCO_3$ and $CaSO_4$. Table 8.4 gives a more extensive list of inverse solubility salts. The relevance of these salts to the components present in river water listed in Table 8.1 is clear. The use of these waters for cooling purposes is very likely therefore, to cause fouling or scaling problems due to the presence of the inverse solubility salts.

TABLE 8.4

Some inverse solubility salts

| Calcium carbonate | $CaCO_3$ |
|---|---|
| Calcium hydroxide | $Ca(OH)_2$ |
| Calcium phosphate | $Ca_3(PO_4)_2$ |
| Calcium silicate | $CaSiO_3$ |
| Calcium sulphate | $CaSO_4$ |
| Lithium carbonate | $Li_2CO_3$ |
| Lithium sulphate | $Li_2SO_4$ |
| Magnesium hydroxide | $Mg(OH)_2$ |
| Magnesium silicate | $MgSiO_3$ |
| Sodium sulphate | $Na_2SO_4$ |

These different characteristics of temperature and solubility can have a pronounced effect on scaling depending on whether or not the saturated solution is being heated or cooled. Inverse solubility salts are likely to form a deposit if the saturated solution is being heated; the reverse is true for normal solubility salts. Problems are likely only to be encountered for sparingly soluble salts where relatively small changes of temperature have a significant effect on solubility.

Duncan and Phillips [1977] have outlined the likely effects of temperature change on the concentration of a normal salt solubility salt solution (see Fig. 8.5). At point A, the solution is undersaturated but, on cooling to point B, it becomes just saturated, i.e. point B represents the solubility at temperature $T_1$. On further cooling, the solution becomes supersaturated, but crystal nucleation does not occur until point C is reached when the temperature is $T_2$. At this point it is assumed that

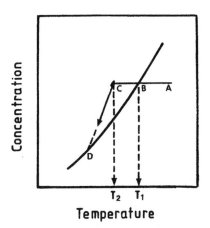

FIGURE 8.5. The cooling of a normal solubility salt solution

crystallisation occurs and the solution yields a swarm of minute crystals. Point C may be regarded as a metastable limit of saturation and the region between B and C as a metastable zone. The temperature difference $(T_1 - T_2)$ is termed the 'maximum allowable undercooling'. As crystallisation and cooling proceed, the solution concentration falls and moves in the direction of point D. In normal solubility conditions, the temperature distribution may be such that saturation can occur in the bulk solution, although in some examples saturation may occur at the heat exchanger surface.

The sequence of events for an inverse solubility salt solution are illustrated on Fig. 8.6 [Bott 1990]. A solution at A is undersaturated; as it is heated it reaches the

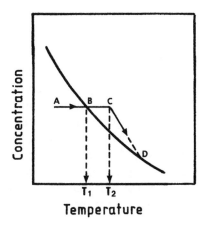

FIGURE 8.6. The heating of an inverse solubility salt solution

solubility limit point B at temperature $T_1$ and under continued heating the solution becomes supersaturated reaching point C at temperature $T_2$ where precipitation occurs. Point C is a metastable point. Further heating continues to remove material from solution and the concentration/temperature equilibrium moves towards point D.

Suitor *et al* [1976] reviewed the temperature effects for inverse solubility salts. An important factor, sometimes overlooked, is that as the deposition process continues the deposit temperature rises at constant heat flux. Under these higher temperature conditions, some process of additional crystallisation and reorientation are likely to occur [Taborek 1972]. The strength of the deposit, and the ease with which it can be removed, may be affected by these additional processes.

The solubility of sparingly soluble salts may also be affected by the presence of impurities. Kemmer [1988] gives as an example the solubility of $CaCO_3$ in the presence of magnesium. If magnesium is precipitated along with $CaCO_3$ the residual concentration of calcium in solution may be increased. The inclusion of other impurities, such as strontium, has also been shown to increase the solubility of $CaCO_3$. The use of empirical data from process plant therefore, has to be treated with caution since the presence of impurities may have a significant effect on the fouling propensity under certain circumstances, and will depend on location.

## 8.3    THE MECHANISM OF CRYSTALLISATION

As discussed in Section 8.2 the attainment of supersaturation is essential for crystallisation to occur. The extent of the supersaturation will in general, determine the rate of the crystallisation or deposition process. Three sequential stages in this process may be visualised:
>    Supersaturation occurs
>    Formation of crystal nuclei and crystallites
>    Growth of crystals

The temperature and flow conditions within the heat exchanger will determine the location at which these various stages occur. For instance the supersaturation and crystallite formation may occur in the bulk fluid with the growing crystals moving towards the wall to form the deposit. The movement of foulant will under these circumstances, follow the processes described in Chapter 7 for particulate deposition. It is possible that due to the level of turbulence within the system, that some (or possibly many) of the crystallites formed are swept into regions where the solution is not supersaturated. Under these conditions the particles will redissolve. On the other hand crystallisation may occur near or at the heat transfer surface. The presence of nucleation sites on a solid surface may encourage the formation of scale on the surface. Under these circumstances the process is largely governed by the mechanics of the crystallisation process.

## 8.3.1 Supersaturation

As already described it is posible for supersaturation to occur if a normal solubility salt is cooled or an inverse solubility salt is heated. Apart from temperature there are other conditions that could lead to supersaturation and hence crystallisation, and may include:

1. Evaporation of water from a solution of salts increases the concentration of the dissolved solids (this occurs for instance in a cooling tower). If the process is carried on for sufficient time it is possible to reach saturation and eventually supersaturation concentrations. If a hot surface is in contact with a solution near saturation, vapourisation of water could lead to direct deposition on the surface.

2. Mixing of solutions can lead to supersaturation conditions. The addition of a soluble salt to a saturated aqueous solution of another salt will usually result in precipitation if the two salts have have a common ion. The so-called "common ion effect" can be explained in terms of the solubility product principle. In this concept the product of the concentration of the ions at saturation is a constant, i.e.

$$\left[M^+\right]_s\left[R^-\right]_s = K_s^1$$

where $M^+$ represents the cation

$R^-$ represents the anion

and $K_s^1$ is the concentration solubility product

If for instance, a solution of $CaCl_2$ is mixed with a solution of $Na_2CO_3$ it is possible for the solubility of $CaCO_3$ to be exceeded and precipitation will occur if the $CaCO_3$ in solution is supersaturated. Again if $\left[Ca^{2+}\right]$ in a solution of $Ca^{2+}$ and $CO_3^{2-}$ is increased then $\left[CO_3^{2-}\right]$ must decrease in accordance with the constant solubility product. In other words the common ion effect acts to depress the saturation concentration of one of the ions forming the precipitate e.g. the excess $Ca^{2+}$ reduces the saturation concentration of $CO_3^{2-}$.

3. The mixing of saturated or near saturated, solutions may also result in supersaturation. Figures 8.7 and 8.8 respectively represent the change in solubility with temperature of a normal solubility salt and a salt displaying inverse solubility respectively. Mixing of solutions represented by points 1 and 2 on either curve might lead to a solution represented by point 3. It will be seen that the mixing results in supersaturation conditions. Had the curves been

convex rather than concave then supersaturation would not have been produced.

4. Gas/liquid equilibria and chemical reactions may produce supersaturation.

If $CO_2$ is dissolved from atmospheric air say in the cooling tower of a cooling water system, equilibrium conditions can be described by:

$$CO_2 \rightleftharpoons CO_2$$
gas    dissolved                                                    (8.1)

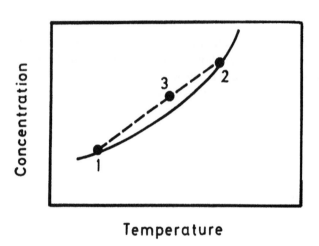

FIGURE 8.7. Normal solubility saturation concentration with temperature

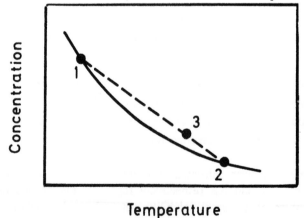

FIGURE 8.8. Inverse solubility saturation concentration with temperature

The presence of the $CO_2$ in solution can give rise to further equilibria that could influence the saturation solubility of related cations, e.g. $Ca^{2+}$.

The following equilibria are possible:

$$H_2O + CO_2 \;\rightleftharpoons\; H_2CO_3 \tag{8.2}$$

$$H_2CO_3 \;\rightleftharpoons\; H^+ + HCO_3^- \tag{8.3}$$

$$HCO_3^- \;\rightleftharpoons\; H^+ + CO_3^{2-} \tag{8.4}$$

The presence of the hydrogen ions suggests that these equilibria are sensitive to *pH*. The quantity of $CO_2$ in solution will also be dependent on the $CO_2$ partial pressure in the air in contact with the water. Figure 8.9 (based on graphs presented by Kemmer [1988] provides data showing the distribution of $CO_2$ related ions and $CO_2$ gas in solution as a function of *pH*.

Combinations of effects within a given system may be responsible for supersaturation when temperature changes are absent, where it might be assumed that supersaturation was unlikely and therefore precipitation would not occur.

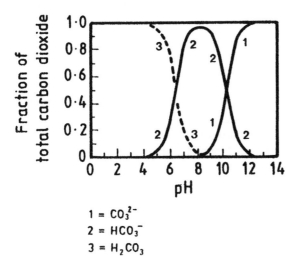

1 = $CO_3^{2-}$
2 = $HCO_3^-$
3 = $H_2CO_3$

FIGURE 8.9. The effect of changing *pH* on the distribution of $CO_2$ related ions and $CO_2$ gas in solution

### 8.3.2   Formation of nuclei and crystallites

Before crystals can grow it is necessary for minute solid particles to be present that can "seed" the crystallisation process. These solids may be foreign bodies or pieces of scale eroded from deposit already present in the system, or spontaneously formed crystallites.   These alternatives give rise to two different definitions depending on the underlying phenomenon.

Where the presence of solid seeds creates the condition for crystallisation to occur, the nucleation is called heterogeneous. Spontaneous nucleation is generally referred to as homogeneous nucleation. An exact understanding of homogeneous nucleation has not yet been achieved, but it is likely to depend on the establishment of a critical cluster of molecules that has some degree of stability, to enable the crystal to be created and to continue to grow. Many reported examples of homogeneous nucleation are found on careful examination, to have been induced in some way [Mullin 1972]. Homogeneous nucleation is unlikely to be responsible for scale formation in industrial heat exchangers since homogeneous nucleation is usually associated with pure solutions in the absence of particles or crystals (see Fig. 8.10). Nucleation sites on surfaces and the presence of particles, e.g. corrosion products or other impurities let alone crystals, are much more likely to create nuclei that result in crystallisation in equipment.

These two mechanisms are usually responsible for what is called primary nucleation. Once crystals appear in the supersaturated solution further crystallites are formed in the vicinity and this establishes the macro-crystallisation process, generally called secondary nucleation. The relationship between the various forms of nucleation are shown on Fig. 8.10.

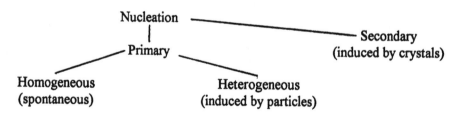

FIGURE 8.10.  Nucleation of crystallisation

### 8.3.3   Crystal growth

Once stable nuclei have been formed within the supersaturated solution they begin to grow into crystals of finite size. Mullin [1972] draws attention to three theories that have been used to explain the process of crystal growth. They include surface energy effects, the presence of an adsorption layer and theories based on diffusion.

In the surface energy theory it is assumed that the shape of a growing crystal is that which has a minimum surface energy. An alternative theory suggests that crystal growth is a discontinuous process, taking place by adsorption where layer upon layer on the crystal surface is responsible for growth. The diffusion concept involves the assumption that material is deposited on the crystal surface at a rate proportional to the difference in concentration between the point of deposition and the bulk solution. The diffusion theory is likely to appeal to chemical engineers since it is a direct application of mass transfer concepts. It may also be imagined that the mass transfer process is a prerequisite for any mechanism that involves further incorporation of material into a growing crystal. It may also be true that the mass transfer process is the rate determining step, particularly where crystallisation is taking place on a surface and an appreciable boundary layer exists between the solid surface and the flowing solution (as might be so in a heat exchanger through which an aqueous solution is flowing). On the other hand, where crystals are forming away from the surface, particularly in regions of high turbulence the resistance to mass transfer to the surface of the crystal may be negligible. Under these circumstances the rate of growth may be controlled by the rate of incorporation of new material into the crystal.

The growth of crystals in terms of diffusion theory can be described by a similar equation to Equation 7.1 used to quantify the mass transfer of particles. The required equation will be of the form:

$$\phi_D = K_t(c_b - c_e) \tag{8.5}$$

where  $\phi_D$ is the mass of solid deposited in unit time

$c_b$ and $c_e$ are the bulk concentration in the bulk (supersaturated) and the equilibrium saturation concentration at the face of the growing crystal

$K_t$ is the mass transfer coefficient

An alternative way of writing Equation 8.5 is in the form

$$\phi_D = \frac{D}{\gamma} A(c_b - c_e) \tag{8.6}$$

where  $D$ is the coefficient of diffusion

$A$ is the area of the crystal receiving the deposit

$\gamma$ is the distance over which the diffusion takes place

In a turbulent flowing system $\gamma$ may be regarded as the thickness of the boundary layer, it being assumed that the concentration at the edge of the boundary layer is the bulk concentration $c_b$.

A weakness of this simple theory is the assumption that once crystal forming material arrives at the liquid/crystal interface it is immediately incorporated into the existing crystal lattice. It is apparent however, that a finite time is likely to be required to enable this process to be carried out, i.e. there is a resistance to the assimilation of new material. This resistance may be expressed in terms of a "rate of reaction" for the solute molecules to arrange themselves into a crystal lattice. If the concentration of solute in solution at the crystal surface/solution interface is $c_i$, Equation 8.5 for diffusion may be written as:

$$\frac{dm}{dt} = K_D A(c_b - c_i) \tag{8.7}$$

where  $m$ is the mass deposition

> $c_i$ is the concentration at the interface

and      $K_D$ is the mass transfer coefficient for diffusion

For the reaction (assuming a first order reaction)

$$\frac{dm}{dt} = k_r A(c_i - c_e) \tag{8.8}$$

where  $k_r$ is the rate constant

In order to simplify the analysis (i.e. avoiding the need to know $c_i$) it is possible to combine Equations 8.7 and 8.8 with the generalised form

$$\frac{dm}{dt} = K_G A(c_b - c_e)^n \tag{8.9}$$

where  $K_G$ is an overall crystal growth coefficient.

As with chemical reaction theory $n$ is usually referred to as the "order" of the overall crystal growth process. Mullin [1972] emphasises that $n$ in crystal dynamics unlike chemical reaction theory, has no fundamental significance.

The magnitude of $K_G$ very much depends on the relative magnitudes of $k_r$ and $K_D$. For instance if $k_r$ is large, i.e. the rate of incorporation of solute into the crystal is high, then

$$K_G \approx K_D \tag{8.10}$$

On the other hand if there is a high resistance to the assimilation of solute into the crystal lattice it may be assumed that

$$K_G \approx k_r \tag{8.11}$$

Nancollas [1983] concludes that the evidence available for sparingly soluble electrolytes, suggests the rate of crystallisation is controlled by the reaction kinetics, rather than the diffusion model. Activation energies for crystal growth are normally considerably higher than those to be expected for simple bulk diffusion. Nancollas [1983] also points out that the rate of crystallisation is normally little affected by changes in the fluid dynamics at the crystal surface; again suggesting that the process is controlled by a mechanism other than diffusion, that is generally sensitive to the flow conditions. He states that for a surface controlled deposition rate for a salt with the generalised formula $M_aX_b$, the rate of crystallisation is given by the following equation:

$$Rate = \frac{d(M_aX_b)}{dt} = -k_r S K_{so}^{n/v} \sigma^n \tag{8.12}$$

where $K_{so}$ is the solubility product

$n$ is the order of reaction

$v = a + b$

$S$ is some function of the surface area

and $\quad \sigma = \left[ (M^{m+})^a (X^{x-})^b \right]^{\frac{1}{v}} - K_{so}^{\frac{1}{v}} \tag{8.13}$

Concentrations or activities of the lattice ions may be used in these equations.

For a number of the common scale forming species such as calcium sulphate and carbonate, the rate of crystallisation follows a parabolic relationship with supersaturation, and the value of the effective order of reaction is 2. An explanation may involve the dehydration of lattice cations at the crystal surface. For other examples when the value of $n > 2$, i.e. where the growth rate varies more strongly with supersaturation, the explanation may involve the interaction between several nuclei and spread of growth [Nielson 1964, O'Hara and Reid 1973].

Nancollas [1983] considers it important to establish not only the dependence of the rate of crystal growth on supersaturation, but also the influence of the ratio of solid to solution, the dynamics of stirring (agitation) and the temperature. He states that for a surface controlled process the activation energy would be of the order of 40 $J$ $mol^{-1}$ as compared to 12 $J$ $mol^{-1}$ for a process controlled by bulk transport.

These relatively simple concepts are often the basis on which models for scaling or crystallisation fouling are formulated.

The analysis of the growth of crystals however, is likely to be much more complex than that described above. Mullin [1972] draws attention to seven processes taking place simultaneously for an ionising solute crystallising from aqueous solution. They include:

1. Bulk diffusion of solvated ions through the diffusion boundary layer.

2. Bulk diffusion of solvated ions through the adsorption layer.

3. Surface diffusion of solvated or unsolvated ions.

4. Partial or total desolvation of ions.

5. Integration of ions into the lattice.

6. Counter-diffusion of released water through the adsorption layer.

7. Counter-diffusion of water through the boundary layer.

It would be extremely difficult to allow for all these processes in a usable expression for crystallisation in a "pure" system. For industrial operations involving fouling, it would be an impossible task to incorporate all these steps into a model, largely because of the large number of unknowns that would be difficult to quantify. Nevertheless the recognition of these potential interacting processes does help to provide an understanding of the complex scaling process.

## 8.4    MODELS OF CRYSTALLISATION FOULING

The previous section has highlighted the complex mechanisms involved in the deposition process associated with crystallisation from solution. For this reason adequate mathematical models for describing the process are difficult to formulate. Reviews of fouling by crystallisation have been made by Hasson [1981] and Bott [1988] that include some basic models.

In order to avoid the problems of taking into account the details of the mechanism by which new material may be incorporated into the crystal lattice it is possible to "lump" these effects together in terms of a "chemical reaction". The rate of reaction under these circumstances will be dependent upon the concentration distribution of ions or crystallising species, in the region of the solid/liquid interface. In general terms the relationship will have the following dependence:

$\phi_D$ is some function of $Ak_r$ (8.14)

where $A$ is the area which is accepting crystal growth

$k_r$ is the rate constant of the "reaction"

$k_r$ may be expressed in the Arrhenius form as

$$k_r = A_1 e^{-\frac{E}{RT}}$$ (8.15)

where $A_1$ is a constant

$E$ the activation energy

$R$ the universal gas constant

$T$ is the absolute temperature

The difficulty of course with this approach is assigning values to $A_1$ and $E$, however it does provide a basis for simple models.

Non-boiling conditions represent the simplest crystallisation fouling process and may be based on the Kern and Seaton [1959] model. Hasson [1981] presents an expression for asymptotic fouling resistance

$$R_{f\infty} = \frac{dm_f}{dt} \cdot \frac{1}{A\rho_f \lambda_s} \cdot \frac{F_A}{F}$$ (8.16)

where $\rho_f$ is the density of the deposit

$F_A$ is some sort of adhesion parameter, i.e. the greater $F_A$ the higher the tenacity of the crystals to stick together on the surface with a lower removal rate

$F$ is the shear force

if $\quad t_c = \frac{F_A}{F}$ (8.17)

$$R_{f\infty} = \frac{dm_f}{dt} \cdot \frac{1}{A\rho_f \lambda_s} \cdot t_c$$ (8.18)

The fouling resistance at any particular time is given by the equation

$$R_f = R_{f\infty}\left(1 - e^{\frac{t-t_D}{t_c}}\right)$$
(8.19)

$t_D$ is the induction time, i.e. the period that elapses before crystallisation commences. For most industrial fouling conditions $t_D$ is negligible so that Equation 8.19 simplifies to

$$R_f = R_{f\infty}\left(1 - e^{\frac{t}{t_c}}\right)$$
(8.20)

By using curve fitting techniques it is possible to derive a value for $t_c$ that will give a good indication of tenacity.

Although this approach goes some way to a model of crystallisation fouling it ignores the important question of the supersaturation mentioned earlier in this chapter.   The basic equation Equation 4.11, has been used to explain the phenomenon of crystallisation fouling, for example by Taborek *et al* [1972] who gave

$$R_f = \frac{K_1}{K_2} e^{\frac{-E}{RT_s}} (1 - e^{-k_2 t})$$
(8.21)

where  $K_1 = C_1 P_d \Omega^n$

and     $K_2 = \dfrac{C_2 \tau \lambda_s}{\Psi}$

and     $C_1$ and $C_2$ are constants

n is an exponent

$P_d$ is the probability of scale deposition

$\lambda_s$ is the thermal conductivity of the scale

$\Omega$ is a water quality factor that, in conjunction with $T_s$, provides a measure of supersaturation

$\tau$ is the shear stress

$\Psi$ is a scale strength factor

For very large values of $t$ and constant operating conditions of water quality, flow velocity and surface temperature, Equation 8.21 can be used to calculate the asymptotic fouling resistance, i.e.

$$R_{f\infty} = \frac{K_1}{K_2} e^{\frac{-E}{RT_s}} \tag{8.22}$$

For constant conditions therefore, the values of $R_{f\infty}$ is an exponential function of temperature of the heat transfer surface.

Various aspects of Equations 8.21 and 8.22 have been studied and the basis verified [Morse and Knudsen [1977], Story and Knudsen [1978], Lee and Knudsen [1979], Coates and Knudsen [1980]]. Data for use with Equation 8.21 have been produced by HTRI (Heat Transfer Research Incorporated) but this is not generally available.

Hasson [1981] quotes the HTRI model simplified to the form

$$\frac{dm_f}{dt} = K' AP \; \Omega \, e^{\frac{-E}{RT_s}} \tag{8.23}$$

where  $m_f$ is the mass of deposit

$K'$ is a constant

$A$ is the area on which the crystallisation is occurring

$P$ is the sticking probability

$T_s$ is the surface temperature

$\Omega$ is some empirical parameter characterising water composition and the supersaturation driving force.

The model is empirical so that any published data that uses this approach is only likely to be applicable to the conditions under which the data were obtained.

Knudsen and Story [1978] have obtained the following equation for $CaCO_3$ deposition from water flowing at 1 $m/s$

$$R_{f\infty} = 3 \times 10^{11} e^{-\frac{42,0000}{RT_s}} \tag{8.24}$$

As the equation given by Knudsen and Story [1978] is empirical the units of $R_{f\infty}$ are $\dfrac{ft \cdot h \cdot {}^\circ F}{BTU}$

Other data quoted by Hasson [1981] for an aqueous system containing $CaCO_3$ $Mg(OH)_2$ and $SiO_2$ may be correlated in the form

$$R_{foo} = 2.5 \times 10^{12} e^{-\frac{44,000}{RT_s}}$$
(8.25)

In this particular example the deposit consisted of $SiO_2$ and $MgO$ in the ratio 1.5 with little $CaCO_3$.

Müller-Steinhagen and Branch [1988] discuss a model originating with Hanson [1969, 1975 and 1981] and modified by Watkinson [1983]. The model considers the diffusion of ionic species from the bulk to the heat transfer surface with chemical reaction taking place at the surface. In terms of $CaCO_3$ deposition (see Section 8.3.1 for the basic chemistry) the development of the equations may be summarised in the following way:

Assuming that there is no removal of deposit taking place the mass flux due to scaling is:

$$= \frac{K_t[Ca^{2+}]((1+4ac/b)^{0.5} - 1)^b}{2a}$$
(8.26)

where $K_t$ is the mass transfer coefficient

$$a = \frac{1 - 4K_2 k_r [Ca^{2+}]}{K_t K_1}$$
(8.27)

$K_1$ and $K_2$ are first and second order dissociation constants.

$k_r$ is reaction constant

$$b = \frac{[CO_2]}{[Ca^{2+}]} + \frac{4K_2 k_r [HCO_3^-]}{K_1 K_t} + \frac{K_{sp} k_r}{K_t [Ca^{2+}]}$$
(8.28)

$K_{sp}$    is the molar solubility product (see Section 8.3.1)

$$c = \frac{K_2 k_r [HCO_3^-]^2}{K_1 K_t [Ca^{2+}]} - \frac{K_{sp} k_r [CO_2]}{K_t [Ca^{2+}]}$$
(8.29)

It is possible to use further equations to determine the constants $k_r$, $K_{sp}$ and $K_1$ and $K_2$, i.e.

$$\ln k_r = 41.04 - \frac{10,417}{T}$$
(8.30)

$$-\log K_{sp} = 0.01183(T - 273.2) + 8.03 \tag{8.31}$$

$$-\log K_1 = \frac{17052}{T} + 215.21 \log T - 0.12675T - 545.56 \tag{8.32}$$

$$-\log K_2 = \frac{2902.39}{T} + 0.023797 - 6.498 \tag{8.33}$$

Müller-Steinhagen and Branch [1988] suggest that the mass transfer coefficient $K_t$ can be calculated from an equation of the form.

$$K_t = 0.023 \, Re^{0.85} \, Sc^{0.33} \, d / d_e \tag{8.34}$$

where $Re$ and $Sc$ are the Reynolds and Schmidt numbers respectively

$d$ and $d_e$ are the outside diameter of tube and the equivalent diameter respectively.

The fouling rate can be calculated from the deposition flux, i.e.

$$\frac{dR_f}{dt} = \frac{dm}{dt} \cdot \frac{1}{(\rho\lambda)_{scale}} \tag{8.35}$$

An average value for $(\rho\lambda)_{scale}$ was found by Watkinson [1983] to be 5000 $KgW/m^4$. Assuming that this mean value can be applied and knowing the *pH*, the temperature, the heat transfer area and the value of $[Ca^{2+}]$, the terms $a,b$ and $c$ may be calculated from Equations 8.27 - 8.29.

Müller-Steinhagen and Branch [1988] conclude that provided the fouling rates are below $10^{-6}$ $m^2$ $K/kJ$ and the fluid velocity is below 0.8 $m/s$ the scaling rate can therefore be estimated from Equation 8.26. Above this velocity the equation will be modified by an unknown removal rate.

The complex crystallisation and scaling phenomena are dependent on a number of interacting variables. The models presented show some limited agreement with laboratory measurements, but it does not appear that these models have been validated in relation to industrial heat exchangers. Furthermore some of the physical property and kinetic data necessary for use in the equations, are not known even for pure systems let alone for industrial conditions where the presence of impurities can influence the chemistry and the tenacity of the scale formed.

Not least of the many difficulties in the modelling of crystallisation fouling is the generally unknown extent of the induction or initiation period. The problem is illustrated by data published by Ritter [1981]. Table 8.5 gives the ranges of experimental induction times he obtained for $CaSO_4$ and $Li_2SO_4$ precipitation. The work reported by Ritter [1981], also suggests that there is a straight line relationship between the fouling resistance due to scale formation and time. The values of the appropriate fouling rates are given in Table 8.5 together with the ranges of some of the operating variables. It is possible however, that had the tests been carried out for a much longer time, the more familiar asymptotic curves might have been obtained.

TABLE 8.5

$CaSO_4$ and $Li_2SO_4$ scale formation [Ritter 1981]

|  | $CaSO_4$ | $Li_2SO_4$ |
|---|---|---|
| Induction period $h$ | 0.15 - 50 | 0.15 - 612 |
| Fouling rate $\dfrac{m^2 K}{Ws}$ | $(0 - 1.4)\ 10^{-7}$ | $(0 - 6.5)\ 10^{-7}$ |
| Reynolds number | $(1.7 - 16)\ 10^4$ | $(0.34 - 6.1)\ 10^4$ |
| Circulating solution concentration wt % | 0.173 - 0.211 | 23.4 - 25.3 |
| Surface temperature °C | 93 - 141 | 38 - 138 |

Bott and Gudmundsson [1978] demonstrated with geothermal water that the fouling trend appeared to be asymptotic. They noted from their experimental work that there did not appear to be an induction period. Indeed the estimates of heat transfer that were made suggested that for an initial period of 700 hours, with silica laden water passing through tubes at Reynolds numbers in the range 23000 - 44000, the fouling resistance appeared to be negative. The reason for this anomaly was attributed to the roughness of the silica deposit which enhanced heat transfer in comparison with the smooth tube correlations used to estimate the heat transfer. After the tests the deposits were found to be rippled as illustrated by the section of fouled tube shown in Fig. 8.11. The increase in heat transfer coefficient was attributed to the formation of ripples on the surface of the deposit causing increased turbulence near the surface. Imdakm and Knudsen [1984] also noted asymptotic fouling curves for forced circulation of $CaCO_3$ solutions under boiling conditions.

Fig. 8.12 is an electron microscope photograph of aragnonite crystals ($CaCO_3$) deposited from a flowing stream. It is possible to recognise a few smaller crystals of calcite (another crystalline form of $CaCO_3$) in among the larger crystals. The complexity of the solid surface presented to the flowing liquid is apparent.

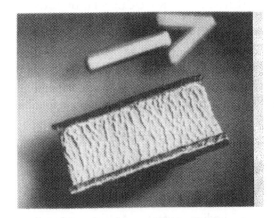

FIGURE 8.11.  Rippled silica deposit (the arrow indicates the direction of flow)

FIGURE 8.12.  Aragonite crystals on a surface (courtesy of Dr. N. Andritsos, CPERI)

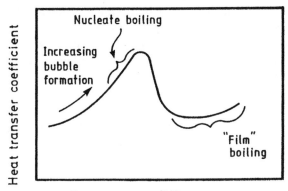

FIGURE 8.13.  The change of boiling heat transfer coefficient with temperature difference

8.5    CRYSTALLISATION AND SCALING UNDER NUCLEATE BOILING
       CONDITIONS

The previous discussion in this chapter has largely centred on crystallisation
fouling under sensible heat transfer, but heat transfer under boiling conditions
occurs in equipment such as steam boilers, evaporators for the concentration of
solutions prior to crystallisation (e.g. sugar or table salt manufacture) or for the
desalination of brackish or sea water.

Fig. 8.13 shows how the boiling heat transfer coefficient changes as the
temperature difference between the heat transfer surface and the boiling liquid is
increased. It will be seen that as the temperature difference is gradually increased
there is a corresponding increase in the heat transfer coefficient, till a peak value is
obtained. The increasing heat transfer coefficient is due to the stirring action of the
bubbles of vapour (steam in the case of boiling water). As the temperature
difference is increased the movement of the bubbles becomes more violent as they
escape into the bulk liquid. The formation of discrete bubbles is generally aided by
nucleation sites, i.e. associated with the roughness of the heat transfer surface,
hence the phenomenon is called nucleate boiling. Eventually the heat flux is such
that the bubbles cannot escape fast enough and a blanket of vapour forms on the
heat transfer surface. Because of the low thermal conductivity of the vapour, the
presence of the vapour layer represents an additional resistance to heat transfer (in
a sense the surface is "fouled" with vapour) the heat transfer coefficient falls from
the peak value as demonstrated by Fig. 8.13. The phenomenon is known as "film
boiling". Further increases in temperature difference accentuate the problem till
radiation heat transfer across the vapour layer offsets some of the thermal
resistance of the vapour.

Because of the high heat transfer rates associated with nucleate boiling, i.e. an
efficient use of the available heat transfer surface, this is the preferred condition in
evaporators and similar equipment and may require values of temperature
difference to be as low as 20°C. Nucleate boiling can bring with it severe problems
of fouling and often as a result, high cleaning costs because the deposits are
difficult to remove. Furthermore Jamialahmadi and Müller-Steinhagen [1991] in a
review of the problem of scale formation under nucleate boiling conditions, suggest
that excessive corrosion of the heat transfer surface can occur under these
conditions.

Port and Herro [1991] have given some data on the qualities of some
components of water-formed deposits that are generally mineral-like in character.
Some of these are reported in Table 8.6.

Steam raising boilers operate at elevated temperatures and pressures. The
increased resistance to heat transfer as a result of deposit or scale formation on the
heat transfer surfaces can lead to metal failure. Fig. 8.14 shows how the yield
stress of carbon steel changes with temperature. It is to be noted how the yield
stress falls rapidly once the metal temperature has reached about 350°C. If the

fouling resistance is high due to scale formation, for a given heat flux the wall temperature could rise and lead to catastrophic metal failure under the prevailing pressure conditions.

TABLE 8.6

Components of water-formed deposits

| Mineral | Chemical formula | Nature of deposit |
|---------|-----------------|-------------------|
| Acmite | $Na_2O.Fe_2O_3.4SiO_2$ | Hard adherent |
| $\alpha$ quartz | $SiO_2$ | Hard adherent |
| Amphibole | $MgO.SiO_2$ | Adherent binder |
| Analcite | $Na_2O.Al_2O_3.4SiO_2.2H_2O$ | Hard adherent |
| Anhydrite | $CaSO_4$ | Hard adherent |
| Araganite | $CaCO_3$ | Hard adherent |
| Bracite | $Mg(OH)_2$ | Flocculant |

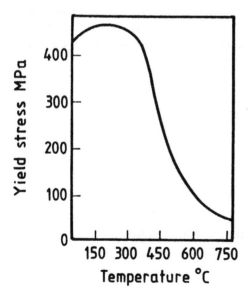

FIGURE 8.14. The variation of yield stress of carbon steel with temperature

According to Mikic and Rohsenow [1969] three mechanisms of heat transfer occur under boiling conditions:

1. Natural convection in the liquid.

2. Conduction through a thin layer of liquid between the bubble and the heat transfer surface.

3. Heat transfer from the superheated liquid boundary layer to the vapour bubble.

Fig. 8.14 schematically shows how deposits can form on a heat transfer surface against which boiling is taking place. As bubbles are generated at the solid surface a triple interface is produced between the solid heat transfer surface, the liquid water and the steam. The triple interface and the bubble dynamics have a profound effect on the deposit morphology. Fig. 8.15 shows a growing bubble in contact with a surface just about to break free from that surface. The bubble formation is initiated at a nucleation site. It may be assumed that the evaporation at the base of the bubble (e.g. at the triple interface), causes supersaturation conditions and hence the possibility of precipitation. Hospeti and Mesler [1965] however concluded from the relative distribution of calcium sulphate deposits beneath bubbles in a nucleate boiling solution, that more evaporation took place beneath the bubble rather than at the triple interface. After the release of the bubble liquid will flow towards the surface to replace the volume originally occupied by the bubble. Depending on the prevailing conditions of temperature and dissolved solid concentration, some of the precipitate may redissolve. If deposits are formed during evaporation, the conditions are unlikely to be such that all the deposit is taken back into solution and a residue will remain.

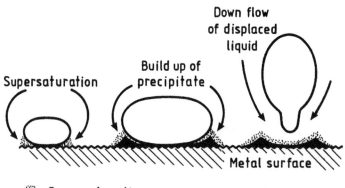

FIGURE 8.15. Bubble growth and scale formation

The sequence is likely to be repeated over a period of time so that gradually the deposit layer builds up on the surface. The morphology of the deposits supports the sequence of events leading to bubble release from the surface. Work by Palethorpe and Bridgwater [1988] clearly shows the formation of rings of a precipitate appearing like "footprints" of the original location of the bubble. The surface roughness appeared to be important in providing suitable nucleation sites for bubble formation.

Freeborn and Lewis [1962] identified five different patterns of scale appearance brought about by the conditions at the triple interface and include:

1. Plain rings of deposit developed at the edge of the bubble that grows in size till the buoyancy forces acting on the bubble are sufficient to enable it to escape from the surface. Under these conditions there is a "clean" breakaway of the bubble from the surface and it travels away from the solid surface.

2. Ladder-like formations occur where the bubble migrates across the surface as it develops. A ring forms initially as the bubble is created and as the bubble moves a short distance (< the bubble diameter) becoming stationary again, the process is repeated till the bubble is released from the surface. As a result of the bubble movement there appears a sequence of intersecting rings with a ladder-like appearance.

3. Continuous bubble migration across the surface will lead to V-tracks till the buoyancy forces allow the bubble to escape from the surface.

4. Contangential ring deposits form as the result of the development of a bubble larger than average so that after detachment smaller bubbles can form within the original ring. As these bubbles grow and detach themselves smaller rings are formed.

5. The converse of contangential rings is concentric rigs. The phenomenon allows small rings to form first with larger rings developing subsequently, rather like the concentric rings form when a stone is dropped into a placid pond.

As the scale continues to develop on the surface the conditions will be modified. For instance, if the deposit layer is porous or contains fissures, it is possible for steam to be generated within the deposit, sometimes referred to as wick boiling, although the phenomenon may be more associated with corrosion deposits. Macbeth [1971] observed steam issuing from cracks in a deposit which he termed "steam chimneys". A schematic diagram is presented as Fig. 8.16. Such a condition would give rise to a different pattern of scale formation.

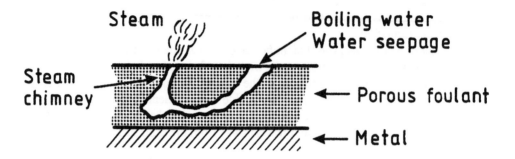

FIGURE 8.16. Wick boiling and steam chimneys

### 8.5.1 Changes in heat transfer and boiling phenomena due to crystallisation fouling

The phenomenon of boiling is noteworthy for the difficulties it presents to achieving a suitable mathematical model that may be used for design purposes. Most design techniques are based on maximum allowable heat flux in conjunction with allowable temperature differences in an attempt to maintain nucleate boiling, thus providing efficient heat transfer to the boiling liquid. Fig. 8.12 is relevant to these requirements. It is clear therefore, with the added effects of crystallisation and deposition of dissolved substances the problem of modelling will be even more complex. It is useful however, to recognise some of the factors that are likely to affect the fouling process under these conditions.

The influence of heat flux on the reduction of heat transfer coefficient for boiling aqueous solutions of calcium sulphate has been demonstrated by Jamialahmadi, Blöchl and Müller-Steinhagen [1989]. Heat transfer at high heat fluxes decreases faster and to a greater extent, than at lower heat fluxes as illustrated by Fig. 8.17. On the figure the ratio of the heat transfer coefficient after 7500 minutes $(\alpha_{7500})$ to the initial heat transfer coefficient $\alpha_o$ decreases linearly with heat flux. A ratio of this kind is sometimes referred to as the cleanliness factor.

The presence of the deposit will radically alter the character of the heat transfer surface and as a consequence, the number of nucleation sites is likely to be markedly changed. Jamialahmadi and Müller-Steinhagen [1989, 1990] measured that variation of nucleation site density with time during the fouling period. The

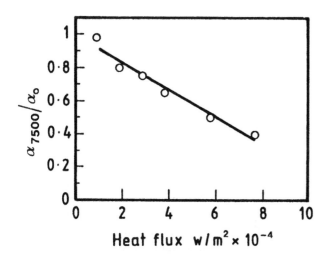

FIGURE 8.17. Reduction rate of heat transfer coefficient as a function of heat flux

number of sites rises towards a maximum after which it falls back to a value close to that at the beginning of the experiment. These data are consistent with an initial porous "open" structure of the deposits that lends itself to bubble formation. As the fouling proceeds the pores of the deposit are likely to become filled with crystals thereby reducing the opportunity for active sites to exist. Similar trends were noted by Palen [1965].

The character of the bubbles (i.e. shape and size) is also likely to be affected by the presence of the deposit. For instance the so-called wick boiling mechanism mentioned earlier, is likely to play an important rôle in the heat transfer process. The evaporation may be considered to take place at the bottom of the steam chimney or on the walls of that channel. If the steam chimneys are absent as might be the case with small pore size or without interconnecting channels, heat transfer is only possible by conduction. It could also be possible to consider that the liquid film was directly on the heating surface. Mass transfer rather than heat transfer might also form the basis of a mathematical model.

## 8.6 CRYSTALLISATION FOULING DUE TO ORGANIC MATERIALS

The deposition of organic crystals onto a heat transfer surface is often associated with "freezing" of the flowing organic fluid on surfaces where the temperature is sufficiently low. It is probably true however, that in reality it is crystals of components of an organic mixture that crystallise on the surface rather than the solvent liquid. For instance many crude oils and liquid hydrocarbons contain dissolved wax molecules that are prone to deposit on cooler surfaces. The transport of crude oil in pipelines under the North Sea has produced some severe problems of solidification on the inside of the pipes. The problem results not only from the low temperatures involved, but also the considerable cooling experienced over the long pipe runs involved, in addition to the chemical composition of the crude oil. The deposits consist mainly of n-paraffins with smaller amounts of branched and cyclic paraffins and aromatics. The process may be regarded as crystallisation from solution rather than freezing fouling described in Chapter 9. Bott [1981 and 1988] has reviewed the mechanism and factors that affect the deposition process using wax deposition from kerosene to illustrate the mechanism.

The concentration of the material contained in the hydrocarbon that leads to the deposition of wax crystals may vary from very small percentages up to substantial amounts. The concentration will affect the so-called "cloud point" which is the temperature at which crystals first appear in the solution that is cooled under standard test procedures. Clearly this temperature will be of considerable significance in fouling by crystallisation of hydrocarbon wax, on cooled heat exchanger surfaces. Of less importance, but still relevant to the fouling process, is the so-called "pour point". The pour point, a temperature lower than the cloud point, is the temperature at which liquid does not flow under prescribed test procedures. Both the cloud point and pour point are empirical measurements useful for comparing the properties of different crude oils.

In common with other crystallisation processes the size and number of crystals formed has a direct relationship to the rate of cooling. Rapid cooling will tend to favour the formation of large quantities of crystals. Under these conditions many centres of crystallisation occur giving rise to a multitude of small crystals. Lower rates of cooling tend to favour larger crystals with fewer centres of crystallisation. Low rates of heat removal will provide an opportunity for the formation of large uniformly packed clusters of crystals.

The process of crystallisation from waxy hydrocarbons may be regarded as similar in some respects to precipitation from a normal solubility salt. The requirement for supersaturation to exist before crystallisation can begin, is less certain than for inorganic solutions, but it is likely to be related to the concentration of the material in solution.

As with scale formation the precipitation of organic compounds will depend on the temperature distribution. Bott and Gudmundsson [1977] proposed a qualitative model of organic precipitation. If a waxy hydrocarbon flowing across a cold surface and the metal/liquid interface is at the cloud point temperature, crystals of wax will form on the surface. If the surface temperature is below the cloud point temperature the cloud point temperature will be located away from the surface towards the bulk liquid. The precise location will depend on the thermal resistance of the laminar sub-layer combined with the resistance of any deposit already on the surface. High thermal resistance between the solid surface and the bulk liquid will cause the cloud point temperature to be located near the solid surface. Low thermal resistances commensurate with high heat transfer coefficients, will mean that the cloud point temperature will be well within the liquid flow, and crystals or crystallites are likely to be produced in the liquid phase. Although solid formation at the interface will be occurring due to the prevailing low temperature, the crystals formed in the bulk may migrate towards the surface thereby adding to the deposit on the surface. The transport of crystals towards the surface will be governed by the mechanisms associated with particle transport described in Chapter 7.

As the deposit develops, the temperature distribution between the original solid surface and the bulk liquid will change. In the limit the outer region of the deposit in contact with the bulk liquid will have a temperature that approximates to the cloud point temperature.

It is probable that the temperature of the deposit near the original surface will approach the pour point temperature. Under these conditions a steady state will exist and there will be little change in deposit thickness with time. Fig. 8.18 illustrates the steady state condition.

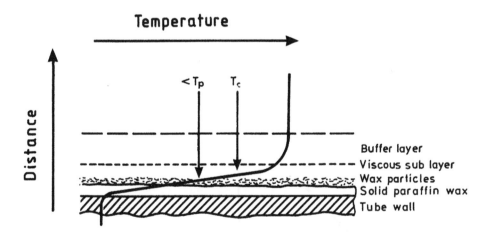

$T_p$ and $T_c$ are the pour point and cloud point temperatures respectively

FIGURE 8.18. The temperature profile across the interface region of wax crystallisation on a solid surface from a flowing liquid

The work by Bott and Gudmundsson [1977] with wax dissolved in kerosene demonstrates that the attainment of the equilibrium condition is likely to be a rapid process. It would appear from Fig. 8.19 that the asymptotic fouling resistance is reached after only 2 hours. Rapid deposition occurs in the first few minutes of exposure of the metal surface to the flowing waxy hydrocarbon.

In the same paper the dependence of the magnitude of the asymptotic fouling on the flow conditions, principally velocity, are demonstrated. For a given geometry, and constant physical properties increasing the velocity (increasing Reynolds number), reduces the asymptotic fouling deposit. Fig. 8.20 in fact shows that the equilibrium fouling resistance is inversely proportional to square of the Reynolds number.

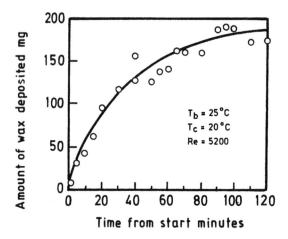

FIGURE 8.19. The development of a wax deposit with time

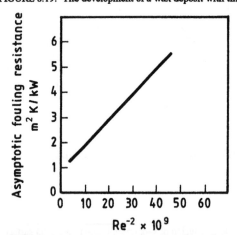

FIGURE 8.20. Asymptotic fouling resistance for deposition from waxy hydrocarbon versus the inverse of Reynolds number squared

The effect of Reynolds number on the fouling resistance is also shown on Fig. 8.21. The effect of higher Reynolds number in lowering the thermal resistance due to the presence of the wax layer is shown. The reduction is in keeping with a reduced wax layer thickness due to the higher shear at a higher Reynolds number.

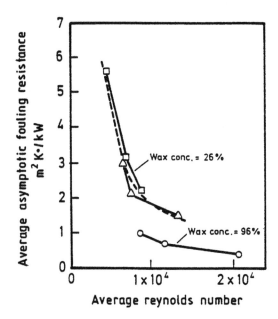

FIGURE 8.21.   Average asymptotic fouling resistance at different Reynolds number for two wax concentrations

Fig. 8.21 also shows the effect of foulant precursor concentration on the magnitude of the fouling resistance. The fact that higher wax concentrations give higher deposit thermal resistance is not necessarily a mass transfer effect due to the higher concentration driving force, but will also be dependent on the different cloud point temperatures at different wax concentrations.

Referring to Fig. 8.18 assuming that the swarm of crystals in the viscous sub-layer are part of the fouling resistance and the deposit thickness remains constant the heat flow rate $q$ from the bulk fluid to the surface of the deposit is given by:

$$q = \alpha_1 (T_1 - T_c) \tag{8.36}$$

where $\alpha_1$ is the heat transfer coefficient on the "hot" side

$T_1$ and $T_c$ are the bulk fluid temperature and the cloud point of the mixture respectively

If $T_2$ is the bulk temperature of the coolant and assuming no fouling on the surface in contact with the coolant.

$$q = \frac{T_c - T_2}{\dfrac{1}{\alpha_2} + \dfrac{x_m}{\lambda_m} + \dfrac{x_{f\infty}}{\lambda_f}} \tag{8.37}$$

where  $\alpha_2$ is the heat transfer coefficient on the coolant side

$x_m$ and $x_{f\infty}$ are the thickness of the metal wall and the asymptotice thickness of the fouling deposit respectively

$\lambda_m$ and $\lambda_f$ are the thermal conductivities of the metal and foulant layer respectively

Combining Equations 8.36 and 8.37 gives

$$\alpha_1 (T_1 - T_c) = \frac{T_c - T_2}{\dfrac{1}{\alpha_2} + \dfrac{x_m}{\lambda_m} + \dfrac{x_{f\infty}}{\lambda_f}} \tag{8.38}$$

and on rearrangement provides an equation from which the thickness of the fouling layer may be calculated:

$$x_{f\infty} = \frac{\lambda_f}{\alpha_1} \frac{(T_c - T_2)}{(T_1 - T_c)} - \lambda_f \left( \frac{1}{\alpha_2} + \frac{x_m}{\lambda_m} \right) \tag{8.39}$$

In theory this equation is simple to use since in general the bulk temperatures will be known along with the cloud point temperature. The physical properties of the system will be known or may be assumed, the geometry will be specified and the heat transfer coefficients could be calculated. In practice however, the use of Equation 8.39 will only give an approximate value of the fouling thickness, which nevertheless may be useful in making an assessment of the fouling potential in a given situation. The limitations of Equation 8.39 are due to the nature of the deposit in relation to the bulk solvent. Unlike inorganic crystallisation from aqueous solution, the liquid/organic crystal deposit is likely to be ill-defined because of the similarities in chemical nature of the deposit and liquid, i.e. the solid and the liquid may both be paraffinic hydrocarbons. In the region of the interface it would be anticipated that the deposit would be "mushy" and easily deformable. Crystals interspersed with liquid could be visualised as the general form of the deposit near the liquid/solid interface; the deposit becoming more robust in the regions close to the metal surface.

Estimation of the heat transfer coefficient at the liquid/deposit interface may be difficult for the following reasons:

1.  As already noted the outer regions of the deposit are likely to be deformable and rough.

2.  In order to calculate the heat transfer coefficient the viscosity of the liquid at the interface is required. The practical value of the viscosity will be modified by the presence of large numbers of crystals and crystallites in the region of the laminar sub-layer.

Equation 8.39 involves the use of $T_c$ the cloud point temperature. Its value depends on the standard method adopted for its determination; principally designed for comparison purposes. The conditions at the interface are unlikely to match those used in the empirical standard method so that $T_c$ may not be the true temeprature at the assumed liquid/solid interface.

Bott [1990] reports data on the deposition of wax from a North Sea crude oil. Using a seven pass plate heat exchanger where measurements of the plate temperatures could be made, the following observations shown in Table 8.7 were obtained:

TABLE 8.7

The deposition of wax in an experimental seven pass plate heat exchanger

| Pass number | Plate temperature range °C | Surface condition |
| --- | --- | --- |
| 1 - 3 | 55 - 73 | No deposit |
| 4 - 6 | 39 - 47 | Some wax deposited |
| 7 | 37 | Heavy wax deposits |

As a result of these experiments six full-sized plate heat exchangers were designed to ensure that no plate temperature was less than 53°C at the same time reducing the crude temperature to 90°C. During operation no wax deposition on the plates was encountered.

The design of these heat exchangers illustrates how careful attention to the significant factors in relation to fouling, can eliminate or reduce the problem.

REFERENCES

Betz Laboratories Inc, 1976, Handbook of industrial water conditioning. 7th Edition. Betz Trevose Pa 19047, USA.

Bott, T.R. and Gudmundsson, J.S., 1977, Deposition of paraffin wax from kerosene in cooled heat exchanger tubes. Can. J. Chem. Eng. 55, 381 - 385.

Bott, T.R. and Gudmundsson, J.S., 1978, Rippled silica deposits in heat exchanger tubes. Proc. 6th Int. Heat Trans. Conf. 4, 373 - 378.

Bott, T.R., 1990, Fouling Notebook Inst. Chem. Engrs., Rugby.

Bott, T.R., 1988, Crystallisation Fouling - Basic Science and Models, in: Melo, L.F., Bott, T.R. and Bernardo, C.A. eds. Fouling Science and Technology. Kluwer Academic Publisher, Dordrecht.

Coates, K.E. and Knudsen, J.G., 1980, Calcium carbonate scaling characteristics of cooling tower water. ASHRAE 86 Trans. 85, Part 2.

Freeborn, J. and Lewis, D., 1962, Initiation of boiler scale formation. J. Med. Sci. 4, 46 - 52.

Gudmundsson, J.S. and Bott, T.R., 1977, Deposition - the geothermal constraint. Symp. Series No. 48. Inst. Chem. Engrs.

Hanlon, L., 1979, Private discussion.

Hasson, D., 1981, Precipitation Fouling, in: Somerscales, E.F.C. and Knudsen, J.G. eds. Fouling of Heat Transfer Equipment. Hemisphere Pub. Corp. Washington, 527 - 568.

Hasson, D., Avriel, M., Resnick, W., Razeman, I. and Shlomo, W., 1968, Mechanism of calcium carbonate deposition on heat transfer surfaces. Ind. Eng. Chem. Fundamentals, 7, 59.

Hasson, D. and Gazit, E., 1975, Scale deposition from an evaporating falling film. Desalination, 17, 339.

Imdakm, A.O. and Knudsen, J.G., 1984, Fouling of heat transfer surfaces under forced circulation boiling conditions. HTD Vol. 35. 22nd Nat. Heat Transfer Conf. Niagara Falls.

Jamialahmadi, M., Blöchl, R. and Müller-Steinhagen, H., 1989, Bubble dynamics and scale formation during boiling of aqueous calcium sulphate solutions. Chem. Eng. Processing, 26, 15 - 26.

Jamialahmadi, M. and Müller-Steinhagen, H., 1990, Effect of surface material and salt concentration on pool boiling heat transfer to aqueous solutions. Proc. 9th Int. Heat Trans. Conf. Jerusalem. 3, 473 - 478.

Kemmer, F.N., 1988, The Nalco Water Handbook. McGraw-Hill Book Co. New York.

Kern, D.O. and Seaton, R.E., 1959, A theoretical analysis of thermal surface fouling. Brit. Chem. Eng. 14, No. 5, 258.

Lee, S.H. and Knudsen, J.G., 1979, Scaling characteristics of cooling tower water. ASHRAE Trans. 85, Part 1.

Macbeth, R.V., 1971, Boiling on surfaces overlaid with a porous deposit: Heat transfer rates obtainable by capillary action. Report AEEW-R711. AERE Winfrith Heath, UK.

Mikic, B.B. and Rohsenow, W.M., 1969, A new correlation of pool boiling data including the effect of heat surface characteristics. J. Heat Trans. 5, 245 - 250.

Morse, R.W. and Knudsen, J.G., 1977, Effect of alkalinity on the scaling of simulated cooling tower water. Can. J. Chem. Eng. 55, 6, 272.

Müller-Steinhagen, H. and Branch, C.A., 1988, Calcium carbonate fouling of heat transfer surfaces. Proc. Chemeca Conf. Sydney, 101 - 106.

Mullin, J.W., 1972, Crystallisation. 2nd Edition. Butterworths, London.

Nancollas, G.H., 1983, The nucleation and growth of scale crystals, in: Bryers, R.W. ed. Fouling of Heat Exchanger Surfaces. United Engineering Trustees Inc. New York.

Nielsen, A.E., 1964, Kinetics of precipitation. Pergamon Press, Oxford.

O'Hara, M. and Reid, R.C. 1973, Modelling crystal growth rates from solution. Prentice Hall, New Jersey.

Palen, J.W., 1965, Fouling of heat transfer surfaces. MSc Thesis, University of Illinois.

Palethorpe, S.J. and Bridgwater, J., 1988, The influence of surface finish on calcium sulphate fouling. Int. Conf. Fouling in Process Plant. Inst. Cor. Sci. and Tech. and Inst. Chem. Engrs. St. Catherine's College, Oxford.

Port, R.D. and Herro, H.M., 1991, The Nalco Guide to Boiler Failure Analysis. McGraw Hill Inc., New York.

Ritter, R.B., 1981, Crystalline fouling studies. HTD Vol. 17. Fouling of heat exchange equipment. 20th ASME/AICHE Heat Transfer Conf. Milwaukee.

Story, M. and Knudsen, J.G., 1978, The effect of heat transfer surface temperature on the scaling behaviour of simulated cooling tower water. Chem. Eng. Prog. Symp. Series No. 174, 74, 25.

Suitor, J.W. *et al*, 1976, 16th Nat. Heat Trans. Conf. 76, CSME/CSChE - 17 St. Louis.

Taborek, J., Aoki, T., Ritter, R.B., Palen, J.W. and Knudsen, J.G., 1972, Fouling: The major unresolved problem in heat transfer. Chem. Eng. Prog. 68, 2, 59 - 67.

Watkinson, A.P., 1983, Water quality effects on fouling from hard waters, in: Hewitt, G.F., Taborek, J. and Afgan, N. Heat Exchangers Theory and Practice. Hemisphere Pub. Corp. Washington.

# CHAPTER 9

## Freezing Fouling or Liquid Solidification

### 9.1    INTRODUCTION

Where a flowing liquid is being cooled and the wall of the channel through which it flows is below the freezing point of the liquid, solidification of the liquid at the surface is likely to take place.  The presence of this solid layer constitutes a resistance to heat removal from the flowing liquid.   In many respects the phenomenon is similar to crystallisation fouling except that in principle, the diffusion or mass transfer of the dissolved solute towards the surface does not apply; the depositing molecules are already at the surface.  The concept of freezing fouling has been applied to organic systems [Bott 1981, 1988] where there is a "spread" of molecular weight, but of essentially the same chemical family, e.g. the deposition paraffin wax during the cooling of waxy hydrocarbons or crude oils (see Chapter 8).

The production of chilled water in the fine chemical manufacture and food processing industries may also give rise to freezing fouling where ice is formed on the cold surface.   The problem may also exist in vapour systems during the recovery of solid products, e.g. the production of phthalic anhydride crystals in so-called "switch condensers".

### 9.2    CONCEPTS AND MATHEMATICAL ANALYSIS

The problem of freezing or liquid (water) solidification, has importance in aspects of climate, conservation in nature and soil mechanics.   In general these represent static systems where the water is not flowing as in an industrial heat exchanger.  Nevertheless, the study of static systems does give some insight into the approach that may be made for liquid systems subject to movement.

#### 9.2.1  Static systems

Considerable experimental and analytical effort has been applied to the problems of phase change during heat removal from non-flowing systems [Riley *et al* 1974, Huang and Shih 1975, Stephen and Holzknecht 1976].   The basic problem is simple in concept (see Fig. 9.1), but difficult to apply to complex geometries.

Considering a flat surface, the temperature of which $(T_s)$ is maintained below 0°C by a large heat sink in contact with stationary water, as illustrated in Fig. 9.1, it is possible to visualise the mechanism of ice formation.

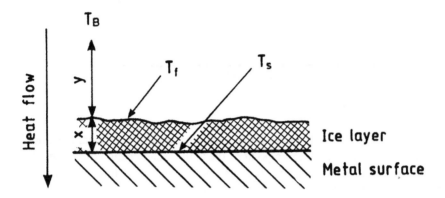

FIGURE 9.1. Ice formation at a surface below 0°C

Sensible heat and the latent heat of freezing are removed from the water at the liquid/solid interface. Under the prevailing static conditions heat will pass from the water to the cold sink by conduction. The resistance to this heat flow initially, will be a combination of the thermal resistances in the liquid and due to the cold solid. Immediately a layer of ice begins to form on the cold surface a further resistance is added to the other thermal resistances. As further heat is extracted from the water, the ice layer thickens representing an advancing boundary between the solid ice and the liquid water, i.e. a transient condition. The transient condition, coupled with complex geometries and different forms of ice structure dependent in turn on the rate of cooling, constitute severe problems of mathematical analysis.

The ice/water interface will advance into the bulk water phase but as the thickness of the ice layer increases with the associated increase in thermal resistance, the rate of heat removal will fall. In turn this will represent an exponentially decreasing rate of growth of the ice layer.

Referring to Fig. 9.1 if at an instant in time when the ice layer thickness is $x$ the thickness of the ice layer increases by $dx$ in time $dt$ the mass of ice formed per unit area is $\rho_{ice}\, dx$ and the heat removed in forming this ice is $h_{ice}\, \rho_{ice}\, dx$ where $h_{ice}$ is the latent heat or enthalpy of fusion of ice and

$$q = -\frac{(T_f - T_s)}{x}\lambda_{ice} \tag{9.1}$$

where $T_f$ is the interface between the water and the ice, i.e. normally 0°C

$$\therefore \frac{dt\,(T_f - T_s)}{x}\lambda_{ice} = h_{ice}\,\rho_{ice}\,dx \tag{9.2}$$

$$\therefore dt = \frac{h_{ice} \rho_{ice} \, x dx}{(T_f - T_s) \lambda_{ice}} \tag{9.3}$$

Assuming that the specific heats of ice and water are small compared with the latent heat of fusion integrating yields:

$$t = Kx^2 \tag{9.4}$$

where $K$ is a constant $= \dfrac{h_{ice} \rho_{ice}}{2(T_f - T_s) \lambda_{ice}}$

It will be seen that the constant $K$ contains terms that are related to the thermal properties of the system namely $\dfrac{h_{ice}}{\Delta T \lambda_{ice}}$ which may be given the collective symbol $\beta$.

where $\Delta T$ represents the temperature difference $(T_i - T_C)$.

Riley *et al* [1974] used this group in a study of inward solidification of spheres and cylinders. The technique assumes constant thermal properties and that the value of $\beta$ is large. Values of $\beta$ used varied from 10 to 20. Stephen and Holzknecht [1976] used a similar parameter that they called the phase conversion factor to solve solidification problems. The magnitude of the errors decreases as the phase conversion factor is increased although values much less than those quoted by Riley *et al* [1974] are claimed.

In a very comprehensive paper Huang and Shih [1975] developed perturbation solutions for liquid solidification in a phase co-ordinate system. The technique adopted involved immobilising the moving solid/liquid boundary by the use of Landau's transformation [Bankoff 1964]. The time variable is replaced by the advancing interface and applying the perturbation technique.

Referring again to Fig. 9.1 at an instant in time when the ice layer thickness is $x$ and assuming that the temperature fall in the liquid from the bulk temperature of $T_B$ to the temperature at the ice/water interface $T_i$ (normally 0°C) occurs over a distance $y$ the rate of heat transfer is given by:

$$q = \frac{T_B - T_s}{\dfrac{x}{\lambda_{ice}} + \dfrac{y}{\lambda_{water}}} \tag{9.5}$$

where $\lambda_{ice}$ and $\lambda_{water}$ are the thermal conductivities of ice and water respectively.

Equation 9.5 assumes that the water is stationary, i.e. no convective heat transfer is present.

An estimation of the heat removed is complex since it not only involves latent heat of fusion, but sensible heat effects that may not be insignificant where large systems are involved. A further complication arises where natural convection in the water at the water/ice interface occurs, i.e. modifying the simple conduction concept implied in Equation 9.5.

The research of Burton and Bowen [1988] into the freezing of water in vertical tubes, clearly shows the complexities of freezing in static conditions. The work was initiated to provide a better understanding of ice plug growth during industrial pipeline maintenance, i.e. potential leakage is controlled by the development of a plug of frozen solid. Fig. 9.2 illustrate three phases of the freezing process identified by Burton and Bowen [1988].

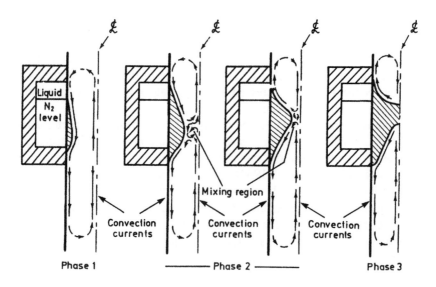

FIGURE 9.2. Three phases of a freezing process

In the first phase freezing is most rapid near the bottom of the cold section. Transition to the second phase may be identified by a reduction in the rate of freezing due to the mixing of the hot and cold components of the water, but at the same time ice continues to form in the upper parts of the freezing section, causing the restricted region to migrate upwards. The third phase occurs when a complete plug has formed when it extends in the axial direction. Changes in the pattern of natural convection are thought to account for these distinct changes. The convective flows will also be influenced by the density inversion of water at 4°C.

Although a study of liquid solidification in static systems is of interest, and gives some insight into the physical behaviour of freezing at a solid surface, the more complex situation that prevails in flowing systems is of more interest to the heat transfer engineer.

## 9.2.2. FLOWING SYSTEMS

Fig. 9.3 illustrates the situation that exists at an instant of time when a flowing fluid (usually a liquid) subject to solidification, is cooled by another fluid flowing counter current on the other side of a metal wall.

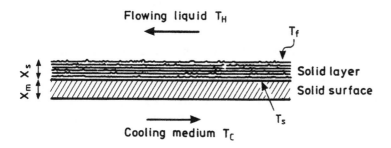

FIGURE 9.3. Solidification in a flowing system

As a result of the temperature difference between the two fluids heat will flow from the liquid subject to freezing, across the metal wall and into the coolant. The rate of heat removal will of course be dependent upon the resistance to heat flow provided by the fluids themselves, the metal wall, the solid frozen layer, and any fouling resistance on the coolant side.

At the beginning of the process, before any solid appears, the layers of warm fluid will flow in contact with the metal wall. In so doing they will lose sensible heat across the metal wall to the coolant. As heat is removed from these layers, the temperature of the liquid in contact with the cold metal surface will eventually fall to the freezing temperature of the liquid. Under these conditions solidification at the metal surface is possible. It is likely that some degree of subcooling will occur before solidification begins, for similar reasons to those mentioned in Chapter 8, for crystallisation of a normal solubility salt.

For turbulent conditions in the fluids with a reasonable finite temperature difference, the onset of solidification is likely to be rapid, since no diffusion of material to the cold metal surface is required before fouling of the surface commences. Immediately solid appears on the surface a further resistance to the transfer of heat is added to the system, thereby reducing the rate of heat transfer.

The process of heat removal will continue till the temperature distribution between the "hot" fluid and the coolant is such that the outer layer of the frozen solid is just at the freezing point. Under these conditions, a steady state is reached when the solid layer remains at constant dimensions. The simple concept will be modified by the practical fouling effects that have already been noted in respect of fouling in general. Principally these effects include changes in velocity for a given flow rate as the deposition reduces the flow area, and the change in roughness presented to the flowing fluid by the deposit surface compared with the original substrate (metal) surface. The combination of these effects on the liquid being cooled, is to change the heat transfer coefficient between the bulk fluid on the solid layer. In turn this will affect the temperature distribution. The effects of the increased shear at the surface may also affect the equilibrium thickness of the solidified layer.

The thickness of the layer is unlikely to be uniform since the temperature difference at equilibrium conditions will vary from the inlet of the freezing liquid to its outlet. The effect will be to change the position of the freezing temperature along the fluid path relative to the wall. Knowing the end temperatures of both streams it would be possible to calculate step changes along the wall and use an interactive procedure to make an estimate of the frozen layer along the wall. The procedure is complex and would most certainly require major simplifying assumptions.

A review of some of the numerical and approximate methods that have been employed to gain mathematical insight into the problem of freezing fouling was made by Siegel and Savino [1966]. As a result they developed a one dimensional solidification model of a liquid flowing over a thin flat plate. The plate is considered sufficiently thin so that the heat removal required to reduce the temperature of the plate is small compared with the heat flow through the wall. The heat flow through the plate is made up of three components; the latent heat of fusion of the solidifying material, the sensible heat removal from the deposited solid and the heat transferred by convection at the frozen layer/liquid interface.

Siegel and Savino [1966] employed three analytical methods. Two of these techniques involve the integration of the transient heat conduction equation which governs the heat flow through the deposit, to provide a generalised form of the growth of the frozen layer with time. These analysts used techniques developed by Adams [1958] and Goodman [1958]. The work of Goodman also provided the basis for the third technique and involves a heat balance at the liquid/frozen layer interface.

For steady state conditions (i.e. constant frozen layer thicknesses) it may be assumed that the temeprature of the liquid/frozen layer interface is the freezing temperature $T_f$ and the only heat removal is the heat transferred across the resistances from the "hot" liquid to the coolant. Referring to Fig. 9.3 the wall temperatures of the hot and cold fluids are $T_H$ and $T_C$ respectively. Under these conditions of steady state the heat flux,

$$q = \alpha_H (T_H - T_f) \tag{9.6}$$

where $\alpha_H$ is the heat transfer coefficient at the "hot" liquid/frozen layer interface.

The resistance to heat flow along the temperature gradient $T_f$ - $T_c$ is:

$$\frac{x_s}{\lambda_s} + \frac{x_m}{\lambda_m} + \frac{1}{\alpha_c}$$

where $\lambda_s$ and $\lambda_m$ are the thermal conductivities of the frozen layer and metal respectively.

$\alpha_c$ is the heat transfer coefficient between the metal/coolant interface.

$$\text{therefore } q = \frac{T_f - T_c}{\dfrac{x_s}{\lambda_s} + \dfrac{x_m}{\lambda_m} + \dfrac{1}{\alpha_c}} \tag{9.7}$$

Combining equations 9.6 and 9.7 yields

$$\alpha_H (T_H - T_f) = \frac{T_f - T_c}{\dfrac{x_s}{\lambda_s} + \dfrac{x_m}{\lambda_m} + \dfrac{1}{\alpha_c}} \tag{9.8}$$

Rearranging gives the value of the equilibrium frozen layer thickness

$$x_s = \frac{\lambda_s}{\alpha_H} \frac{(T_f - T_c)}{(T_H - T_f)} - \lambda_s \left( \frac{x_m}{\lambda_m} + \frac{1}{\alpha_c} \right) \tag{9.9}$$

Equation 9.9 indicates the effect of the variables on the thickness of the solid layer and confirms some intuitive observations. Increasing the liquid and coolant temperatures $T_H$ and $T_c$ and the heat transfer coefficient $\alpha_c$ will also reduce the frozen layer thickness. Opposite changes in the variables will have the converse effect.

When no frozen layer exists $x_s = 0$ so that equation 9.9 reduces to

$$\frac{\lambda_s(T_f - T_c)}{\alpha_H(T_H - T_f)} = \lambda_s\left(\frac{x_m}{\lambda_m} + \frac{1}{\alpha_c}\right)$$

(9.10)

i.e. $T_H = T_f + \left(\dfrac{T_f - T_c}{\alpha_H}\right)\cdot\dfrac{1}{\left(\dfrac{x_m}{\lambda_m} + \dfrac{1}{\alpha_c}\right)}$

(9.11)

For freezing fouling to be prevented the bulk temperature of the hot fluid $T_H$ must be $\left(\dfrac{T_f - T_c}{\alpha_H}\right)\cdot\dfrac{1}{\left(\dfrac{x_m}{\lambda_m} + \dfrac{1}{\alpha_c}\right)}$ higher than the freezing temperature.

Ignoring the thermal resistance of the substrate (e.g. metals are good conductors of heat)

$$T_H = T_f + \frac{\alpha_c}{\alpha_H}(T_f - T_c)$$

(9.12)

Two parameters can be introduced to facilitate the analysis of freezing fouling problems. Siegel and Sarino [1966] include the sub-cooling parameter $S$ sometimes known as the Stefan number where

$$S = \frac{c_{ps}(T_f - T_c)}{h_s}$$

(9.13)

$c_{ps}$ is the specific heat of the solid layer

and the ratio of the heat transfer resistances $R$ where

$$R = \frac{\left(\dfrac{x_s}{\lambda_s}\right)}{\left(\dfrac{1}{\alpha_c} + \dfrac{x_m}{\lambda_m}\right)}$$

(9.14)

It may be noted that $S$ is the inverse of the factor introduced by Riley *et al* [1974] for the analysis of static systems.

The term $c_{ps}(T_f - T_c)$ in the Stefan number represents the maximum internal energy it is possible to remove from unit mass of the solidified liquid in reducing its temperature from the freezing point down to the coolant temperature. If the coolant temperature is reduced towards absolute zero, $S$ will increase towards a maximum value. Siegel and Savino [1966] observe that for many materials, the maximum value of $S$ is approximately 3.

For thin layers of frozen solid, $\dfrac{x_s}{\lambda_s}$ tends to zero so that the value of the parameter $R$ will tend to zero. For steady state conditions $\alpha_c$ will be constant, $x_m$ is fixed and the physical properties $\lambda_s$ and $\lambda_m$ will also be constant, so that the value of $R$ is a measure of the solid layer thickness $x_s$, under steady state conditions.

The heat of fusion to be removed to produce a frozen layer thickness of $x_s$ is $\rho_s h_s x_s$. If this freezing process takes place in time $t$ the total heat transferred by convection from the hot fluid to the solid is $t\alpha_H (T_H - T_f)$ per unit area.

The ratio of these two quantities of heat may be used to define a dimensionless time $t^1$ so that

$$t^1 = \frac{t\alpha_1(T_H - T_f)}{\rho_s h_s x_s} \tag{9.15}$$

It is also possible to define an dimensionless frozen layer thickness as

$$x^1 = \frac{x}{x_s} \tag{9.16}$$

where $x$ is the instantaneous frozen layer thickness.

$x^1$ will vary from zero when the metal surface is clean to unity when equilibrium has been established.

A large value of dimensionless time $t^1$ suggests that the heat transfer by convection is large compared to the heat removed to create the foulant layer.

For different sets of conditions within a system with a fixed value of the dimensionless thickness $x^1$, it is possible to plot a graph of dimensionless time $t^1$ against $R$ the ratio of the heat transfer resistances for different values of $S$ the Stefan number, as shown on Fig. 9.4, originally suggested by Siegel and Savino [1966].

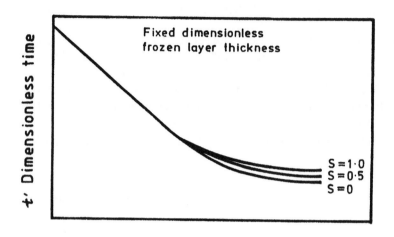

FIGURE 9.4. Dimensionless time $t^1$ plotted against $R$ for different values of $S$

The usefulness of figures similar to Fig. 9.4 for particular systems is that they provide a means of estimating the frozen layer thickness at any time $t$. For a given set of conditions, where all the properties of the system are known. Temperatures $T_H$ and $T_C$ are specified, and the values of $\alpha_H$ and $\alpha_C$ may be calculated from the standard equations, the values of $R$, $S$ and $t^1$ may be estimated. Some inaccuracy in the calculation of $\alpha_H$ may be introduced since the roughness of the deposit may not be readily assessed. Increased roughness will increase the value of $\alpha_H$. Having plotted the graphs along the lines of Fig. 9.4 the relationship between $x$ and $t$ can be calculated from the equations for $t^1$ and $x^1$.

Using a jet of warm water directed at a plate maintained at a low temperature Savino *et al* [1970] have compared the data obtained with those from the mathematical analysis. The comparison made on Fig. 9.5 shows excellent agreement between the experimental data and the theory.

It is noteworthy that the equilibrium ice layer is reached after a relatively short time. The fall in wall temperature mirrors the increase in equilibrium ice layer thickness. The general shape of the accumulation of ice on the surface follows the exponential curve of deposit thickness with time often associated with fouling of heat exchanger surfaces.

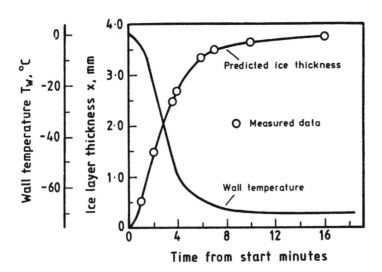

FIGURE 9.5 Experimental data compared to the mathematical model

REFERENCES

Bankoff, S.G., 1964, Heat conduction or diffusion with change of phase. Advances in Chemical Engineering. Academic Press, New York.

Bott, T.R., 1981, Fouling due to liquid solidification, in: Somerscales, E.F.C. and Knudsen, J.G. eds. Fouling of Heat Transfer Equipment. Hemisphere Publishing Corp. Washington.

Bott, T.R., 1988, Crystallisation of organic materials, in: Melo, L.F., Bott, T.R. and Bernardo, C.A. eds. Fouling Science and Technology. Kluwer Academic Publishers, Dordrecht.

Burton, M.J. and Bowen, R.J., 1988, Effect of convection on plug formation during cryogenic pipe freezing. 2nd UK Nat. Heat Transfer Conf. Vol. 1, 465 - 476, Glasgow. IMechE/IChemE.

Goodman, T.R., 1958, The heat balance integral and its application to problems involving a change of phase. ASME, 80, 335 - 342.

Huang, C.L. and Shih, S.Y., 1975, Perturbation solutions of planar diffusion controlled moving boundary problems. Int. J. Heat and Mass Transfer, 18, 689 - 695.

Riley, D.S. *et al*, 1974, The inward solidification of spheres and cylinders. Int. J. Heat and Mass Transfer, 17, 1507 - 1516.

Savino, J.M. *et al*, 1970, Experimental study of freezing and melting of flowing warm water at a stagnation point on a cold plate. 4th Int. Heat Transfer Conf. 1, 2 - 11.

Siegel, R. and Savino, J.M., 1966, An analysis of the transient solidification of a flowing warm liquid on a convectively cooled wall. 3rd Int. Heat Transfer Conf., 4, 141 - 151.

Stephen, K. and Holzkrecht, B, 1976, Die asymptotischen lösungen für vorgänge des erstarrens. Int. J. Heat and Mass Transfer, 19, 597 - 602.

# CHAPTER 10

## Fouling Due to Corrosion

### 10.1 INTRODUCTION

Corrosion in process plant is an ever present problem and the engineers responsible for design, operation and maintenance need to be vigilant in the control of corrosion. Corrosion may be defined as the deterioration and loss of material due to some form of chemical attack. The origin of the corrosion may be the process fluid itself, or a constituent of it, or corrosion may be the result of the presence of impurities, perhaps in trace quantities carried in the fluid stream. In one sense corrosion may be regarded as reaction fouling, with the chemical reactions involving the surface rather than the constituents of the process fluid. Corrosion is often accelerated by the presence of other deposits such as scale or biofilms. On the other hand corrosion protection is often afforded by the presence of metal oxides on surfaces. If the oxide layer is removed by chemical action or erosion then the underlying metal may be seriously affected. It is possible to limit corrosion or eliminate the problem altogether, by the correct choice of material of construction. Unfortunately in many applications, the cost penalty will be prohibitive. For products with a high added value however, the first cost (investment) may be entirely justified particularly where contamination from products of corrosion cannot be entertained.

There is a wide variety of different materials of construction for the fabrication of corrosion resistant heat exchangers. For liquid systems associated with shell and tube heat heat exchangers, tubes are often of various copper base alloys, principally Admiralty brass, aluminium brass, cupro nickel (70:30) and aluminium bronze together with stainless steels, Monel and other high quality corrosion resistant alloys. Mild steel tubes are also used for low cost applications. In certain operating conditions a combination of environments involved on the inside and outside of the tubes may require the use of "duplex" or bimetal tubes. One example is the use of steel/aluminium brass tubes for applications of ammoniacal liquor on one side with salt water on the other. Lake [1990] has given an excellent review of the properties of duplex steels. Tube plates may be of mild steel where the tubes are of this material, but otherwise they are normally of non ferrous metals. The shells themselves are often of mild steel (on account of cost), and the other components of shell and tube heat exchangers, e.g. shell covers, baffles, floating heads may employ special materials for corrosion resistance.

Corrosion technology is a vast subject in its own right and therefore the discussion must be limited. Consideration will be given only to the situation where the heat transfer surface is under attack. The presence of the corrosion products on the surface represents a resistance to heat transfer and therefore constitutes a fouling deposit. Corrosion in other parts of the process plant may give rise to heat exchanger fouling, but here the mechanism of deposition is not generally related to chemical reactions, but rather to particulate deposition.

## 10.2   GENERAL LIQUID CORROSION THEORY

The well known brown rust (ferric hydroxide) is formed by the combination of metallic iron with oxygen and water. The overall relationship is:

$$4Fe + 3O_2 + 6H_2O \rightarrow 4Fe(OH)_3 \qquad (10.1)$$

Equation 10.1 illustrates the most common corrosion mechanism involving an electrochemical process essentially metal oxidation, and necessitates the removal of electrons from a metal. Equations 10.2 and 10.3 illustrate the release of electrons from iron metal to produce first ferrous ions followed by conversion to ferric ions by further oxidation.

$$Fe \rightarrow Fe^{2+} + 2e \qquad (10.2)$$
$$\text{Ferrous ion}$$

$$\text{and} \quad Fe^{2+} \rightarrow Fe^{3+} + e \qquad (10.3)$$
$$\text{Ferric ion}$$

The removal of a metal atom from an anodic site (Equations 10.2 and 10.3) on the metal surface gives an ion in solution and an excess of electrons on the metal surface. Utilisation of the electrons on a nearby cathodic site gives a balancing reaction which in solutions that have a near neutral *pH*, usually involves the reduction of dissovled oxygen to hydroxyl ions according to

$$O_2 + 2H_2O + 4e \rightarrow 4OH^- \qquad (10.4)$$

Other reduction reactions are possible for solutions with a *pH* lower than 7 (acidic environments) i.e. reduction of hydrogen ions to hydrogen gas as shown in Equation 10.5.

$$2H^+ + 2e \rightarrow 2H \rightarrow H_2 \qquad (10.5)$$

Fig. 10.1 illustrates how the oxidation/reduction reactions occur on the surface of a piece of iron resulting in corrosion. Other electron consuming reactions may be important in other specific examples.

Different metals have different oxidation potentials, i.e. the energy required to remove electrons varies from metal to metal. Electron removal may be facilitated in some solutions compared to others. Both anodic and cathodic reactions occur evenly over a metal surface in simple examples of corrosion. The location of the individual sites will depend very much on the surface characteristics, i.e. discontinuities in a protective oxide layer, grain boundaries and crevices.

When a metal is immersed in a solution containing ions of that metal and equilibrium is allowed to be established, a potential difference is developed at the metal/solution interface generally called the electrode potential of the metal. If the potential is measured using a molar solution of the metal ions at 25°C it is referred to as the standard electrode potential.

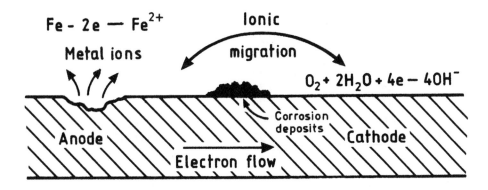

FIGURE 10.1. The electrochemical corrosion cell

The electrode potential of a single half cell cannot be directly measured, but must be obtained using pairs of half cells and measuring the potential difference of the combination.

Rewriting Equation 10.2 as a reversible reaction

$$Fe \rightleftarrows Fe^{2+} + 2e \qquad (10.6)$$

may be considered as one half cell.

For comparison purposes it is necessary to determine each electrode potential in relation to an arbitrary standard. It is usual to assume the standard hydrogen electrode potential to be zero, i.e. the combination of hydrogen ions and electrons as given by Equation 10.2 simplified to

$$2H^+ + 2e \rightarrow H_2 \qquad (10.7)$$

Individual half cell potentials on the hydrogen scale may be compared to all other half cells referred to the hydrogen scale. Table 10.1 lists some standard electrode potentials for some common metals. The sign allocated to the standard potentials is positive but the opposite sign is often used by corrosion engineers. As long as consistency is observed these differences present no difficulties.

TABLE 10.1

Standard electrode potentials at 25°C and molar solutions of metal ions

| | Metal ion | Potential (volts) | |
|---|---|---|---|
| NOBLE | $Au^+$ | -1.50 | CATHODIC |
| ↑ | $Cu^{2+}$ | -0.34 | ↑ |
| | $H^+$ | ZERO | |
| | $Fe^{3+}$ | +0.05 | |
| | $Ni^{2+}$ | +0.23 | |
| | $Fe^{2+}$ | +0.44 | |
| | $Cr^{2+}$ | +0.56 | |
| | $Zn^{2+}$ | +0.76 | |
| | $Al^{3+}$ | +1.70 | |
| ↓ | $Mg^{2+}$ | +2.40 | ↓ |
| ACTIVE | $K^+$ | +2.92 | ANODIC |

The alkaline earth and alkali metals hold their outer shell electrons rather loosely and demonstrate a greater potential difference than iron. At the other extreme noble metals such as gold produce fewer electrons than does hydrogen and are therefore low down on the potential scale. Table 10.2 gives a galvanic series rating [Fontana and Greene 1967] in terms of noble or cathodic to active or anodic metals and alloys in contact with sea water.

An alternative way of expressing the potential corrosion is to consider the rate of corrosion over the range of *pH*. Figs. 10.2 - 10.5 illustrate the differences between groups of metals with similar properties.

When a metal is in contact with a solution containing its own ions, it attains a state of dynamic equilibrium where the rate of dissolution of metals from the surface is equal to the rate of deposition from solution. If the reaction is to proceed in one direction or another, the equilibrium must be displaced in an anodic sense to promote the dissolution reaction (oxidation, see Equations 10.2 and 10.3) and in a negative sense to promote the deposition reaction (reduction). The extent to which the potential of an electrode is displaced from its equilibrium value is termed the polarisation of the electrode. In the corrosion situation the difference between the equilibrium potential for metal dissolution and for the cathodic reaction on the same metal provides the driving force for the dissolution of the metal.

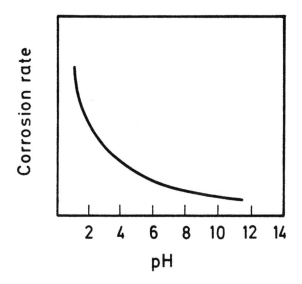

FIGURE 10.2.  Corrosion rate at different *pH* for acid affected metals, e.g. *Ni, Cu, Mn, Mg, Co, Cr, Cd,* (i.e. anodic to hydrogen)

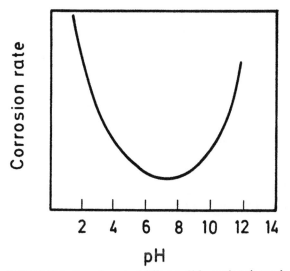

FIGURE 10.3.  Corrosion rate at different *pH* for amphoteric metals (or oxides) *Al, Zn, Sn*

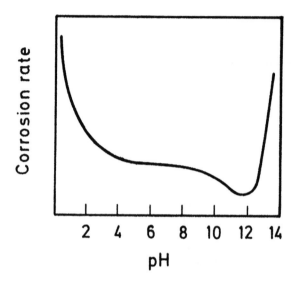

FIGURE 10.4. Corrosion rate at different *pH* for iron and steel

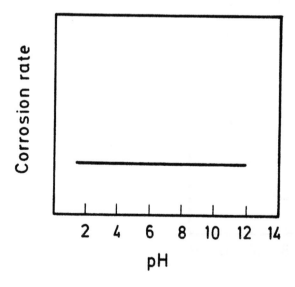

FIGURE 10.5. The passivity of noble metals in relation to *pH*

TABLE 10.2

Galvanic series of some commercial metals and alloys in contact with sea water

| | |
|---|---|
| NOBLE | Platinum |
| ↑ | Gold |
| | Graphite |
| | Titanium |
| | Silver |
| | $\left\{\begin{array}{l}\text{Chlorimet 3} (62\,Ni,18\,Cr,18\,Mo) \\ \text{Hastelloy C} (62\,Ni,17\,Cr,15\,Mo)\end{array}\right.$ |
| | $\left\{\begin{array}{l}18-8\,Mo\ \text{s/steel (passive)} \\ 18-8\,\text{s/steel (passive)} \\ \text{Chromium s/steel } 11\text{-}30\%\ Cr \text{ (passive)}\end{array}\right.$ |
| | $\left\{\begin{array}{l}\text{Inconel (passive)} (80\,Ni,13\,Cr,7\,Fe) \\ \text{Nickel (passive)}\end{array}\right.$ |
| | Silver solder |
| | $\left\{\begin{array}{l}\text{Monel} (70\,Ni,30\,Cu) \\ \text{Cupro-nickels } (60\text{-}90\ Cu,\ 40\text{-}10\ Ni) \\ \text{Bronzes} (Cu-Sn) \\ \text{Copper} \\ \text{Brasses } (Cu\text{-}Zn)\end{array}\right.$ |
| | $\left\{\begin{array}{l}\text{Chlorimet 2} (66\,Ni,32\,Mo,1\,Fe) \\ \text{Hastelloy B } (60\ Ni,\ 30\ Mo,\ 6\ Fe,\ 1\ Mn)\end{array}\right.$ |
| | $\left\{\begin{array}{l}\text{Inconel (active)} \\ \text{Nickel (active)}\end{array}\right.$ |
| | Tin |
| | Lead |
| | Lead-tin solders |
| | $\left\{\begin{array}{l}18\text{-}8\,Mo\ \text{s/steel (active)} \\ 18-8\,\text{s/steel (active)}\end{array}\right.$ |
| | *Ni*-Resist (high *Ni* cast iron) |
| | Chromium s/steel, 13% *Cr* (active) |
| | $\left\{\begin{array}{l}\text{Cast iron} \\ \text{Steel or iron}\end{array}\right.$ |
| | Cadmium |
| | Commercially pure aluminium |
| ↓ | Zinc |
| ACTIVE | Magnesium and magnesium alloys |

Although most metals display an active or activation controlled region, when polarised anodically from the equilibrium potential, many metals and perhaps even more so alloys developed for engineering applications, produce a solid corrosion product. In many examples the solid is an oxide that is the stable phase rather than the ion in solution. If this solid product is formed at the metal surface and has good intimate contact with the metal, and features low ion-conductivity, the dissolution rate of the metal is limited to the rate at which metal ions can migrate through the film. The layer of corrosion product acts as a barrier to further ion movement across the interface. The resistance afforded by this corrosion layer is generally referred to as the passivity. Alloys such as the stainless steels, nickel alloys and metals like titanium owe their corrosion resistance to this passive layer.

The discussion so far has concentrated on the surface effects that produce corrosion but there are other corrosion cells that can lead to metal wastage.

### 10.2.1 Galvanic corrosion involving two dissimilar metals

When two dissimilar metals are immersed in an electrolyte they usually develop different potentials in accordance with the theory already presented. If the metals are in contact the potential difference provides the driving force for corrosion. Severe corrosion often occurs as a result of the contact between two metals. In shell and tube heat exchangers where the tubes are fabricated from a corrosion resistant alloy, and the shell is made from mild steel for instance to reduce the capital cost, corrosion is very likely unless adequate protection is made. The less resistant of the two metals is caused to corrode, or to corrode more rapidly, while the resistant metal or alloy corrodes much less or may be even completely protected. The basis for galvanic corrosion is illustrated on Fig. 10.6. Metal A has a lower electrode potential than metal B. Ions migrate in the conducting solution while electrons flow across the junction of the two metals, as a result metal A is corroded at C.

The ranking given in Table 10.2 and similar tables are a useful guide to the potential corrosion (or resistance to corrosion) by different alloys and metals. Although such tables give indications of the anticipated corrosion they give no indication of the actual corrosion rate to be encountered. The conditions within the exchanger, e.g. the temperature and velocity will affect the corrosion rate.

The conductivity of the solution in contact with the metal will also affect the rate of corrosion. In a highly conductive environment such as sea water, the galvanic currents can travel over relatively long distances which has the effect of spreading the corrosion over a wide area. Where the solution is a relatively poor conductor of electricity the corrosion is likely to be more localised.

The geometry of the system and in particular the relative properties of the dissimilar metals in contact will also affect the intensity of corrosion in particular locations. The situation is illustrated on Fig. 10.7. The small rivets joining the

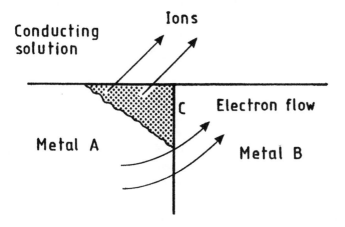

FIGURE 10.6. Corrosion at the junction of two dissimilar metals

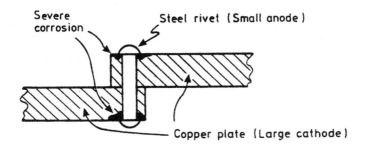

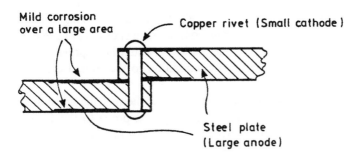

FIGURE 10.7. Copper and steel in contact in sea water

copper plates (large cathode) represent small anodes and cause severe corrosion attack around the rivets. If the situation is reversed and steel plates are joined by copper rivets (small cathode), the steel plates representing a large anode are mildly attacked over a large area.

Two dissimilar metals in contact in an aqueous environment does not necessarily give rise to galvanic corrosion. Turner [1990] cites an example of a shell and tube heat exchanger (steam condenser), containing aluminium brass tubes expanded into carbon steel tube plates. It could be anticipated that with the relatively large area of the noble aluminium brass, severe galvanic corrosion would occur on the smaller area of exposed steel resulting in a short service life. In fact the condenser had been in service for 26 years when the examination took place!

On the water side the tube plates had corroded fairly uniformally to a depth of about 3 *mm*. On the steam side the only severe corrosion was in a narrow zone about 2 - 3 *mm* wide around each tube where the depth of corrosion was something like 6 - 10 *mm*. As the tube plates were 100 *mm* thick, a further long period of service could be anticipated without problems resulting from corrosion.

Turner [1990] put forward the following explanation for the good corrosion resistance of the condenser. The condensing steam from a turbine was on the shell side of the exchanger under vacuum. The boiler plant supplying the steam was well maintained and operated. Dissolved oxygen and carbon dioxide in the condensate were low as was the conductivity. The thin film of electrolyte providing the ionic pathway between the brass and the steel therefore had a high electrical resistance. Due to the low level of dissolved oxygen the cathode reaction was strongly polarised. The low conductivity ensured that the activity at both anode and cathode was confined to a narrow zone in the immediate vicinity of the junction between the brass and copper.

Cooling water from a recirculating evaporative system flowed through the inside of the condenser tubes. The *pH* of the water was controlled and an inhibitor used to reduce corrosion. The water had a moderately high conductivity so that corrosion was uniformly distributed over the face of the tube sheet. Turner [1990] suggests that water would have been near saturation with respect to oxygen and cathodic polarisation would not have been very pronounced. At the same time the cathodic area was not likely to be large. The conductivity was not so high that the bore of the tubes beyond about 100 - 200 *mm* from the tube ends could have acted as a cathode. The resistance along the ionic pathway would have prevented the brass surface making more than a marginal contribution to the overall corrosion current.

Alloys sometimes establish corrosion cells if grains of the metallic constituents exist on the surface. Brass an alloy of zinc and copper, is prone to this form of corrosion as the zinc can be removed to leave an open porous matrix of copper.

A corrosion cell may be set up between a metal and the surface oxide film. Cracks in the oxide film on the surface of iron may give rise to the brown coloured *Fe (OH)$_3$*. The process is illustrated on Fig. 10.8.

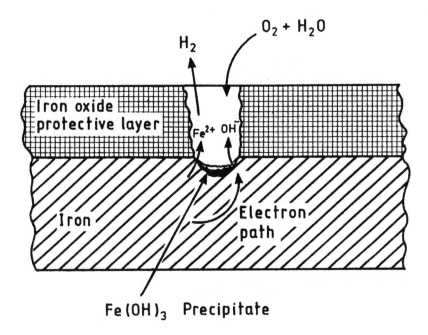

FIGURE 10.8. Ferric hydroxide formation at a crack in an iron oxide film on an iron surface

In the water in contact with the oxide layer residing on the metal surface some dissociation of the water molecules takes place, i.e.

$$H_2O \quad \rightarrow \quad H^+ + OH^- \tag{10.8}$$

The oxide layer acts as an electron donating area, neutralises the hydrogen ions so that hydrogen gas is formed. The equilibrium of Equation 10.8 is disturbed so that there is an excess of $OH^-$ ions in the region of the oxide layer. A crack in the oxide layer will provide the opportunity for $Fe^{2+}$ ions to pass into solution. The co-existence of $Fe^{2+}$ and $OH^-$ ions near the surface will enable precipitation of the hydroxide to occur.

$$Fe^{2+} + 2OH^- \rightarrow Fe(OH)_2 \tag{10.9}$$

The ferrous hydroxide ($Fe(OH)_2$) is rapidly converted to ferric hydroxide ($Fe(OH)_3$) in the presence of dissolved oxygen according to Equation 10.10.

$$4Fe(OH)_2 + 2H_2O + O_2 \rightarrow 4Fe(OH)_3 \tag{10.10}$$

The reactions are likely to occur near the crack in the oxide layer so that deposition will be in this region (see Fig. 10.8). Under these circumstances the metal below the crack in the oxide will remain unprotected and the corrosion will continue.

If the ion formation can occur only at a few cracks in the oxide layer, the total loss of metal will be relatively small. The remainder of the large surface area of oxide will provide efficient ion neutralisation, but the wastage of metal at the few exposed regions will be extremely severe.

Corrosion near a crack in an iron oxide layer can give rise to the phenomenon of "tuberculation". Fig. 10.9 shows a cross section of a tubercle consisting of various

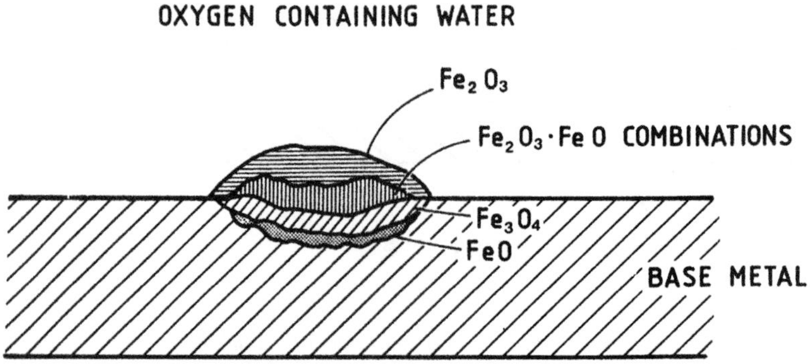

FIGURE 10.9. A cross section of a tubercle

forms of iron oxide and other corrosion products. The mechanism described in connection with corrosion at the bottom of a crack in an iron oxide layer continues till the pit is filled with porous iron compounds. The attachment of oxygen atoms among others, to the original iron makes the corrosion products more voluminous than the original metal so that a characteristic mound is formed. The porous structure of the tubercle generally contains an aqueous solution with relatively high concentrations of chlorides, sulphates and low concentrations of oxygen that can lead to the establishment of further corrosion cells.

The presence of tubercles generally increases the roughness of the metal surface that in turn increases the resistance to fluid flow. Large tubercles may break loose from the surface as a result of shear stress, and become lodged in downstream equipment such as heat exchanger header boxes.

The neutralisation of $H^+$ ions by electrons to form hydrogen gas is likely to affect the rate of corrosion since the presence of the gas will restrict the movement of ions and electrons. The build up of ions at the blocking (or polarising) layer will affect the electrode potential. Hydrogen bubbled over a metal surface can alter its electrode potention by as much as 0.7 V [Lewis and Secker 1965].

The presence of dissolved salts in the water will accelerate the corrosion process by the provision of ions available to effect a more rapid neutralisation of $Fe^{2+}$ and $OH^-$ ions. For example if the water contains dissolved sodium chloride the following reactions at the electrodes are possible.

$$Na^+ + OH^- \rightarrow NaOH \qquad\qquad (10.11)$$
$$\text{soluble}$$

$$Fe^{2+} + 2Cl^- \rightarrow FeCl_2 \qquad\qquad (10.12)$$
$$\text{soluble}$$

The presence of $FeCl_2$ and $NaOH$ in solution leads to the precipitation of ferrous hydroxide ($Fe(OH)_2$) when the solubility limit is exceeded (see Chapter 8).

## 10.2.2 Crevice corrosion

Cracks in an oxide layer can give rise to localised attack of the metal as described in the previous section. Localised corrosion can also occur due to crevices at the metal surface produced during fabrication, e.g. under bolt or rivet heads or lap joints or may even be initiated by deposits such as biofilms or particulate matter. Under these conditions the crevice solution becomes more aggressive than the bulk solution. Rapid metal wastage in these regions is often referred to as crevice corrosion. Turner [1989] has outlined the problems associated with crevice corrosion.

Crevice corrosion occurs as a result of a non-uniform concentration of eletrolyte solution. The local electrode potential varies with the concentration of electrolyte corrosion currents (electron and ion migration) that occur in the metal and solution. For instance a piece of copper immersed in copper sulphate solution of varying composition, tends to be more positive in the region of high copper sulphate concentration due to the enhanced rate of deposition of $Cu^{2+}$ ions. A flow of electrons through the metal towards this region occurs and the ion deposition is a continuous process. As a result corrosion occurs in the metal where the solution concentration is at its lowest, and $Cu^{2+}$ ions pass into solution.

Where there is an oxygen gradient in a electrolyte a corrosion cell can be established. The presence of a crevice can lead to poor oxygen availability resulting in the cessation of the cathodic reaction. As the dissolution of the metal proceeds, an excess positive charge builds up due to the presence of the metal ions in solution in the crevice and as a result, negatively charged ions diffuse into the crevice from the bulk solution. As the process continues a point is reached at which the solubility limit of the metal hydroxide is exceeded and deposition occurs. The result is an excess of $H^+$ ions in the crevice with an attendent lowering of the

local *pH*, making the crevice environment more aggressive.    Under certain
conditions the localised corrosion can be extremely severe, for instance with
stainless steels.

### 10.2.3 Pitting corrosion

Pitting corrosion is also a localised attack of a metal surface.  It may give rise to
serious damage to industrial equipment even when the total metal loss is extremely
small.  Initiation of pit formation may be extremely long but once the attack has
begun, it can proceed by the mechanisms previously described in connection with
crevice and crack corrosion.  Initiation is likely to occur as a result of occasional
fluctuations in conditions.  Once pits have started to develop it is unusual for the
rate of attack to decrease again.  Oxygen depletion will aid the corrosion process
and as the oxygen content of the electrolyte at the bottom of the pit becomes
progressively less, the corrosion cell becomes more active.  Fig. 10.10 provides an
idealised picture of the pitting phenomenon.

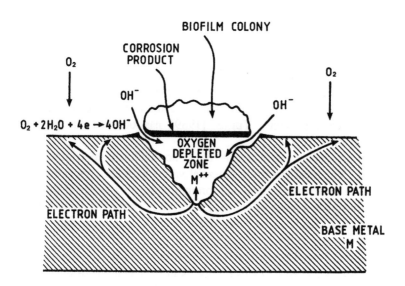

FIGURE 10.10.  The mechanism of pitting

### 10.2.4 "Hot spot" corrosion

Localised and rapid pitting may develop in response to local high temperatures
in the metal wall.  Cupro-nickel (70:30) appears to be rather susceptible to this
form of attack [Gilbert 1960].  A characteristic feature is that the pits are generally
filled with metallic copper.

10.2.5 Stress corrosion cracking

Stress corrosion cracking occurs as a combination of corrosion and mechanical conditions. It would appear that a number of alloys fail in very specific environments, under tensile stresses significantly lower than the tensile strength of the material, and frequently lower than the yield strength. The metal stress may be applied by any kind of external force, e.g. bending or stretching. It may be due to internal stresses in the metal set up during fabrication processes, e.g. rolling, drawing, shaping or welding. Combinations of stress and environment may be high stress and mild corrosion potential or low stress and high corrosive potential. If the combined effects are conducive to stress cracking, both intergranular and transgranular cracks develop at right angles to the prevailing force. Highly corrosive waters tend to encourage randomly oriented cracking.

Careful design, good fabrication techniques and annealing processes will do much to reduce or eliminate, the problem of stress corrosion cracking.

Caustic embrittlement is a particular version of stress corrosion cracking that sometimes occurs in boilers. Three contributory factors are considered to be involved in the problem [Kemmer 1988] and include:

1. Some mechanism by which a boiler water could produce high concentrations of sodium hydroxide.

2. Where the concentration occurs the metal is under high stress conditions, for instance where the water tubes are rolled into the steam drum.

3. The boiler water contains silica which directs the attack to grain boundaries leading to intercrystalline attack of an embrittling nature.

Another specific version of stress corrosion cracking results from a chloride concentration cell. It is sometimes referred to as chloride embrittlement and can affect stainless steels. For the problem to appear a combination of local chloride concentration and transgranular cracking occurs, leading ultimately to metal failure.

10.2.6 Selective leaching

Selective leaching applies to the situation where one element of an alloy is removed preferentially to another metallic constituent. The most common example of this phenomenon is the so-called dezincification of brass (see Section 10.2.1), but it may also occur through the selective removal of nickel from cupro nickel alloys and aluminium from aluminium bronze.

### 10.2.7 Corrosion fatigue cracking

A combination of corrosive environments and repeated working of the metal may result in metal failure. The fatigue is brought about by a routine cyclic application of stress: the vibration of tubes in a shell and tube heat exchanger for instance. The nature of the aggressive conditions and the metal involved are not specific; it is this combination that gives rise to cracking. The fatigue cracks are usually at right angles to the applied stress and the propagation of the crack will be a function of the corrosivity of the environment, the stress level and the number of cycles per unit time. Corrosion fatigue cracking is more common than stress corrosion cracking [Kemmer 1988].

### 10.2.8 Impingement attack

It is possible for the passivation (oxide) layer on the surface of a metal to be continuously removed or not allowed to develop, by erosion from particulate matter or gas bubbles. Not only is the surface eroded but the removal of the protective oxide layer allows corrosion to take place. The problem is accentuated by the presence of an obstruction or debris, on the metal surface that diverts and accelerates the flow near the surface along a defined path. Sato *et al* [1977] report experimental data on erosion-corrosion resistance of condenser tubes fabricated from various cupro nickel alloys. They suggest that high iron bearing cupro nickels are superior in respect of erosion corrosion by clean sea water.

## 10.3    CORROSION FOULING IN GAS SYSTEMS

Corrosion of heat transfer surfaces subject to gaseous environments can occur at almost any temperature. It is useful however to divide the discussion into two parts, namely low and high temperature corrosion, since in general, the mechanisms involved are different. As already described the presence of a metal oxide on a surface affords protection against further oxidation and metal wastage, but where corrosion occurs in gaseous atmospheres, the oxide layer may become involved in chemical reactions thereby removing the protection, and allow corrosion to proceed. The effects may be severe, even in the presence of an oxidising environment.

### 10 3.1  Low temperature corrosion

At relatively low temperatures the problem of corrosion in heat exchangers is likely to arise from condensation from the gas stream of aggressive agents, usually involving the presence of moisture. Under these circumstances the corrosion is likely to follow the mechanisms described under Section 10.2. An example is condensation from flue gases in combustion plant, where the flue gas temperature is reduced by the removal of the heat energy contained in the gases. The final heat extraction is often used to preheat the air for the combustion process and consequently air preheaters are prone to corrosion.

In addition to oxides of sulphur other oxides and chemical compounds, flue gases contain water vapour originally as moisture in the fuel or produced from the combustion process by converting the hydrogen content of the fuel to water. As the flue gas temperature is progressively reduced, condensation will occur often at some specific temperature. The resulting condensate is usually acidic in character since it will generally contain sulphur oxides in solution, i.e. $SO_2$ and $SO_3$ that combine with water to lower the $pH$. The reactions involved include:

$$SO_2 + H_2O \rightarrow SO_3^{2-} + 2H^+ \tag{10.13}$$

$$\text{and } SO_3 + H_2O \rightarrow SO_4^{2-} + 2H^+ \tag{10.14}$$

Oxides of nitrogen produced from nitrogen-containing fuel may also give rise to acidic condensate conditions.

The temperature at which condensate appears in combustion gases is usually called the acid dew point and is generally around 120°C although the particular value will depend to a large extent on the character of the fuel burnt. Because of the corrosion problems associated with these low temperatures, it is usual to operate the heat exchanger above the recognised dew point temperature. Modern combustion technology is being developed so that the exit flue gas temeprature can be lower than the acid dew point temperature by using specially designed heat exchangers and corrosion resistant materials of construction. It is claimed [Anon 1985] that bulk fluid temperatures as low as 38°C can be obtained with Teflon® covered tubes. Improvements in thermal efficiency are obtained by the technique. Domestic central heating boilers are also benefiting from the improved technology.

10.3.2 High temperature corrosion in gas systems

Although high temperature corrosion can occur in many systems, given the conditions that promote the problem, it is usually in combustion systems that the problem is manifest. ESDU [1992] reviewed corrosion associated with combustion. The nature of the fuel involved, particularly its chemical composition, will affect the potential corrosion. In general given good combustion control, it is usually the impurities in the fuel that give rise to the problem. The impurities are often minerals that were included in the fuel when it was formed or, in the case of waste material, from the constituents of the waste.

A more comprehensive disucssion of fouling and slagging in combustion systems as it affects mitigation, is to be found in Chapter 16, but for completeness, implications in respect of corrosion are discussed in this chapter.

Corrosion fouling may be associated with the deposition of other foulants such as particulate matter or molten fly ash. When corrosion occurs under deposits at high temperature alkali sulphates are often identified as the cause of the problem. Corrosion under deposits may be apparent during coal combustion. It is possible for sulphates such as $Na_2SO_4$ or $K_2SO_4$ or polysulphates such as $Na_2S_2O_7$ to form ionic melts on surfaces when the temperature is higher than the corresponding melting point. The presence of molten material provides the opportunity for electrochemical corrosion cells to be established in a similar way to that described in Section 10.2. The melting point of the deposits may be lower than expected due to the formation of fluxing agents that serve to lower the melting point, thereby enhancing the corrosion process.

The melting points of some of the mineral compounds associated with deposition and corrosion during combustion of fuel are presented in Table 10.3 [Tippler 1964].

TABLE 10.3

Melting points of some deposits

| Compound | Chemical formula | Melting point °C |
|---|---|---|
| Aluminium oxide | $Al_2O_3$ | 2049 |
| Calcium sulphate | $CaSO_4$ | 1450 |
| Ferric oxide | $Fe_2O_3$ | 1566 |
| Potassium sulphate | $K_2SO_4$ | 1069 |
| Potassium pyrosulphate | $K_2S_2O_7$ | 300 |
| Silicon oxide | $SiO_2$ | 1721 |
| Sodium sulphate | $Na_2SO_4$ | 882 |
| Sodium pyrosulphate | $Na_2S_2O_7$ | 401 |
| Sodium ferric sulphate | $Na_3Fe(SO_4)_3$ | 504 |

The relatively low melting points of sodium and potassium pyrosulphates and sodium ferric sulphate, demonstrate how the liquid electrical pathways can be readily produced.

An idealised sketch of the mechanism is given on Fig. 10.11. It represents a heat exchanger tube exposed to flue gas that contains a depositing foulant material. It is at the anodic site that corrosion occurs.

$$Fe \rightarrow Fe^{3+} + 3e \qquad (10.15)$$

In the cathodic region electrons are consumed by reactions such as:

$$O_2 + 4e \rightarrow 2O^{2-} \qquad (10.16)$$

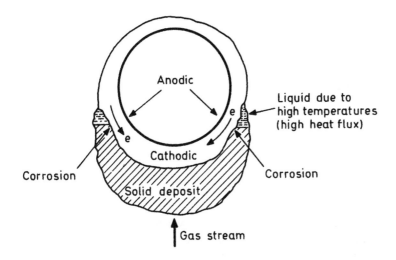

FIGURE 10.11. Corrosion of a heat exchanger tube in association with the deposition of particulate matter

The regions of molten ash deposits serve as ionic conduction pathways to complete the electrical circuit. In practical examples of high temperature corrosion under deposits, the mechanism is likely to be far more complex than this simple explanation.

High temperature corrosion of oil fired and other combustion equipment, probably proceeds by similar mechanisms to those in coal fired boilers. Differences will depend on the mineral content in the particular fuel and its chemical composition.

It is convenient to discuss corrosion fouling from combustion gases under the heading of the particular fuel.

*Coal combustion.* The extent and chemical composition of the mineral content of coal depends very much on the geology of the site from where the coal was obtained. Metal atoms are diverse and may include *Fe, Al, Ca, Na, K, Ti* and *Mg* and non metals usually include *O, Cl, S, P* and *Si.* The extent and combination of the mineral matter will affect the corrosion potential of a particular coal. An extensive review of the mineral content of coal was provided by Raask [1985].

Sulphur is present in coal either organically bound or as part of some mineral matter such as pyrite *(FeS$_2$)*. In general the inorganic sulphur content is higher than that associated with organic molecules. Under the high temperature conditions of a flame, the pyrite undergoes decomposition to *FeS* and elemental sulphur. The reaction involved is:

$$2FeS_2 \rightarrow 2FeS + S_2 \qquad (10.17)$$

Further reactions are possible according to the following equations:

$$2FeS + 3O_2 \rightarrow 2FeO + 2SO_2 \tag{10.18}$$

and the ferrous oxide reacts according to:

$$5FeO + O_2 \rightarrow Fe_2O_3 + Fe_3O_4 \tag{10.19}$$

The evolution of the acid gas $SO_2$ in the course of these reactions, has implications for further reaction to produce $SO_3$.

$$SO_2 + \frac{1}{2}O_2 \rightarrow SO_3 \tag{10.20}$$

The corrosive potential of $SO_2$ and $SO_3$ for low temperature conditions have already been described (Section 10.3.1) but these compounds can also become involved in reactions that lead to corrosion at higher temperatures.

Because of the high temperatures involved it is possible for some of the mineral matter to go forward in the flue gases as droplets (i.e. in the molten state). Such conditions are conducive to enhanced corrosion.

The rate of corrosion of the usual construction metals depends to a large extent on the integrity of the protective oxide layer. The formation of the oxides on the surface of a trivalent metal ($M$) can be accomplished by one or two consecutive chemical reactions:

$$M + \frac{1}{2}O_2 \rightarrow MO \tag{10.21}$$

and $$2MO + \frac{1}{2}O_2 \rightarrow M_2O_3 \tag{10.22}$$

Typical reactions with acid gases and subsequent chemistry include:

$$M + SO_3 \rightarrow MO\,SO_2 \tag{10.23}$$

Reactions with the protective oxide layer $M_2O_3$ may involve a molten sulphate for example:

$$3M_2O_3 + 3X\,SO_4 \rightarrow 3M_2SO_4 + 3X\,O_2 \tag{10.24}$$

where $X$ is any suitable cation and may be a complex combination of elements.

Under reducing conditions that may occur in some parts of the combustion space where the combustion is not complete and $CO$ is present, the protective oxide layer may be partially destroyed according to the equation

$$M_2O_3 + CO \rightarrow 2MO + CO_2 \qquad (10.25)$$

In the presence of sulphur or sulphur bearing compounds sulphides may be produced

$$M + S \rightarrow MS \qquad (10.26)$$

or in the presence of chlorine or its compounds

$$M + 2Cl \rightarrow MCl_2 \qquad (10.27)$$

A large temperature difference is often present across the protective oxide layer, and this could influence the effectiveness of the protection against further corrosion.

The method of firing will influence the chemical reactions that take place in the hotter regions of the combustion space because each of the techniques will have different temperature/time characteristics and hence the possibility of subsequent corrosive reactions (see also Chapter 16).

*Oil combustion.* Much higher proportions of sodium are encountered in fuel oils than in coal, and there are relatively high levels of vanadium, an element not usually found in coal. Furthermore the total non combustible fraction of oil is likely to be as low as 0.1%, whereas for poor quality coal it might be as high as 50%.

Vanadium, nickel and iron are indigenous in crude oil but sodium occurs in a brine phase. The silicon and calcium are present as a result of the oil refining process.

It is the sodium-vanadate-sulphur system that is often responsible for the corrosion in oil fired equipment. Niles and Sanders [1962] suggested that reactions between sodium sulphate and vanadium pentoxide are important. The three principal reactions are:

$$Na_2SO_4 + V_2O_5 \rightarrow 2NaVO_3 + SO_3 \qquad (10.28)$$

$$Na_2SO_4 + 3V_2O_5 \rightarrow Na_2O.3V_2O_5 + SO_3 \qquad (10.29)$$

and $Na_2SO_4 + 6V_2O_5 \rightarrow Na_2O6V_2O_5 + SO_3 \qquad (10.30)$

The relatively low melting points of vanadium petoxide ($V_2O_5$) and sodium vanadate (690 and 630°C respectively) may sustain electro chemical activity and corrosion, if the metal surface temperature exceeds these values.

The presence of $V_2O_5$ in the deposit/corrosion product matrix may provide a catalyst for the oxidation of $SO_2$ to $SO_3$ with the implications for the acid dew point of the system.

A more extensive discussion of fouling problems in fuel oil combustion is to be found in Chapter 16.

*Natural gas combustion.* Provided proper control is exercised the combustion of natural gas does not generally produce problems of corrosion since it is virtually completely combustible with no mineral impurities.

*Waste combustion.* In general the problems associated with coal combustion may be present during the combustion of waste materials, but in detail there are likely to be substantial differences. Each fuel will have its own characteristics and corrosion potential.

## 10.4   MATERIALS OF CONSTRUCTION TO RESIST CORROSION

Although glass and plastic materials have been used for the fabrication of heat exchangers there are limitations to their use. In the former the problem of fragility exists and in the latter there are temperature restrictions. At the present time the majority of heat exchangers are usually fabricated from metallic alloys where corrosion is anticipated.

Bendall and Guha [1990] in a concise paper have presented the options open to the designers of chemical process plant, including heat exchangers. They give the following useful summary of advice on metallic alloys generally for use in liquid systems.

1.   Standard austenitic stainless steels such as type 316 (18 *Cr*: 10 *Ni*: 3 *Mo*) have useful if limited resistance, to acids and reasonable resistance to pitting corrosion. Type 304 (18 *Cr*: 10 *Ni*) stainless steel has a good resistance to nitric acid. Austenitic stainless steels have relatively low strength, poor anti-erosion and abrasion properties and do not possess the ability to resist stress corrosion cracking.

2.   Super austenitic stainless steels with relatively high nickel content (approx. 20 *Cr*: 29 - 34 *Ni*), sometimes referred to as "alloy 20" are more costly than standard austenitic steels but provide excellent resistance to acids and some acid chlorides.

3.   Duplex stainless steels offer high strength, coupled with resistance to abrasion and erosion and to stress corrosion cracking. It is claimed that Ferralium alloy

255 (with 25% *Cr*), has excellent pitting and crevice corrosion with good resistance to acids.

4.  Nickel based alloys, such as Hastelloy, have outstanding corrosion resistance in reducing acids, mixed acids and acids at high temperatures. The principal restriction in their use is the high cost.

Apart from the corrosion resistance additionally any limitations on fabrication must be properly considered, i.e. the ability to be machined, rolled, welded, or cast.

## 10.5   FACTORS AFFECTING CORROSION FOULING

Although there are certain similarities between corrosion in contact with aggressive liquids and corrosion in the presence of gas streams, it is useful to discuss separately the influence of system variables in these two phases.

### 10.5.1 Liquid systems

As with most fouling mechanisms of prime importance in the corrosion of heat exchanger surfaces in contact with liquids is the flow velocity across the surface. Since corrosion of metals depends on the availability of chemical agents to effect the solution of metal, the rate of diffusion of these agents across the boundary layers will influence the corrosion rate. For instance in Section 10.2 the utilisation of electrons on nearby cathodic sites was illustrated with reference to the reduction of dissolved oxygen according to Equation 10.4. Limited availability of $O_2$ due to reduced diffusion will restrict the extent of the corrosion involving $O_2$. It could be anticipated that an increase of flow rate will only enhance the corrosion rate till the reaction rate at the surface becomes the limiting factor. The likely relationship between corrosion rate and velocity is schematically illustrated in Fig. 10.12. Even under stagnant conditions (zero flow) some corrosion is likely to occur that may be regarded as equivalent to some effective liquid flow. For relatively low velocities (i.e. in the laminar regime) the rate of corrosion may be directly related to velocity by an equation of the form

$$r_c = K_o u^n \tag{10.31}$$

where  $r_c$ is the rate of corrosion

   $K_o$ is a constant dependent upon the metal and the geometry

   $n$ is a constant

   $u$ is the velocity

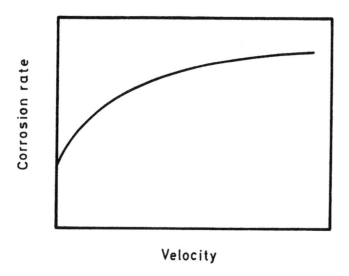

FIGURE 10.12. The change in corrosion rate with velocity

In the laminar conditions $n$ is $< 1$ but as turbulence is increased $n$ tends towards unity.

Increased availability of oxygen may provide an opportunity for some metals to change from an active mode to a passive condition. Mild steel corrodes actively and as already discussed, the oxide layer does not afford much protection. It is possible however, under certain conditions with enhanced oxygen availability, for the oxide layer to become more stable as passivation occurs. The resistance to corrosion under these circumstances is very dependent upon the flow rate, and active corrosion may recommence if the flow rate is signfiicantly reduced. Although Fig. 10.12 suggests that high velocities stabilise the corrosion rate, too high a velocity could destroy the protective layer due to erosion.

Since corrosion is the result of chemical reactions it would be anticipated that an increase in temperature would give rise to an increase in the rate of corrosion. The effect of temperature however, is complex since it will affect different components of the corrosion system in different ways. An increase of temperature may:

1. Modify the chemical composition and possibly the physical properties of the aggressive agent. The equilibrium within a liquid system is often dictated by the temperature so that a shift of temperature will change the equilibrium concentrations and hence the chemical composition of the liquid; for instance the reduction of oxygen solubility with temperature. Another example is the relationship between $HCO_3^-$ $CO_3^{2-}$ and dissolved $CO_2$ (see Chapter 9).

2. Affect the electrochemical behaviour of the metal/liquid system. A galvanic cell may be established between areas on the same heat exchanger that are at different temperatures. The hotter regions may be anodic or cathodic to cooler areas. Problems may arise when the thermogalvanic cell gives rise to activity in one area and passivity in another, that may already exist or is in the process of being developed.

3. Affect the nature and properties of oxide layers and corrosion deposits.

In addition the actual corrosion rate may be limited by the availability (diffusion to the surface) of the reactants, even though the raised temperature could increase the rate of reaction

Although the rate of corrosion may be regarded as high at the initiation of the process, for instance Mott and Bott [1993] noticed that corrosion sites on mild steel in contact with a simulated cooling water appeared within 3 - 4 hours exposure, it is likely that as the corrosion layer thickens the rate of reaction will decrease due to the barrier imposed by the deposit. Much will depend on the structure of the corrosion products, i.e. whether they are porous or compact.

### 10.5.2 Gas systems

Factors that influence corrosion fouling in liquid systems are also relevant to gas systems but there can be differences. In some combustion systems, such as coal combustion and refuse incineration, and to a lesser extent oil burning, particulate material is carried forward (see Section 10.3.2) with the gas stream. The particles may affect corrosion rates. At high velocities it is possible that particulate matter may erode protective corrosion layers as they form to expose fresh metal that may be subject to further corrosion. The presence of particulate material deposited on the metal surface, will affect the temperature of the metal/deposit interface. In combustion systems where heat is being removed from the combustion zone, as the deposit grows the metal temperature will decrease and consequently the rate of corrosion of the metal will be affected. The growth of the deposit due to particulate deposition and corrosion may in time, lead to spalling and foulant shedding and thus allowing the deposition/corrosion process to be repeated on a cyclic basis. Lister [1981] draws attention to this phenomenon of spalling, pointing out that the compactness of an oxide film determines the degree of protection it will afford. He suggests that the amount of oxide on a surface $m_o$ is proportional to the square root of time, i.e.

$$m_o \, \alpha \, t^{\frac{1}{2}} \tag{10.32}$$

If the deposit layer spalls at a certain thickness a succession of parabolic growths can lead to a deposit thickness versus time curve as illustrated in Fig. 10.13.

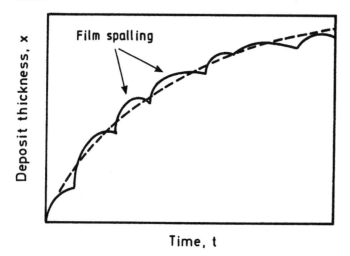

FIGURE 10.13. The development of an oxide layer under repeated spalling conditions

The spalling phenomenon may, due to thermal effects, be brought about by changes in temperature that affect the tenacity of the film, or weaknesses in the deposit as a result of gradual changes in composition.

## 10.6   CORROSION FOULING MODELS

There has been little published work on the mathematical modelling of corrosion fouling although some pioneering work has been carried out by Somerscales and his co-workers [1981, 1983, 1984 and 1988].

Somerscales [1983] generalises Equations 10.2 and 10.4 so that they apply to any metal using the elemental valency $Z$.

At the anodic surface Equation 10.2 becomes:

$$M \rightarrow M^{Z+} + Ze^-$$                                          (10.33)

where $M$ represents a metal

At the cathodic surface Equation 10.4 becomes:

$$\frac{Z}{4}O_2 + \frac{Z}{2}H_2O + Ze^- \rightarrow ZOH^-$$               (10.34)

and Equation 10.9 becomes:

$$M^{2+} + Z\,OH^- \rightarrow M(OH)_z \tag{10.35}$$

For aqueous systems the overall reaction becomes:

$$O_2 + 2\,H_2O + \frac{4}{Z}\,M \rightarrow \frac{4}{Z}\,M(OH)_z \tag{10.36}$$

$\frac{4}{Z}\,M(OH)_z$ represents the general corrosion product in an aqueous sytem involving the presence of oxygen, and a metal $M$ such as iron, steel, copper or aluminium.

The corrosion rate $r_c$ of metals that degenerate according to Equations 10.32 to 10.35 depends on the rate of supply of oxygen to the cathodic portions of the surface $r_{O_2}$, i.e.

$$r_c = K_1 r_{O_2} \tag{10.37}$$

where $\quad K_1 = \dfrac{4}{Z} \cdot \dfrac{M_m}{M_{O_2}}$ \hfill (10.38)

and $\quad M_m$ and $M_{O_2}$ are the molar mass of metal and oxygen respectively.

Somerscales [1983] points out that the *pH* of the water and impurities will also affect the rate of corrosion but, in order to develop the mathematics it is necessary to make suitable simplifying assumptions. It is assumed that the *pH* is in the range 4 - 7 although it is recognised that high or low *pH* values could be of importance. It is assumed further, that the presence of other chemical species can be neglected.

Fig. 10.14 is a schematic illustration of the oxygen, velocity and temperature profiles in a cooling medium in contact with a metal that is corroding according to equations 10.32 - 10.35.

The oxygen transport can be seen as involving three steps:

1. Convective transport from the bulk of the water to the water/corrosion product interface.

2. Diffusion through the deposit of corrosion product.

3. Reaction with the metal according to equations 10.32 - 10.35.

It is customary to assume that the corrosion reaction rate is so high that it plays no part in controlling the corrosion rate. As a result it is possible to postulate that the oxygen concentration at the interface is zero.

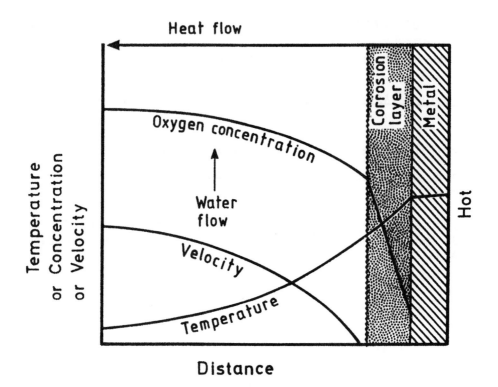

FIGURE 10.14. Oxygen, velocity and temperature profiles at a corroding surface

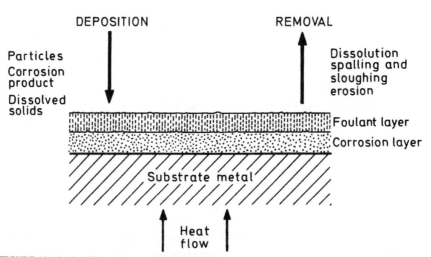

FIGURE 10.15. Conditions at a corroding metal surface

It is likely that removal of the corrosion layer will occur depending on the flow conditions, the solubility, the cohesive strength of the products of corrosion and their adhesion to the underlying metal. Fig. 10.15 provides an idealised concept of the conditions in the interface region.

Somerscales [1981] states that the removal process may include one or more of the following mechanisms:

1. Dissolution, i.e. the material leaves the surface in ionic form.

2. Erosion, i.e. particles of corrosion product are removed.

3. Spalling, i.e. material is removed in relatively large pieces.

Each of these mechanisms is likely to be influenced by the prevailing velocity, either through shear effects or the convective transport of material away from the solid surface.

Controlling mechanisms considered by different workers are given in Table 10.4.

TABLE 10.4

Controlling mechanisms in corrosion fouling [Somerscales 1981]

| Application | Control | Reference |
|---|---|---|
| Initial stages of corrosion fouling | Mass transfer | Butler [1966] |
| Corrosion fouling where there is a significant fouling deposit | Diffusion | Dillon [1959] Butler [1966] |
| Corrosion fouling where mass transfer and diffusion apply. Not originally devised for non-ferrous metals. | Mass transfer and diffusion | Mahato *et al* [1968] Galloway [1973] |
| Diffusion controlled growth and constant rate of removal | Diffusion | Dillon [1959] |
| Depends on the processes controlling formation and removal of corrosion products and heat transfer | Probably mass transfer | Khailov *et al* [1975] |

Assuming that the corrosion process is controlled by mass transfer to and diffusion through the existing layer it is possible to write:

$$r_c = \frac{K_1 c_{bo_2}}{\dfrac{1}{K_{o_2}} + \dfrac{x}{D_{o_2}}} \tag{10.39}$$

where $c_{bo_2}$ is the bulk concentration of oxygen

$K_{o_2}$ is the mass transfer coefficient of oxygen

$D_{o_2}$ is the diffusivity of oxygen

$x$ is the instantaneous thickness of the corrosion layer

It is not possible to use Equation 10.39 to estimate the rate of corrosion at an instant of time because $x$ is generally unknown.

The mass of metal lost by corrosion per unit area $m_c$ is given by

$$m_c = \rho_m (x + \varepsilon) \tag{10.40}$$

where $\varepsilon$ is the "thickness" of metal lost from the surface by some removal mechanism.

and     $\rho_m$ is the metal density

Combining Equations 10.38 and 10.39 produces:

$$r_c = \frac{K_1 D_{o_2} \rho_m c_{Bo_2}}{m_c - \varepsilon \rho_m + \dfrac{D_{o_2} \rho_m}{K_{o_2}}} \tag{10.41}$$

If the removal process is by dissolution it is possible to write an equation of the form:

$$\rho_m \frac{d\varepsilon}{dt} = k_2 (c_{i_2} - c_{b_2}) \tag{10.42}$$

where $c_{i_2}$ and $c_{b_2}$ are the concentrations of the dissolving species 2 at the liquid/solid interface and in the bulk liquid respectively.

$K_2$ is the mass transfer coefficient.

It is not possible to write equivalent equations for erosion and spalling since there is insufficient theory to describe these processes.

The concentration of the dissolving species at the interface will be the equilibrium value in contact with the solid. Under these circumstances it is possible to use the solubility product $K_{s_o}$ as a basis for determining $c_{i_2}$, i.e.

$$c_{i_2} = \frac{K_{s_o}}{[c_{OH}]^z f_M f_{OH}}$$

(10.43)

where $f_M$ and $f_{OH}$ are the activities of $M^{z+}$ and $OH^-$ respectively.

Combining Equations 10.41 and 10.42 yields:

$$\rho_m \frac{d\varepsilon}{dt} = K_2 \left\{ \frac{K_{s_o}}{[c_{OH}]^z f_M f_{OH}} - c_{b_2} \right\}$$

(10.44)

In order to determine the value of $\varepsilon$ the amount of corrosion product removed from the surface, it is necessary to integrate Equation 10.44.

$$\rho_m \varepsilon = k_2 \left\{ \frac{K_{s_o}}{[c_{OH}]^z f_m f_{OH}} - c_{b_2} \right\} \int_{t=o}^{t=t} dt$$

(10.45)

Strictly Equation 10.41 only applies to a single point on a corroding surface at an instant of time. In a practical example conditions will not be uniform across the surface, however it could be assumed that the equation refers to average values. Equation 10.41 in combination with an equation describing the removal of corrosion products, could therefore be used to describe the overall process.

First considering isothermal conditions Somerscales [1983] attempted to resolve an extremely difficult mathematical problem and devised an equation in terms of dimensionless groups, i.e:

$$\frac{\Gamma \mu_c}{K_1} = f\left( \frac{\tau \Gamma^2}{K_1} \right)$$

(10.46)

where $\Gamma$ (dimensionless) is defined by

$$\Gamma = \frac{\phi_R \rho_m l_m}{c_{bo_2} D_{o_2}}$$

(10.47)

$\phi_R$ is the rate of removal

$l_m$ is a characteristic dimension of the system

$\mu_c$ (dimensionless) is defined by

$$\mu_c = \frac{m_c}{\rho_m l_m} \tag{10.48}$$

and    $\tau$ is dimensionless time

$$= \frac{t.c_{bo_2} D_{o_2}}{\rho_m l_m^{\,2}}$$

Somerscales [1983] applied the model (Equation 10.46) to the data published by English *et al* [1961] and the results are plotted on Figs. 10.16 and 10.17. There are limitations with these data since they do not relate to a uniform cross section for flow. They do however, demonstrate the applicability of the model.

Further development of the concepts to take into account the effects of heat transfer led Somerscales [1983] to put forward further equations.

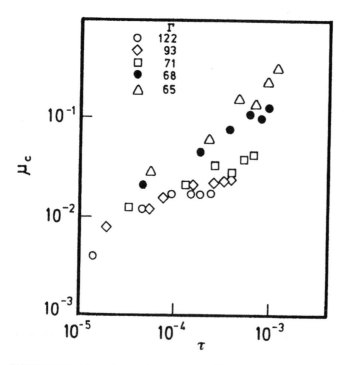

FIGURE 10.16. Dimensionless plot $\mu_c$ versus $\tau$ of data for the corrosion of aluminium

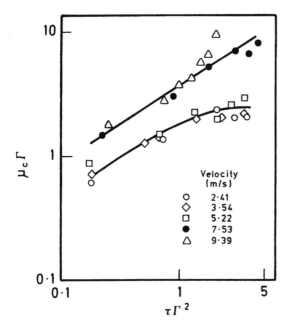

FIGURE 10.17. Dimensionless plot $\Gamma \mu_c$ versus $\tau \Gamma^2$ of data for the corrosion of aluminium

A dimensionless grouping $E$ may be based on these temperatures where

$$E = \frac{T_s - T_{so}}{T_{so} - T_b} \qquad (10.49)$$

and  $T_{so}$ the initial temperature of the metal/water interface (i.e. before corrosion commences) is identical to $T_i$ provided the heat transfer coefficient remains constant.

Other dimensionless groupings may be defined

$$D = Sh_2 \frac{D_2}{D_{o_2}} \frac{c_{b_2}}{c_{bo_2}} \cdot \left[ \frac{A}{c_{b_2}} e^{-\frac{B}{T_i}} - c_{b2} \right] \qquad (10.50)$$

where $Sh_2$ is the Sherwood number for mass transfer of the dissolved corrosion defined as

$$\frac{k_2 l_m}{D_2}$$

$D_2$ is the diffusivity of the corrosion products

$c_{b_2}$ is the concentration of corrosion products in the bulk water

$A$ and $B$ are constants

and a *Biot* number $Bi$ for the deposit may be defined as

$$\frac{q l_m}{\lambda(T_{so} - T_b)}$$

where $q$ is the heat flux

and $\lambda$ the thermal conductivity of the corrosion layer

The relationship involving these dimensionless groups suggested by Somerscales [1983] is:

$$\frac{\tau Bi\, D^2}{K_1} = f\left(\frac{DE}{K_1}\right) \tag{10.51}$$

The model represented by Equation 10.51 is extremely complex and does involve a number of assumptions and in practice could be difficult to employ with any degree of confidence. However the basis and development of the equations does assist a better understanding of what is, after all a complex process.

## REFERENCES

Anon., 1985, Subdew condensing heat exchangers. Montair Andersen BV, Sevenum.

Bendall, K. and Guha, P., 1990, Balancing the cost of corrosion resistance. Proc. Ind. J. March 31 - 34.

Butler, G., 1966, Proc. Int. Conf. on Metallic Corrosion, 1, 271, Moscow.

Dillon, R.L., 1959, Corrosion, 15, 13.

English, J.L., Rice, L., Griess, J.C., 1961, AEC Report No. ORNL, 3063.

ESDU, 1992, Data Item 92012 Fouling and Slagging in Combustion Plant. ESDU International, London.

Fontana, M.G. and Greene, N.D., 1967. Corrosion Engineering. McGraw Book Co., New York.

Galloway, T.R., 1973, Int. J. Heat Mass Transfer, 16, 443.

Gilbert, P.T., 1960, Corrosion problems arising in tubular heat exchanger equipment in corrosion problems of the petroleum industry. Monograph 10. Soc. Chem. Ind., London.

Kemmer, F.N., 1988, Ed. The Nalco Water Handbook. McGraw Hill Book Co., New York.

Khailov, V.S. *et al*, 1975, J. App. Chem. USSR, 48, 2059.

Lake, T., 1990, Duplex steels resist corrosion in heat transfer. Proc. Ind. J. July 15 - 17.

Lewis, T.J. and Secker, P.E., 1965, Science of Materials. George G. Harrap & Co. Ltd., London.

Lister, D.H., 1981, Corrosion products in power generating systems, in: Somerscales, E.F.C. and Knudsen, J.G. eds. Fouling of Heat Transfer Equipment. Hemisphere Publishing Corp., Washington.

Mahato, B.K. *et al*, 1968, Corrosion Science, Vol. 8, 173.

Mott, I.E.C. and Bott, T.R., 1993, Biofilm adhesion under annular and tubular flow regimes using a simulated cooling water system. 29th Nat. Heat Trans. Conf. Atlanta, August.

Niles, W.D., and Sanders, H.R., 1962, Reactions of magnesium with inorganic constituents of heavy fuel oil and characteristics of compounds formed. Trans. ASME J. Engng. Power, 178.

Raask, E., 1985, Mineral impurities in coal combustion. Hemisphere Publishing Corp., Washington.

Sato, S., Nagata, K. and Yamaha, T., 1977, Re-evaluation of corrosion resistance of copper-nickel based alloys to sea water. INCO Power Conference, Lousanne.

Somerscales, E.F.C., 1981, Corrosion fouling, in: Chenoweth, J.M. and Impagliazzo, M. eds. Fouling in Heat Exchange Equipment. 20th ASME/AIChE Heat Transfer Conference, Milwaukee HTD, Vol. 17.

Somerscales, E.F.C., 1983, Fundamental ideas in corrosion testing in the presence of heat transfer. Proceedings Conference on Corrosion in Heat Transfer Conditions. Soc. Chem. Ind., London.

Somerscales, E.F.C. and M. Kassemi, 1984, Fouling due to in-situ corrosion products, in: Suiter, J.W. and Pritchard, A.M. eds. Fouling in Heat Exchanger Equipment. 22nd Nat. Heat Transfer Conf. Niagara Falls, ASME HTD, Vol. 35.

Somerscales, E.F.C., 1988, Corrosion fouling: liquid side, in: Melo, L.F., Bott, T.R. and Bernardo, C.A. eds. Fouling Science and Technology, Kluwer Academic Publishers, Dordrecht.

Turner, M., 1990, What every chemical engineer should know about galvanic corrosion. Chem. Engr. January 28 - 32.

Turner, M., 1989, What every chemical engineer should know about crevice corrosion. Chem. Engr. September 41 - 43.

# CHAPTER 11

## Chemical Reaction Fouling

### 11.1  INTRODUCTION

Chemical reaction fouling is usually associated with organic chemicals as opposed to reactions of metals with aggressive agents as discussed in Chapter 10. The mechanisms and problems have been reviewed and discussed by a number of authors among them: Froment [1981], Crittenden [1988], Watkinson [1988], and Murphy and Campbell [1992].

Under the influence of the temperatures present in the heat exchanger chemical reactions can take place that do not involve the heat transfer surface as a reactant, although in some circumstances the metal surface may act as a catalyst. In others the metal surface may reduce or inhibit, the extent of the potential chemical reaction.  The chemical reactions are usually complex and may involve such mechanisms as autoxidation, polymerisation, cracking or coke formation. Processes that are susceptible to fouling of this kind are often found in oil refining, petrochemical manufacture and food processing.  Generally the problems arise in liquid streams or in mixtures of gases and liquids, and sometimes in streams of vapours.

In the preheating of crude oils prior to the primary distillation process, the temperature may be progressively raised from ambient to around 250°C in the preheaters and subsequently to as high as 400°C usually in an oil fired furnace. Other refinery operations where chemical reaction fouling may occur are in thermal cracking of heavy hydrocarbon fractions for the production of lighter fractions. Modern technologies use catalysts to effect the cracking reactions (catalytic cracking) but chemical reaction fouling may still be a problem.  In the petrochemical industry it is usual to "crack" hydrocarbons varying from ethane to gas oil or even crude oils, to satisfy the demand for olefins and aromatics which are the principal feedstocks for further processing.  In contrast some petrochemical operations are designed to produce coke or carbon as the primary product, to be used as filler material for rubber and plastic products, e.g. motor tyres. In the food industry many of the large organic molecules that make up the food substance are heat sensitive so that exposure even to relatively low temperatures, can produce breakdown products that will foul heat transfer surfaces.  The pasteurisation of milk in plate heat exchangers for example, usually results in chemically complex deposits after some hours operation so that cleaning of the heat exchanger is required to maintain product quality and heat transfer efficiency [Delplace et al 1994].

In addition to carbon and hydrogen, deposits may contain significant amounts of inorganic matter which may aid further chemical degradation by catalysing the reactions or participating in or encouraging corrosion. Inorganic constituents of the deposit may be there as a result of:

1. Salts dissolved in an aqueous fraction associated with the organic liquid. For instance in crude oil processing steps are taken to "desalt" the hydrocarbon but there may be some "carry over" in the organic liquid stream.

2. Organic molecules containing hetero atoms. For example sulphur is often a constituent of crude oil in the form of mercaptans or other organo-sulphur compounds.

3. Organo-metallic compounds. Crude oil for instance usually contains vanadium and nickel that may be present in combination with complex organic structures.

4. Corrosion products that have become detached from upstream equipment and are trapped within the deposits produced by chemical reactions.

The reduction of heat transfer coefficient due to reaction fouling in an experimental recycle apparatus through which a light Arabian crude oil (containing 10% waxy residue from a crude oil tank) was flowing at 0.5 *m/s* is given in Fig. 11.1 [Crittenden *et al* 1993].

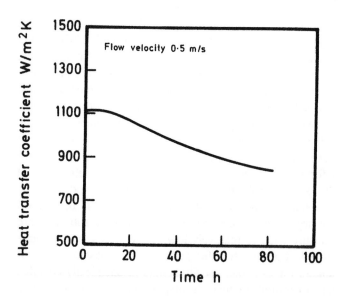

FIGURE 11.1. The change in heat transfer coefficient with time attributed to chemical reaction fouling

The chemical reaction either forms the deposit directly on the heat transfer surface or is involved in forming deposit precursors, which subsequently result in the formation of deposits. Precursors may be formed in the bulk liquid, in the boundary layers near the heat exchanger surfaces or directly on the surface. The precursor may be soluble in the bulk fluid and only give rise to deposition when it is carried by diffusion or eddy transport to the wall region. It is entirely possible that if the precursor precipitates or reacts to cause a solid to be formed remote from the wall, then the deposition process will be particulate as described in Chapter 7. If there is a reaction with the wall itself then the mechanism could be regarded as corrosion fouling.

At the present state of knowledge it is impossible to separate the creation of precursors from deposit appearance on the heat exchanger surface. It is the overall result that is apparent. Furthermore the presence of precursors, let alone their chemical character may not be known, and they may appear as a result of the very small concentrations of trace substances in the process liquid. Many alternative chemical routes and reactions to the deposit are possible. The precursors themselves may already be present in the liquid stream or they may be the result of chemical reactions within the heat exchanger.

The complexities of chemical reaction fouling may be compared to crystallisation or particulate fouling, where the depositing species is generally known and does not undergo chemical change during the deposition process. There are similarities in that the character of the deposit is strongly dependent on the conditions within the heat exchanger and the length of time the deposit has been in place. It is extremely unlikely that the deposit due to chemical reactions, will be uniform throughout its thickness since, as the deposit thickens there will be a redistribution of temperature that could change the appearance of the deposit. For instance if a heavy oil is being heated in a coil and deposit formed on the inner surface of the coil, the interface temperature between the metal and the deposit will gradually increase. The result may be the formation of a dense "coke-like" material on the wall with outer layers of a more porous nature. Trapped liquid within the pores of the initial deposit may thermally degrade under the influence of the increasing temperature, to form further dense coke within the pores of the original structure. The release of gaseous products may produce crevices and pores within the deposit. As time passes the liquid filled porous "front" gradually moves away from the metal surface so that the deposit becomes more robust and densified.

The discussion of chemical reaction fouling will be restricted to that associated with hydrocarbon processing. Problems with food processing are covered in Chapter 16.

## 11.2 BASIC CHEMISTRY

It is not possible within the necessary limitations of this book, to discuss all the potential chemical reactions that may lead to residues and coke formation on heat transfer surfaces, but some of the principal chemical mechanisms will be presented.

The structure as well as the concentration of species present even in trace quantities, can have pronounced effects on fouling rates.

## 11.2.1 Background

Before considering some detailed chemistry it is useful to record general observations that have a bearing on the discussion. Crittenden [1984] has listed some relevant facts:

1.  Liquids containing molecules of high molecular weight have a greater tendency to form deposits compared with lighter fractions. In contradiction however, for pure n-paraffins at moderate temperatures (366 - 505 $K$), the autoxidation reactions leading to deposition decreases with increasing numbers of carbon atoms in the chain in the range 10 - 16.

2.  Branched chains have a greater tendency to deposition (i.e. increased deposition rates).

3.  Thermally decomposed liquid streams are particularly susceptible to gum formation and deposition. Deposition from n-decane in the range 366 - 450 $K$ is inhibited in the presence of aromatics and naphthenes but encouraged in the presence of olefins.

4.  The presence of oxygen (and its concentration) has a marked effect on fouling rates.

5.  Some sulphur and nitrogen containing compounds, particularly those that break down under the conditions prevailing near the heat exchanger surfaces to produce free radicals, promote deposition.

6.  Dissolved metal complexes and metals can increase deposition through catalytic action. Copper is said to be particularly deleterious.

7.  Antioxidants, and corrosion inhibitors may increase deposition rates.

8.  Desalting of crude oil streams does not always eliminate deposition of inorganic material.

## 11.2.2 Possible reactions leading to deposition

Stephenson and Row [1993] have outlined the chemistry of chemical reaction fouling and Watkinson [1988] has provided a very useful review of the literature relating to some mechanisms that can lead to insoluble foulant materials. They include:

1. *Gum formation.* Heating of paraffinic hydrocarbon mixtures found in jet fuels, gas oils and similar products may result in the precipitation of gum-like material. A sequence of events has been proposed by Taylor [1967, 1968]. Autoxidation forms a soluble oxidation product, with further oxidation to an insoluble polymer. The polymer may be formed on the wall or transported as particles to the wall.

$$\text{HYDROCARBON} \xrightarrow[\text{OF O}_2]{\text{TRACES}} \text{SOLUBLE OXIDATION PRODUCT} \xrightarrow{\text{O}_2} \begin{array}{c}\text{INSOLUBLE POLYMER}\\ \text{M. Wt. 400 - 600}\end{array}$$

2. *Deposition of asphaltenes from crude oil.* A series of degradation stages as crude oil is heated has been described by Eaton and Lux [1984]. The process is similar to (1), i.e. oxidation and polymerisation.

$$\begin{array}{c}\text{SATURATED}\\ \text{HYDROCARBONS}\end{array} \xrightarrow[\text{ACIDS}]{\text{INORGANIC}} \begin{array}{c}\text{UNSATURATED}\\ \text{HYDROCARBONS}\end{array} \xrightarrow{\text{O}_2} \begin{array}{c}\text{ORGANIC}\\ \text{ACIDS}\end{array}$$

and

$$\begin{array}{c}\text{ORGANIC}\\ \text{ACIDS}\end{array} \xrightarrow{\text{METALS}} \begin{array}{c}\text{RESINS AND}\\ \text{ASPHALTENES}\end{array} \xrightarrow[\text{(WALL)}]{\text{HOT METALS}} \begin{array}{c}\text{COKE-LIKE}\\ \text{DEPOSIT}\end{array}$$

Time is also necessary to allow these reactions to proceed to any degree [Crittenden 1988]. Dickakian and Seay [1988] suggested that asphaltene precipitation is the major contribution to crude oil heat exchanger fouling. The incompatibility of asphaltenes with crude oil, either through reactions or insolubility, causes precipitation and adherence to heat transfer surfaces. Residence on the hot exchanger wall allows carbonisation to take place with the formation of coke. A similar mechanism involving resins as well as asphaltenes has been suggested by Eaton and Lux [1984].

3. *Oxidation of lubricating oil.* A scheme for deposition following the oxidation of lubricating oils occurs through the formation of low molecular weight precursors and polymerisation, has been suggested by Tseregounes *et al* [1987]. The similarity with scheme (1) can be seen.

$$\begin{array}{c}\text{LUBRICATING}\\ \text{OIL}\end{array} \xrightarrow{\text{O}_2} \begin{array}{c}\text{LOW MOLECULAR WEIGHT}\\ \text{PRIMARY OXIDATION}\\ \text{PRODUCT}\end{array} \xrightarrow{\text{O}_2} \begin{array}{c}\text{HIGH MOLECULAR}\\ \text{WEIGHT PRODUCT}\end{array}$$

and

$$\begin{array}{c}\text{HIGH MOLECULAR}\\ \text{WEIGHT PRODUCT}\end{array} \rightarrow \text{DEPOSIT}$$

4. *Coke formation during cracking of light hydrocarbons.* Ethylene and propylene are strong coke precursors, probably not by polymerisation but rather by cyclisation with higher diolefins and aromatics [Froment 1981]. Based on these observations it is possible to offer the following scheme:

$$\text{OLEFINS} \rightarrow \begin{array}{c}\text{HIGHER}\\\text{MOLECULAR WEIGHT}\\\text{HYDROCARBONS}\end{array} \xrightarrow{\text{CYCLISATION}} \begin{array}{c}\text{CYCLIC}\\\text{INTERMEDIATES}\end{array}$$

and

$$\begin{array}{c}\text{CYCLIC}\\\text{INTERMEDIATES}\end{array} \xrightarrow{\text{POLYMERISATION}} \begin{array}{c}\text{HIGH}\\\text{MOLECULAR WEIGHT}\\\text{HYDROCARBONS}\end{array} \rightarrow \begin{array}{c}\text{TARS}\\\text{AND}\\\text{CARBON}\end{array}$$

Watkinson [1988] observes that the last stages of this mechanism also occur in heating of heavy organics in the absence of oxygen.

5. *Sludge promoted fouling.* Carry over of sludge from storage tanks has been known to cause severe heat exchanger fouling problems in refinery operations. Nixon [1962] suggests that heteroatomic species are involved. Pyrroles are examples of nitrogen-containing molecules that are deposit forming through oxidation or co-oxidation followed by condensation.

$$\text{PYRROLES} \xrightarrow[\begin{array}{c}\text{OLEFINS}\\\text{SULPHIDES}\end{array}]{O_2} \begin{array}{c}\text{DIMERS}\\\text{AND}\\\text{TRIMERS}\end{array} \xrightarrow{\text{CONDENSATION}} \text{POLYMERS}$$

6. *Deposition from coal products.* Deposits could form in heat exchangers and have been identified as being in part at least, products of phenolic coupling reactions [Gayla *et al* 1986]

$$\text{PHENOL} \xrightarrow{O_2} \begin{array}{c}\text{HIGH MOLECULAR}\\\text{WEIGHT PHENOLIC}\\\text{SUBSTANCES}\end{array} + \text{WATER}$$

Watkinson [1988] observes that polymerisation reactions of some kind are the key to a better appreciation of organic chemical reaction fouling. Polymerisable species may already be contained in the feedstock, or they could be produced as the result of pyrolysis (thermal decomposition) or by autoxidation.

11.2.3 Pyrolysis, cracking and dehydration

At temperatures in excess of 650 $K$ the primary reactions are rapid and in the gas phase the mechanisms are well understood [Fitzer *et al* 1971].
Cracking may occur according to the equation

$$2R_1 - CH_2 - CH_2 - CH_2 - R_2 \rightarrow R_1CH_2 - CH_3 + R_2 - CH = CH_2 \tag{11.1}$$

and/or

$$R_1 - CH = CH_2 + R_2 - CH_2 - CH_3$$

where $R_1$ and $R_2$ are organic radicals.

Dehydration could follow the equation

$$CH_3 - CH_2 - CH_2 - CH_3 \rightarrow CH_3 - CH = CH - CH_3 + H_2 \tag{11.2}$$

The pyrolytic rupture of the *C-C* or *C-H* bond can occur via radical or ion formation.

i.e. $\quad R_1 - R_2 \rightarrow R_1^{\cdot} + R_2^{\cdot} \qquad$ radical

or $\quad R_1 - R_2 \rightarrow R_1^{+} + R_2^{-} \qquad$ ionic $\tag{11.3}$

Ionic decomposition requires more energy than decomposition via radical formation, so that thermal cracking proceeds through the free radical route [Crittenden 1988] and where a catalyst is involved, the chemical reaction pathway involves ions.

Secondary synthesis reactions involving the products of primary reactions from mechanisms such as those given in Equations 11.1 and 11.2 involve cyclisation to form aromatic compounds followed by condensation to form high molecular weight compounds.

### 11.2.4 Autoxidation

Full radical autoxidation reactions are initiated by the removal of hydrogen from an organic molecule by a free radical ($X^{\cdot}$), i.e.

$$R - H + X^{\cdot} \rightarrow R^{\cdot} + XH \tag{11.4}$$

Reaction of the organic free radical with molecular oxygen together with further hydrogen removal results in a chain reaction:

$$R^{\cdot} + O_2 \rightarrow ROO^{\cdot} \tag{11.5}$$

$$ROO^{\cdot} + RH \rightarrow ROOH + R^{\cdot} \tag{11.6}$$

Termination may occur through the chance combination of free radicals, i.e.

$$R + R \rightarrow R - R \tag{11.7}$$

$$R + ROO \rightarrow ROOR \tag{11.8}$$

$$ROO + ROO \rightarrow ROOR + O_2 \tag{11.9}$$

The rupture of *0-0* bonds in the organic acid produced by reaction 11.6 can give rise to further free radicals according to

$$ROOH \rightarrow RO + OH \tag{11.10}$$

In the presence of oxygen free radicals are bred through mechanisms depicted in Equations 11.4, 11.5, 11.6 and 11.10 and so the process is self sustaining.

11.2.5 Polymerisation

Polymerisation, the final stages of deposit formation may involve three basic steps [Crittenden 1988]. They include:

$$XH \rightarrow X + H \tag{11.11}$$

$$\text{and } R - CH - CH_2 + X \rightarrow R - CH - CH_2X \tag{11.12}$$

followed by

$$R - CH - CH_2X + n(R - CH = CH_2) \rightarrow R - CH - \left[ CH_2 - \overset{R}{CH} \right]_n CH_2 - \overset{R}{CH} \tag{11.13}$$

which propagates the free radical formation

Chain termination may occur by a mechanism such as:

$$R - \underset{CH_2}{C} - \left[ -CH_2 - \overset{R}{CH} \right]_n - CH_2 - \overset{R}{C} + H \rightarrow R - \underset{CH_2}{CH} - \left[ -CH_2 - \overset{R}{CH} - \right]_n - CH_2 - \overset{R}{CH_2} \tag{11.14}$$

The rate at which each step proceeds depends on a number of interacting factors. Oxygen, halides, sulphides, nitrogen compounds, some metals and metallic compounds are able to initiate polymer formation. The temperature and the stability of the individual radicals determine the extent to which they participate in the fouling process.

## 11.3   DEPOSIT COMPOSITION

Although the composition of deposits will be very specific to the particular flow stream from which they were derived, and therefore cannot be taken as general, it is useful to examine some published data.

In a comprehensive study of fouling in crude oil preheat exchangers on a refinery, Crittenden *et al* [1992] were able to make an analysis of some of the deposits. These data are reproduced in Table 11.1 and are related to the processing of "light" crudes.

TABLE 11.1

Analysis of deposits from crude oil preheaters [Crittenden *et al* 1992]

| Fraction wt% | 1 | 2 | 3 | 4 |
|---|---|---|---|---|
| *n*-heptane-soluble | 49.8 | 22.6 | 56.2 | 57.4 |
| Toluene-soluble | 1.9 | 1.1 | 1.6 | 1.2 |
| Loss on ignition at 820 $K$ | 32.8 | 37.2 | 24.6 | 25.3 |
| Remaining ash | 15.5 | 39.1 | 17.6 | 16.1 |
| TOTAL | 100.0 | 100.0 | 100.0 | 100.0 |
| *Components in ash wt%* | | | | |
| Iron | 35.5 | 28.1 | 37.1 | 42.2 |
| Sulphur | 29.0 | 18.3 | 27.4 | 28.0 |
| Sodium | 20.0 | 29.6 | 21.7 | 18.0 |
| Calcium | 7.7 | 3.3 | 4.1 | 5.6 |
| Zinc | 2.6 | 1.0 | 2.8 | 3.1 |
| Magnesium | 1.3 | 0.5 | 0.6 | - |
| Chlorine | - | 14.1 | 1.1 | 0.6 |
| Others | 3.9 | 5.1 | 5.2 | 2.5 |
| TOTAL | 100.0 | 100.0 | 100.0 | 100.0 |

Columns 1 - 4 in Table 11.1 refer to different heat exchangers. The heptane soluble fraction is assumed to be resins and free oil.
The toluene soluble fraction is assumed to be asphaltenes.
The coke fraction is defined as the loss on ignition at 820 $K$ of the toluene insoluble fraction.

The authors comment that the absolute numbers cannot be given too much importance because of the necessary shutdown procedure before the exchangers could be opened up for inspection, but they do show general similarities. The relatively high inorganic content may be noted. The presence of iron suggests corrosion reactions (see Chapter 10). The sulphur is likely to originate from the crude oil. It is suggested that the high sodium content is probably due to the practice of injecting caustic to control *pH* and hence corrosion. The high ash sodium and chlorine contents of Exchanger 2 are attributed to carry over from the associated desalter.

Some limited data on deposits isolated in different refinery processing units were quoted by Watkinson [1988] and are reproduced in Table 11.2. In many respects these data reflect those of Crittenden *et al* [1992]. Data also relating to deposition in crude oil processing were reported by Lambourne and Durrieu [1983], and are included in Table 11.3.

Although it is difficult to make precise comparisons because of the unknown differences in heat exchanger operation and the differences in the analytical techniques, certain observations may be made in respect of these data and those of Crittenden *et al* [1988].

TABLE 11.2

Deposits in refinery processing units [Watkinson 1988]

| Unit | Crude | Hydro-desulphur-iser | Reformer | Coker-visbreaker | Alkylation |
|------|-------|----------------------|----------|------------------|------------|
| Max. fluid temp. °C | 360 | 370 | 510 | 500 | 150 |
| Max. metal temp. °C | 400 | 440 | 510 | 540 | 240 |
| Fraction wt % | | | | | |
| Organics | 80 | 86-93 | 34-65 | 98 | 36-97 |
| $Fe_2O_3$ | 20 | 3-11 | 25-30 | | 3-63 |
| $Fe\,S$ | | 3- 5 | 5 | | |
| Sulphur | | | 20 | 2 | |
| | | | $NH_4Cl$ | | |
| Others | | | 20 | | |

The striking difference is in terms of the asphaltene content. The data in Table 11.1 are roughly half the value of the lower limit quoted in Table 11.3 for "light" crudes. The relatively high asphaltene fractions in deposits associated with "heavy" crudes are worthy of note. The "organics" mentioned in Table 11.3 at a weight fraction of 80% of the deposit, are of a roughly similar order to the total organic components quoted by Crittenden *et al* [1992].

TABLE 11.3

Deposits in crude oil processing

| Deposit fraction wt % | Various crudes API ~ 34° | Light crudes API > 40 |
|---|---|---|
| Asphaltenes | 60-75 | 3-10 |
| Water soluble salts | 1- 5 | 1- 5 |
| Iron salts | 20-35 | 75-90 |

It is also interesting that the iron salts reported in Table 11.3 are much greater than even the total inorganic ash recorded in Table 11.1. High levels of iron in deposits taken from desulphuriser unit preheaters, contining up to 38.6 wt% as $Fe_2O_3$ were reported by Canapary [1961]. In the radiant section of the unit furnace the $Fe_2O_3$ wt fraction was estimated as 6.6%. It is not surprising that relatively high levels of sulphur were recorded in deposits in a unit of this kind.

Bott [1990] reported some qualitative data on the composition of deposits taken from the "tar side" of a tar oil heater. The appearance of the deposit was that of a very thick dark coloured paste that included fibres, iron (as rust), carbon, sand and glass particles. The deformable nature of the deposit was likely to be due to the flexible bonding of the particles by the process oil. The presence of components in the deposit other than organic related material, confirms the observation that parallel to chemical reaction fouling on process plant, there are other mechanisms involved.

Coggins [1968] made a survey of deposits in gas-making equipment. The survey comes from a time in the UK when petroleum fractions with reforming processes were used to produce towns gas. The necessity for these units disappeared when natural gas from the North Sea became available. Problems of fouling arose largely as a result of subjecting the petroleum feedstocks to high temperatures, very often in the search for thermal efficiency. Bott [1990] reports an example of a vaporiser installed in the flue gas system of a furnace that was subject to considerable fouling problems (the problem was resolved completely by using steam to vaporise the organic liquid). The data for preheaters, vaporisers and superheaters, together with the timescale for the deposition to become a serious problem are given in Table 11.4.

It is clear from the table that there is considerable variation between units not only in the quality of the deposit, but also the extent of the problems as far as satisfactory operation is concerned. In the same report Coggins provided some data on the composition of the deposits taken from naphtha vaporisers (see Table 11.5). The main constituents of the ash appeared to be iron, sulphur and lead together with many other inorganic elements and compounds including vanadium and sulphur.

It is clear that the formation of deposits from what might be considered to be basically due to chemical reactions, is an extremely complex process and will very much depend on the nature of the process liquid (i.e. whether or not it is a mixture of compounds and its trace components) and the temperature regime it encounters.

The analysis of the deposit may give some indication of the processes by which the deposit was formed. The presence of oxygen or otherwise, suggests that autoxidation may or may not be involved in the mechanism. The presence of sulphur or nitrogen in the deposit could demonstrate that a free radical mechanism had been involved in the formation of deposits.

TABLE 11.4

Deposits in gas-making units [Coggins 1968]

|  | Preheaters | Vaporisers | Superheaters |
|---|---|---|---|
|  | % | % | % |
| Type of deposits: |  |  |  |
| Hard, brittle, scaly | 57 | 19 | 26 |
| Massive, coke-like, spongy | 17 | 7 | - |
| Finely divided powder | 17 | 9 | 36 |
| Scaly + spongy | 5 | 28 | 16 |
| Spongy and powder | - | 4 | - |
| Powder + scaly | - | 24 | 11 |
| Scaly, spongy + powder | 4 | 9 | 11 |
| Frequency of deposits becoming serious |  |  |  |
| 0- 5 weeks | 36 | 24 | 40 |
| 5-10 weeks | 9 | 34 | 30 |
| 10-20 weeks | 27 | 22 | 20 |
| 20-52 weeks | - | 9 | 10 |
| More than 52 weeks | 28 | 11 | - |

TABLE 11.5

The analysis of deposits from naphtha vaporisers [Coggins 1968]

| Constituent | Composition % weight |
|---|---|
| Carbon | 73 |
| Hydrogen | 4 |
| Ash | 23 |

Khater [1982] noted differences in the microscopic and macroscopic appearance of deposits at low and high temperature during the vaporisation of kerosene. He concluded that at the lower furnace temperatures the fouling was the result of autoxidation in the liquid phase. At the higher furnace temperatures vapour phase thermal cracking was considered to be responsible for deposit formation. Such changes in the chemical reaction origins of the deposit may occur at right angles to the flow of the fouling liquid. On the clean surface at high temperature, thermal cracking may be responsible for the initial layers of deposit, but as the solid/fluid interface temperature falls due to the insulating effect of the deposit itself, the lower temperature may be conducive to autoxidation rather than thermal cracking.

In all but the simplest of systems the chemistry is complex and is not likely to be fully understood.

## 11.4   THE EFFECTS OF SYSTEM OPERATING VARIABLES

In addition to the composition of the process fluid (see Section 11.2) that produces the deposit, the conditions under which the equipment is operated will also influence the deposition process. Crittenden [1984] has reviewed some of the effects of these variables.

### 11.4.1 Temperature

Already some mention of the effects of temperature have been made: the higher the temperature the more likely are deposition problems associated with chemical reactions.

Generally an increase of temperature favours chemical reaction with an exponential increase in the rate constant $(r_c)$ with absolute temperature $(T)$ according to the Arrhenius equation, i.e.

$$r_c \alpha e^{\frac{-E}{RT}} \tag{11.15}$$

where  $E$ is the activation energy

$R$ is the universal gas constant

Deposition rates will be a function of the rate constant $r_c$.

Fig. 11.2 gives a plot of experimental data [Vranos *et al* 1981] for the coking rate of jet fuel in terms of $\frac{1}{T}$. The value of $E$ the activation energy is 41.6 *kJ/mol*. Crittenden [1984] stated that for relatively moderate temperatures in the range 366 - 505 *K* deposition rates from pure hydrocarbon streams correlate well with $E = 42$ *kJ/mol*. Watkinson [1988] suggests that values around 40 *kJ/mol* appear to be typical of antioxidation processes. The data from Fig. 11.2 agree with these observations. Watkinson and Epstein [1969] observed that the initial deposition

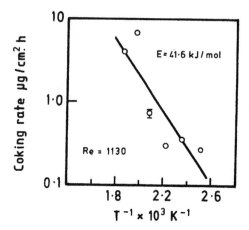

FIGURE 11.2. Arrhenius plot of coking rate of jet fuel

rate in a sensible heat exchanger operating at constant heat flux, was also correlated by an Arrhenius type equation, modified by a flow rate term, where $E = 121$ *kJ/mol*. The initial surface temperature was in the range 419 - 478 *K*.

Watkinson [1988] provides a summary of published activation emergencies and these are reproduced in Table 11.6. The data show a wide range of values.

TABLE 11.6

Some activation energies from the literature [Watkinson 1988]

| Fouling fluid | Activation energy kJ/mol | Reference |
|---|---|---|
| Hydrocarbons | 40 - 120 | Braun and Hansler [1976] |
| Kerosene | 70 | Crittenden and Khater [1984] |
| Gas oil | 120 | Watkinson and Epstein [1969] |
| Jet fuels | 42 | Taylor *et al* [1967] [1968] |
| Crude oil | 53 | Scarborough *et al* [1979] |
| Crude oil | 36 | Eaton and Lux [1984] |
| Shale oil | 15 - 20 | Latos and Franke [1982] |
| Lubricating oils | 74 - 97 | Steele *et al* [1981] |
| Styrene polymerisation | 39 | Crittenden *et al* [1987] |

Crittenden [1984] also observes that there is generally a minimum temperature below which fouling will not proceed, and upper temperatures above which the deposition rate drops rapidly with increasing temperature due to changes in the underlying chemical reactions. The experimental data on Fig. 11.3 for deposition during the processing of lubricating oil [Steele *et al* 1981] illustrate that at the lower temperatures little coke formation is experienced.

Data reported by Braun [1977] reproduced on Fig. 11.4 show that above a critical temperature fouling falls with increasing temperature. The figure also shows how the concentration of oxygen affects the rate of deposition.

An empirical relationship for the rate of fouling of plain and indented tubes for surface temperatures $(T_s)$ over the range 408 - 490 $K$ for solutions of indene in kerosene was obtained by Zhang *et al* [1993].

$$\frac{dR_f}{dt} = 2100e^{\frac{-4675}{T_s}}$$ (11.16)

Indene was used in the experiments since it has a tendency to form polymeric peroxides under oxidising conditions [Asomaning and Watkinson 1992].

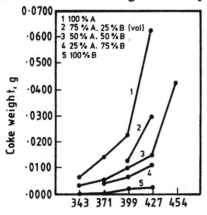

FIGURE 11.3. Gain in tube weight as a function of temperature for lubricating oil coking

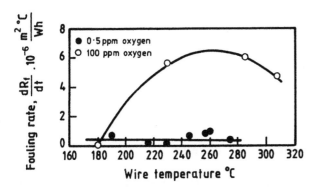

FIGURE 11.4. The effect of temperature and oxygen concentration on fouling rate (Illinois feed)

## 11.4.2 Pressure

Since pressure can affect the physical properties of a fluid it is to be expected that it will affect fouling tendencies. For instance where vaporisation occurs the boiling point curve of an organic liquid mixture may be affected. The changed relationship between associated vapour and liquid compositions could affect the extent of chemical reactions in the different phases. Pressure may increase the solubility of gaseous impurities (e.g. oxygen) that through autoxidation, may enhance chemical reactions leading to deposition. Additionally pressure may favour reactions in the gas phase if polymerisation is involved.

## 11.4.3 Flow rate or velocity

Two aspects of velocity that need to be kept in mind in a consideration of the effects of velocity in connection with chemical reaction fouling are the effects on mass transfer and heat transfer. The transfer of mass and heat will be increased with increased turbulence. If the fouling deposition is solely controlled by the reaction rate then enhanced mass transfer to the surface will not change the situation. On the other hand if some of the products of reaction or intermediates even, remain in the reaction zone then the rate of production of deposits could be reduced. Increased turbulence (i.e. increased flow rate for a given geometry) will assist not only the rate of arrival of reactants but also the removal of products of reaction from the reaction zone. It is possible that in part, this may account for some of the contradictions apparent in the literature in respect of flow rate.

Another cause of confusion over the effects of velocity arise as a result of the effects of velocity on the temperature of the wall receiving the deposit. For a given heat flux $q$ increasing the heat transfer coefficient $\alpha$ by increasing the velocity reduces the wall temperature. If the temperature at the interface between solid and fluid is $T_i$ then:

$$q = \alpha(T_i - T_b) \tag{11.17}$$

where $T_b$ is the bulk fluid temperature

$$\therefore T_i = \frac{q}{\alpha} + T_b \tag{11.18}$$

It follows therefore that as $\alpha$ increases $T_i$ decreases.

Another aspect of increasing flow rate is to decrease the thickness of the laminar or viscous sub-layer. In effect this reduces the volume of fluid that has a temperature that approximates to the interface temperature, i.e. temperatures in excess of the fluid bulk temperature.

Shalhi [1993] constructed a table from literature data that qualitatively demonstrates the effects of flow velocity on fouling in certain processing operations. Part of this table has been used to produce Table 11.7. It can be seen that there is no consistent trend.

TABLE 11.7

The effects of velocity on chemical reaction fouling

| System | Reference |
|---|---|
| Reduced deposition rate with increased velocity | |
| Oil refining | Nelson [1958] |
| Forced circulation reboilers | Chantry and Church [1958] |
| Sour gas oils | Watkinson and Epstein [1969] |
| | |
| Increased deposition rate with increased velocity | |
| Aviation kerosene | Smith [1969] |
| Ethylene cracking | Chen and Maddock [1973] |
| Vapour phase cracking of naphthia and light gas oil | Fernandez-Baujin and Solomon [1976] |
| Jet fuels | Vranos *et al* [1981] |
| | |
| Increased and decreased deposition rate with increased temperature | |
| Vapour phase cracking of n-Octane | Shah *et al* [1976] |
| Polymerisation of styrene | Crittenden *et al* [1987] |

## 11.4.4 The effects of impurities

In Section 11.2 it was stated that trace impurities can affect the extent of a reaction fouling problem. Figs. 11.4 and 11.5 illustrate the effects of relatively small amounts of oxygen in the fouling rate of crude oil processing.

In Fig. 11.4 [Braun 1977] it is shown that an oxygen concentration as low as 5 *mg/l* has a pronounced effect on the rate of fouling. For a different feedstock changes in oxygen content from 0.5 to 100 *mg/l* of oxygen (Fig. 11.5) are required to produce changes in rates of deposition of the same order as those in Fig. 11.4. It is anticipated that the basis of the chemical reactions will be autoxidation according to Equations 11.4 - 11.9.

## 11.5    MODELS FOR CHEMICAL REACTION FOULING

Previous sections (11.2 and 11.4) have illustrated the dependence of chemical reaction fouling on the chemistry of the system and the operating variables. The

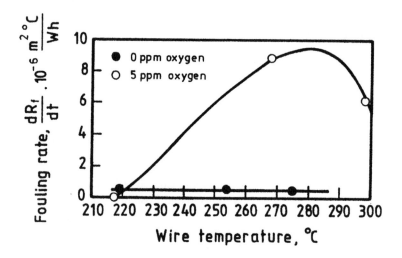

FIGURE 11.5. Fouling rate and dissolved oxygen content (Texas feed)

deposition of foulant material on the heating surface as a result of chemical reactions involves:

1. Transport of the foulant precursors, present in the fluid normal to the tube surface.

2. Chemical decomposition and/or polymerisation close to or at the heating surface.

3. Possible transport of the products including foulant, away from the reaction zone.

Ideally any model of chemical reaction fouling should take account of these three aspects. Because in many industrial processes, the precise chemical mechanism may not be understood, particularly where the fouling is dependent on impurities, it will be necessary to make simplifying assumptions in order that resultant model has limited complexity. In some respects this approach may be regarded as empirical or semi-empirical.

Table 11.8 lists some of the models that have been discussed in the literature [Shalhi 1993] in chronological order. The increasing model complexity with time is illustrated by Table 11.8 with gradual recognition of the importance of hydrodynamics, mass transfer and reaction kinetics. In some of the models no account of removal is taken, and indeed in some extreme examples of chemical reaction fouling, removal of a deposit once formed is indeed negligible.

TABLE 11.8

Some chemical reaction fouling models (table reproduced by kind permission of Dr. A.M. Shalhi)

| Authors (year) | Application | Deposition term | Removal term | Remarks |
|---|---|---|---|---|
| Nelson [1934] | Oil refining | Rate is directly dependent upon thickness of thermal boundary layer | None considered | Fouling rate can be reduced by increasing fluid velocity |
| Atkins [1962] | Fired heaters in oil industry | Constant monthly increase in coke resistance for various refinery steams | None considered | Two layer concept porous coke adjacent to fluid and hard coke adjacent to wall |
| Nijsing [1964] | Organic coolant in nuclear reactors | Hydrodynamic boundary layer and diffusion partial differential equations | Product diffusion back to the fluid bulk | Solution with diffusion control fits plant data, fouling rate predicted to increase with velocity |
| Watkinson and Epstein [1969] | Liquid phase fouling from gas oil | Mass transfer and adhesion of suspended particles | First order Kern and Seaton shear removal term* | (1) Correct prediction of initial rate dependence on velocity (2) Incorrect prediction of asymptotic resistance on velocity |
| Fernandez-Baujin and Solomon [1976] | Vapour phase pyrolysis | Kinetics and/or mass transfer control with first order reaction | None considered | Solution with mass transfer control fits plant run-time date |
| Crittenden and Kolaczkowski [1979] | Hydrocarbons in general | Kinetics and/or mass transfer control with first order reaction | (1) Diffusion of foulant back into fluid bulk (2) First order Kern and Seaton shear removal term* | (1) Limited testing (2) Complex - many parameters (3) Extended to two layer concept proposed by Atkins [1962] |

(cont'd)

TABLE 11.8 (continued)

| Authors (year) | Application | Deposition term | Removal term | Remarks |
|---|---|---|---|---|
| Crittenden et al [1987] | Dilute solution polymerisation of styrene | Non-zero order kinetics | (1) Diffusion of foulant back into fluid bulk (2) First order Kern and Seaton shear removal term* | |
| Kolaczkowski et al [1988] | Crude oil | Kinetics control | First order Kern and Seaton shear removal term* | Demonstrated interactive nature of fouling in process networks |
| Yen et al [1988] | Fluid catalytic cracking | Coke formation is kinetics control with fourth order reaction | None considered | (1) Complex - many parameters (2) Tested against actual coke data from both pilot and commercial fluid catalytic cracking units |
| Takatsuka et al [1989] | Residual oil | Coke formation is kinetics control | None considered | Predicts the effects of major operation conditions, i.e. velocity, surface temperature, etc. of cracking furnaces on tubular fouling |
| Crittenden et al [1992] | Crude oils | Kinetics control | None considered | Plant data used to obtain coefficients in predictive model; good fits obtained |

* See Equation 4.12

It is not productive to discuss all the models in detail but to select appropriately to illustrate certain aspects of development. The very first model included in Table 11.8, that due to Nelson [1934], assumes that the rate of coke deposition in oil processing is dependent on the thickness of the thermal boundary layer, i.e. the thicker the boundary layer the greater the volume of oil subject to high temperature (see Section 11.4.3).

An important observation made by Atkins [1962] was that in fired heater tubes fouling deposits appeared as two layers; an outer porous (or tarry layer) and a hard crust of deposit next to the tube wall (see Fig. 11.6).

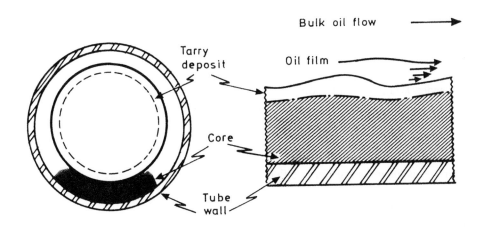

FIGURE 11.6. Idealised diagram of deposit layers in a fired heater tube [Atkins 1962]

Atkins' perception of the fouling process was that a primary tarry deposit was subject to further chemical reaction (or decomposition) resulting in the hard coke layer. Vapours or gases resulting from the decomposition process left a porous deposit structure remote from the tube wall. The overall fouling resistance $R_{fo}$ on the tubeside under these circumstances would be given by:

$$R_{fo} = \frac{1}{\alpha_i} + R_{f(coke)} + R_{f(tarry)} \tag{11.19}$$

where  $\alpha_i$ is the heat transfer coefficient on the inside of the tube

$R_{f(coke)}$ and $R_{f(tarry)}$ are the thermal resistances due to the coke and tar layers respectively

Nijsing [1964] assumed that fouling from an organic coolant in a nuclear reactor was caused by an instantaneous, irreversible reaction of a precursor to a product that deposits rapidly in comparison to its diffusion to the deposition surface. Applying simplifying assumptions the underlying equations for flow and diffusion were solved to give an average rate of deposition, i.e:

$$\phi_D \, \alpha \frac{(c_b D (Re))^{0.875} (Sc)^{0.33}}{d} \tag{11.20}$$

$c_b$ is the bulk concentration of the precursor

$D$ is the diffusion coefficient for the precursor

$Re$ and $Sc$ are the appropriate Reynolds and Schmidt numbers respectively

$d$ is the tube diameter

Fernandez-Baujin and Solomon [1976] incorporated film theory into their chemical reaction fouling model. The mass flux of reactant $N_r$ to the surface is given by:

$$N_r = K_t (c_b - c_w) \tag{11.21}$$

where $K_t$ is the mass transfer coefficient

$c_w$ is the reactant concentration at the wall

Assuming that all the material that arrives at the wall reacts then

$$N_r = k c_w^n \tag{11.22}$$

where $k$ is the reaction velocity constant for the reaction order $n$.

and $\quad K_t (c_b - c_w) = k c_w^n$ \hfill (11.23)

The simple concepts of Nelson [1934] and Atkins [1962], the analysis provided by Kern and Seaton [1959] and the use of film mass transfer and simple chemical kinetics led Crittenden and Kolaczkowski [1979] to develop a comprehensive fouling model that allowed chemical reaction and transport to and from a surface to be taken into account.

Diffusional transport and location of the reaction zone are strongly influenced by the hydrodynamic conditions. The relative roles of diffusional and kinetic phenomena may be illustrated by considering and idealised exchanger tube illustrated on Fig. 11.7 which is the basis of a mathematical analysis attempted by Crittenden and Kolaczkowski [1979].

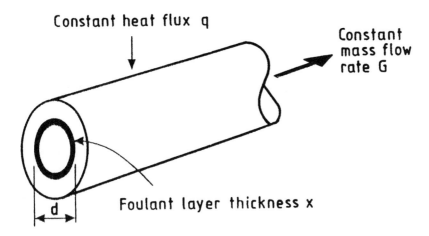

FIGURE 11.7. Heat exchanger tube in which chemical reaction is occurring [Crittenden and Kolaczkowski 1979]

In order to make the mathematics manageable it is necessary to make some simplifying assumptions, i.e.
1.  The heat flux and the mass transfer are maintained constant.

2.  Foulant material is formed by an irreversible reaction.

    Precursor → Deposit + light compounds

    Occurring at temperature $T$ somewhere between bulk temperature, $T_b$ and interface temperature $T_i$.

3.  The precursor concentration in the bulk fluid $C_{pb}$ may be considered constant in view of the relatively low ratio of deposition flux to the flow through the exchanger.

4.  The precursor diffuses to the reaction zone but the foulant may diffuse away from the zone into the main flow as well as to the heating surface on which it deposits.

5.  There is no induction or initiation period.

6. The properties of the foulant do not alter as the fouling process proceeds and the properties of the fluid are constant over the temperature range covered by the conditions.

7. In the first instance it may be assumed that the reaction zone is the clean wall/fluid interface (and subsequently the deposit/fluid interface), i.e. $T = T_i$.

8. Mass transfer of light products (other than foulant) away from the reaction zone is not likely to be deposition rate limiting owing to the relatively high diffusion rates of these relatively small molecules and the existence of potentially high driving forces.

The rate of increase in foulant deposit thickness $x$ is the difference between the precursor mass flux to the reaction zone $N_r$ and the foulant mass flux away from the zone $N_D$, i.e.

$$\frac{dx}{dt} = \frac{1}{\rho_f}(N_r - N_D) \tag{11.24}$$

$\rho_f$ is the deposit density

Using the film theory of mass transfer the mass fluxes may be written down in terms of a mass transfer coefficient $K_{tr}$ and a concentration driving force

$$N_r = K_{tr}(c_{rb} - c_{ri}) \tag{11.25}$$

and $N_D = K_{tD}(c_{Di} - c_{Db}) \tag{11.26}$

$c$ is the concentration

$b$ refers to bulk

$i$ to interface

$r$ refers to precursors or reactants

$D$ to deposit

$c_{Db}$ could in most situations, be considered = ZERO

The mass flux of precursors can be balanced by the rate of reaction.

$$N_r = kc_{ri}^n \tag{11.27}$$

*n* is the reaction order

*k* is the velocity constant

$c_{ri}$ the concentration of precursors at the interface is unknown - an indeterminate quantity. Therefore eliminating between Equations 10.25 and 10.27 yields.

$$c_{rb} = \left(\frac{N_r}{K_{tr}}\right) + \left(\frac{N_r}{k}\right)^{\frac{1}{n}}$$

(11.28)

$N_r$ the mass flux of the precursors may be estimated from this equation and assumptions regarding the order of the reaction may be made, eg. assuming first order, i.e. n=1.

It is possible to continue the analysis further without making assumptions as to the order of reaction, by applying the theory relating to kinetic or diffusion controlled transfer (deposition).

$$N_r = \frac{c_{rb}}{\frac{1}{K_{tr}} + \frac{1}{k}}$$

(11.29)

The rate of change of fouling resistance is given by

$$\frac{dR_f}{dt} = \frac{d\left(\frac{x}{\lambda_f}\right)}{dt} = \frac{1}{\rho_f \lambda_f}\left\{\frac{c_{rb}}{\frac{1}{K_{tr}} + \frac{1}{k}} - K_{iD}c_{Di}\right\}$$

(11.30)

where $\lambda_f$ is the thermal conductivity of the deposit

and    $x$ is the deposit thickness

In general mass transfer coefficients will be unknown but they may be estimated from physical properties, hydrodynamic considerations, and the analogy between heat and mass transfer.

The temperature dependence of the reaction velocity constant may be expressed by the Arrhenius equation as discussed in Section 11.4.1.

Crittenden and Kolaczkowski [1979] arrive at an extremely complex equation taking these factors into account; an equation which would be extremely difficult to use in design or in the analysis of an operating system.

$$\frac{dR_f}{dt} = \frac{1}{\rho_f \lambda_f} \left\{ \frac{c_{rb}}{\dfrac{\rho(d-2x)^{1.8}(Sc_r)^{0.67}}{1.213\lambda_1 \eta^{0.2}G^{0.8}} + \dfrac{1}{Ae^{\frac{-E}{RT}}}} - \frac{1.213\lambda_1 \eta^{0.2}G^{0.8}c_{Di}}{\rho(d-2x)^{1.8}(Sc_D)^{0.67}} \right\}$$

(11.31)

where  $\lambda_1$  is a function of surface roughness

$Sc_r$ and $Sc_D$   are the Schmidt numbers of precursors and deposit respectively $\dfrac{\eta}{\rho D}$

$D$          is the diffusivity

$E$          is the activation energy of the chemical reaction

$A$          is a constant

$$T = T_b + \left\{ \frac{q}{0.023(Re)^{0.8}(Pr)^{0.4}} \right\}$$

(11.32)

Diffusion of foulant back into the bulk fluid may be important when the deposit contains a relatively mobile species, i.e. when $Sc_D$ is small. the concentration $c_{ri}$ is then the solubility of the species in the bulk fluid and likely to reduce as the molecular weight increases, but increases as the temperature rises. At the same time higher interfacial temperatures generally tend to give further degradation to form higher molecular weight deposits. There is evidence of the importance of back diffusion of deposit in some process plant, away from the heat transfer site to other parts of the plant. The following observations may be made on examination of the equations.

The magnitude of the mass flow rate through the tube $G$ is important.

1.  For relatively low temperatures, small diameters and high mass flow rates deposition is likely to be controlled by reaction kinetics (i.e. $k \ll K_{tr}$).

2.  If foulant back diffusion is important then $\dfrac{dR_f}{dt}$ decreases as $t$ (or $x$) increases and asymptotic fouling can occur.

3.  The initial rate of fouling (when there is no deposit, i.e. $x = o$) decreases as the mass flowrate increases.

4. If deposition is kinetically controlled but foulant back diffusion is negligible then the rate of fouling $\dfrac{dR_f}{dt}$ still decreases as $t$ increases but asymptotic fouling cannot result unless there is an alternative deposit removal mechanism which is dependent upon deposit thickness.

5. Reduction of fouling, if kinetically controlled, may be achieved by the use of smaller diameter tubes and high mass flow rates; unfortunately this could lead to high pressure drop.

6. At higher temperatures deposition may be controlled by diffusion, i.e. $K_{tr} \ll k$ (because the reaction rate is higher at higher temperatures). Inspection of the equation suggests that the fouling resistance cannot tend towards asymptotic.

7. At very high temperatures severe thermal degradation to hard coke occurs and thus the interfacial foulant concentration $C_{Di}$ may be assumed to be zero and equation 11.31 reduces to

$$\frac{dR_f}{dt} = \frac{1.213\lambda_1 \eta^{0.2} G^{0.8} C_{rb}}{k_f \rho_f \rho (d-2x)^{1.8} (Sc_r)^{0.67}} \tag{11.33}$$

The model put forward by Fernandez-Baujin and Solomon [1976] for the deposition of coke in industrial olefin furnaces under diffusion control shows a similarity to Equation 11.33. The rate of coke formation is given by:

$$\frac{K^* G^{0.8}}{(d-2x)^{1.8}} \tag{11.34}$$

$K^*$ is a function of feedstock, cracking severity, dilution steam ratio and other system variables; it is an empirical factor.

Fig. 11.8 illustrates qualitatively the trends in the initial fouling rate with mass flow rate based on the Crittenden and Kolaczkowski [1979] model.

Increased temperatures generally result in increased rates of fouling but the effects of tube diameter and mass flow rate, are more complex and are not yet fully understood. Surface roughness, that changes once the deposition mechanism has commenced, will influence deposition, but the precise effect will depend on the controlling mechanism. Increased roughness can lead either to reduced or increased deposition rates depending on whether the deposition is kinetically or diffusion controlled.

A model was developed [Onifer and Knudsen 1993] for reaction fouling under sub-cooled boiling conditions for the polymerisation of styrene. It is similar to the

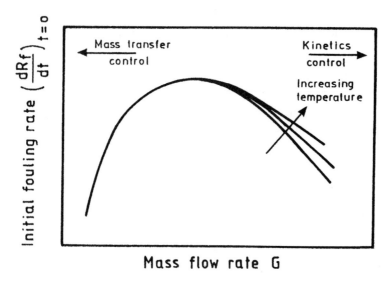

FIGURE 11.8.  Variation of initial fouling rate with mass flow rate and temperature

model presented by Crittenden [1988].  The equation which describes the transport of precursor mass flux to the surface $N_r$ is:

$$N_r = \rho_f \lambda_f \frac{dR_f}{dt} + C_3 \frac{\tau}{\psi} R_f \qquad (11.35)$$

where $\rho_f$ and $\lambda_f$ are the deposit density and thermal conductivity respectively

    $\tau$ is the shear stress

    $\psi$ is the deposit strength

    $C_3$ is a constant

The asymptotic fouling resistance is given by

$$R_{f\infty} = \frac{1}{C_3}\left(\frac{\psi N_r}{\tau}\right) \qquad (11.36)$$

The difficulty in the use of these equations is again assigning values to the variables.

Comparisons between predicted and experimental values were not good, but it was concluded that the model might be improved by using a different equation to describe the boiling conditions.

Panchal and Watkinson [1993] have presented the basis for a more sophisticated model. It is relies on the assumption that the key chemical reaction leading to fouling can be expressed in a two-step reaction involving first the generation of a soluble precursor followed for the formation of an insoluble foulant. The rates of reaction are $r_1$ and $r_2$ respectively. The scheme is.

REACTANTS (A) $\rightarrow$ PRECURSOR (B) $\rightarrow$ FOULANT (C)
(SOLUBLE) (SPARINGLY SOLUBLE) (INSOLUBLE)

Panchal and Watkinson [1993] visualise a number of different reaction mechanisms that can account for the fouling, but dependent on where the reactions take place in relation to the surface that is subject to fouling deposits, i.e. in the bulk fluid, within the thermal boundary layer, or on the surface. They suggest three possible cases:

Case 1          Precursor generation in the bulk liquid

Case 1a         $A \rightarrow B$ molecular transport to wall where $B \rightarrow C$

Case 1b         $B \rightarrow C$ particle transport to wall

Case 2          Precursor generation in the boundary layer
                $A \rightarrow B$ molecular transport to wall
                $A \rightarrow B \rightarrow C$ particle transport to wall

Case 3          Precursor generation at the surface
                $A \rightarrow B \rightarrow C$ occurs on the wall

In cases 2 and 3 it is possible for precursor B to "back diffuse" into the bulk.

An expression for the reaction rate $r$, may be described in terms of two reactants with reaction order $m$ and $n$. Such an expression is applicable to the autoxidation of olefins.

$$r_1 = k_1 c_{r1}^m c_{r2}^n \tag{11.37}$$

$k_1$ is the rate constant

$c_r$ and $c_{r2}$ are the concentrations of the reactants

The second stage in the formation of foulant may be expressed

$$r_2 = k_2 c_P \tag{11.38}$$

$k_2$ is the rate constant

$c_P$ is the concentration of the precursor

Considering now the three alternative cases Panchal and Watkinson [1993] suggest it is possible to write equations that are specific.

*Case 1.  Precursor generation in the bulk.*  In Case 1a where the foulant formation takes place at the wall, the transfer of the precursor may be expressed in terms of a mass transfer coefficient $K_{tP}$ for the transfer of the soluble precursor to the surface.

$$N_r = K_{tP}(c_{Pb} - c_{Pi}) \tag{11.39}$$

$N_r$ is the mass flux of precursor to the surface

$c_{Pb}$ and $c_{Pi}$ are the precursor concentrations in the bulk and at the interface respectively

The rate of foulant formation ($r_f$) may then be expressed as

$$r_f = \tilde{a}\, k_{2i} c_{Pi}^l \tag{11.40}$$

where $\tilde{a}$ is a conversion factor to change the rate from per unit volume to per unit area

$k_{2i}$ the rate constant is evaluated at the interface temperature

$l$ is the order of reaction for the precursor

In Case 1b Equation 11.37 is used to calculate the rate of foulant formation at the bulk temperature.  A similar expression to Equation 11.38 may be used to calculate the rate of transport of insoluble foulant to the wall but using a mass transfer coefficient applicable to particulate transport.  The equation quoted by Panchal and Watkinson [1993] for the mass transfer coefficient $K_t$ is the one suggested by Nijsing [1964] namely:

$$K_t = \frac{u\left(\dfrac{f}{8}\right)^{\frac{1}{2}} (Sc)^{\frac{2}{3}}}{11.8} \tag{11.41}$$

where $u$ is the fluid velocity

$f$ is the friction factor

*Sc* is the Schmidt number

The Stokes-Einstein equation may be used to calculate the diffusion coefficient for the particles $D_p$, i.e.

$$D_p = \frac{1.38 \; 10^{-23} \, T}{3 \pi \eta d_p}$$ 

(11.42)

where  *T* is the absolute temperature

$\eta$ is the viscosity

$d_p$ is the particle diameter

*Case 2. Precursor generation in the boundary layer.*  Panchal and Watkinson [1993] suggest two possible approaches may be taken to model the simultaneous mass transfer and reaction in the thermal boundary layer.
1.  The use of the simple film theory, but allowing for different film thicknesses for heat and mass transfer.

2.  Applying a two-layer turbulent flow model, in which the turbulent flow structures are modelled differently for the bulk and wall regions.

For a stagnant boundary layer the equation for reactant 1 is:

$$\rho_c D_{c1} \frac{d^2 c_{r1}}{dx^2} = k_1 c_{r1}{}^m c_{r2}{}^n$$

(11.43)

where  *x* is the *x* co-ordinate in the boundary layer

$D_{c1}$ is the diffusivity of the dissolved component 1

A similar equation exists for reactant 2.

At the fluid/wall interface

$$\frac{dc_{r1}}{dx} = \frac{dc_r{}^2}{dx} = \frac{dc_P}{dx} = 0$$

(11.44)

At the edge of the diffusion boundary layer $c_{r1}$, $c_{r2}$ and $c_P$ are the bulk concentration values.

The thermal boundary layer thickness $\delta$ may be calculated from an equation of the form:

$$\delta_{ml} = \delta_T (Pr / Sc_1)^{\frac{2}{3}} \tag{11.45}$$

where $\delta_T = \lambda / \alpha$

where $\alpha$ is the heat transfer coefficient

Simplifying assumptions may be made to the analysis depending on the conditions.

*Case 3. Precursor generation at the wall.* Panchal and Watkinson observe that Case 3 is the simplest of the three fouling cases and has been used by other researchers, e.g. Crittenden and Koloczkowski [1979].
    The diffusion of reactants to the surface may be expressed as:

$$N_1 = K_{ml}(c_{r1b} - c_{r1i}) \tag{11.46}$$

$K_{ml}$ is the mass transfer coefficient for reactant 1

$c_{r1b}$ and $c_{r1i}$ are the concentrations of reactant 1 in the bulk and at the interface respectively, the rate of reaction $r$, is given by

$$r_1 = \tilde{a}\, k_{1i} c_{r1i}{}^m c_{r2i}{}^n \tag{11.47}$$

where $k_{1i}$ is the rate constant

$c_{r1i}$ and $c_{r2i}$ are the interface concentrations of reactants 1 and 2 respectively.

Similar equations may be obtained for reactant 2.

An overall expression to take account of precursor generation, consumption and back diffusion has the form:

$$\tilde{a}\,\beta K_{1i} c_{r1i}{}^m c_{r2i}{}^n - k_{2i} c_{Pi}{}^l = K_{mp}(c_{pi} - c_{pb}) \tag{11.48}$$

where $\beta$ is a stochiometric parameter.

Panchal and Watkinson [1993] made a comparison between the models and some experimental data. Although the conclusions were tentative it would appear that these proposed models show promise. The work is ongoing.

Using the Equations 11.25 and 11.27 as a basis Epstein [1994] has developed a mathematical model to describe the initial rate of chemical reaction fouling.
Combining Equations 11.25 and 11.27 gives

$$N_r = \frac{c_{rb}}{\dfrac{1}{K_{tr}} + \dfrac{1}{Kc_{ri}}^{n-1}} \tag{11.49}$$

If the initial fouling rate $\left(\dfrac{dR_f}{dt}\right)_o$ is related to $N_r$

$$\left(\frac{dR_f}{dt}\right)_o = \frac{mN_r}{\lambda_f \rho_f} \tag{11.50}$$

where the stoichiometric factor $m$ represents the mass of fouling deposit per mass of precursor transported to and reacted at the wall.

Epstein [1994] suggests that the mass trnasfer coefficient for well developed turbulent flow may be written in terms of the friction velocity $u*$

i.e. $K_t = \dfrac{u*}{K_1}$ \hfill (11.51)

where $K_1$ is a constant and may be taken as a function of the prevailing Schmidt number.

It is argued that the rate constant $k$

$$= \frac{\eta e^{\frac{-E}{RT_i}}}{K^{11} \rho u*^2} \tag{11.52}$$

where $K^{11}$ is a constant

For a given fluid and a fixed surface temperature Equation 11.52 reduces to

$$k = \frac{1}{K_2 u*^2} \tag{11.53}$$

where $K_2 = \dfrac{K^{11} \rho e^{\frac{E}{RT}}}{\eta}$ \hfill (11.54)

Epstein [1994] makes the important point that instead of modifying Equation 11.25 for mass transfer, with $c_i$ = zero, by the application of a sticking probability, attachment is instead, treated as a process in series with the mass transfer.

Manipulation of the equations yields

$$N = \frac{c_{rb}}{K_1 u^{*-1} + K_2 u^{*2} c_i^{(1-n)}}$$ (11.55)

and $K_2 = \left(\frac{nK_1}{2}\right)^n \frac{N_{max}^{n-1}}{(u^*)_{crit}^{n+2}}$ (11.56)

where $N_{max}$ is the maximum value of $N_r$ at constant $T_i$

and    *crit* indicates a critical condition corresponding to $N_{max}$.

For a first order reaction Equation 11.56 simplifies to:

$$(u^*)_{crit} = (K_1 / 2K_2)^{\frac{1}{3}}$$ (11.57)

In terms of concentration

$$(c_i)_{crit} = n\, c_b / (n+2)$$ (11.58)

Equation 11.58 shows that the value of $c_i$ at the point of maximum deposition flux for a given interface temperature, is the same for all interface temperatures.

Epstein [1994] was able to demonstrate the validity of his model using data published by Crittenden *et al* [1987] for the polymerisation of styrene. He claims that the model is flexible enough to be adaptable to other examples of chemical reaction fouling as well as other mechanisms of fouling.

As with many model developments for fouling mechanisms the mathematical analysis is valuable since it draws attention to the salient effects of the variables. The present state-of-the-art models put forward, only go so far towards a complete mathematical solution that may be incorporated in the design of heat exchangers operating under chemical reaction fouling conditions. They are useful however, in suggesting a basis for an empirical formula to correlate experimental data that might so be used.

REFERENCES

Asomaning, S. and Watkinson, A.P., 1992, Can. J. Chem. Eng. 70, 444 - 451.
Atkins, G.T., 1962, What to do about high coking rates. Petro/Chem. Eng. 34, April 20 - 25.

Bott, T.R., 1990, Fouling Notebook. Instn. Chem. Engrs., Rugby.

Braun, R. and Hausler, R.H., 1976, Contribution to understanding fouling phenomena in the petroleum industry. Paper 76 - CSME/CSChE - 23. 16th Nat. Heat Trans. Conf., St. Louis.

Braun, R., 1977, The nature of petroleum process fouling. Materials performance, 16, No. 11, 35 - 41.

Canapary, R.C., 1961, How to control refinery fouling. Oil and Gas J. 9th October, 114 - 118.

Chantry, W.A., Church, D.M., 1958, Chem. Eng. Prog. 74, No. 5, 53 - 57.

Chen, J. and Maddock, M.J., 1973, Hydrocarbon Processing, 52, No. 5, 147 - 150.

Coggins, J.R., 1968, Report No. ER117. The Gas Council (British Gas), Midlands Research Station.

Crittenden, B.D., and Kolaczkowski, S.T., 1979, Mass transfer and chemical kinetics in hydrocarbon fouling. Proc. Conf. Fouling - Science or Art? Inst. Cor. Sci. Tech. and I. Chem. E. University of Surrey, Guildford, 169 - 187.

Crittenden, B.D., 1984, Chemical reaction fouling in fouling and heat exchanger efficiency. Inst. Chem. Engrs. Course, University of Leeds, 78.

Crittenden, B.D. and Khater, E.M.H., 1984, Fouling in a hydrocarbon vapouriser. 1st UK Nat. Heat Trans. Conf. I.Mech.E./I.Chem.E. Symp. Series 86, 401 - 414.

Crittenden, B.D., Hout, S.A. and Alderman, N.J., 1987, Model experiments of chemical reaction fouling. Chem. Eng. Res. Des. 65, 165 - 170.

Crittenden, B.D., Kolaczkowski, S.T. and Hout, S.A., 1987, Modelling hydrocarbon fouling. Chem. Eng. Res. Des., 65, 171 - 179.

Crittenden, B.C., 1988, Basic science and models of reaction fouling, in: Melo, L.F., Bott, T.R. and Bernardo, C.A. eds. Fouling Science and Technology, Kluwer Academic Publishers, Dordrecht.

Crittenden, B.D., Kolaczkowski, S.T. and Downey, I.L., 1992, Fouling in crude oil preheat exchangers. Trans. Inst. Chem. Engrs. 70, Part A, 547 - 557.

Crittenden, B.D. Kolaczkowski, S.T. and Takemoto, T., 1993, Use of in tube inserts to reduce fouling from crude oils. Heat Transfer - Atlanta A.I.Ch.E. Symposium Series, Vol. 89, pages 300 - 307.

Delplace, F., Leuliet, J.C. and Tissier, J.P., 1994, Fouling experiments of a plate heat exchanger by whey proteins solutions. Trans. I.Chem.E., Vol. 72, Part C, 163 - 169.

Dickakian, G. and Seay, S., 1988, Oil and Gas J. 86 (March 7), 47 - 50.

Eaton, P. and Lux, R., 1984, Laboratory fouling test apparatus for hydrocarbon feedstocks, in: Suitor, J.W. and Pritchard, A.M. eds. Fouling in Heat Exchange Equipment, HTD No. 35, ASME.

Epstein, N., 1994, A model of the initial chemical reaction fouling rate for flow within a heated tube, and its verification. Proc. 10th Int. Heat Trans. Conf. Brighton, I.Chem.E. 4, 225 - 229.

Fernandez-Baujin, J.M. and Solomon, S.M., 1976, in: Albright, L.F. and Crynes, B.L. eds. Industrial and Laboratory Pyrolysis. A.Ch.E. Symp. Ser. 32, 345 - 372.

Fitzer, E., Mueller, K. and Schaeffer, W., 1971, The chemistry of pyrolytic conversion of organic compounds to carbon, in: Walker, P.L. ed. Chemistry and Physics of Carbon, Vol. 7, Marcel Dekker, New York.

Froment, G.F., 1981, Fouling of heat transfer surfaces by coke formation in petrochemical reactors, in: Somerscales, E.F.C. and Knudsen, J.G. eds. Fouling of Heat Transfer Equipment. Hemisphere Publishing Corp., Washington.

Gayla, L.G., Cronauer, D.C., Painter, P.C., Li, N.C., 1986, Thermal stability of coal derived naphthanes. Ind. Eng. Chem. Found., 25, 129 - 135.

Khater, E.M.M., 1982, Fouling in a hydrocarbon vaporizer, PhD Thesis University of Bath.

Kern, D.Q., and Seaton, R.E., 1959, A theoretical analysis of thermal surface fouling. Brit. Chem. Eng., 4, No. 5, 258 - 262.

Kolaczkowski, S.T., Crittenden, B.D. and Varley, R., 1988, Operability of crude oil preheat exchangers - computer software tools for managing the problems of heat exchanger fouling in oil refining and petrochemical plants. In. Conf. Fouling in Process Plant. Inst. Cor. Sci. Tech. and I. Chem. E. University of Oxford, Oxford, 52 - 67.

Lambourne, G.A. and Durrieu, M., 1986, Fouling in crude oil preheat trains, in: Taborek, J., Hewitt, G.H. and Afgan, M. eds. Heat Exchangers Theory and Practice. Hemisphere Pub. Corp., Washington.

Latos, E.J. and Franke, F.H. Relative thermal fouling rates of petroleum and shale oils. Corrosion '82, Paper 103, NACE.

Murphy, G. and Campbell, J., 1992, Fouling in refinery heat exchangers: Causes, effects measurement and control, in: Bohnet, M. *et al*, eds. Fouling Mechanisms. Editions Europeénes Thermique et Industrie, Paris.

Nelson, W.L., 1934, Fouling of Heat Exchangers. Refining and Natural Gasoline Manufacturer. 13, No. 7, 271 - 276 and No. 8, 292 - 298.

Nelson, W.L., 1958, Petroleum Refining Engineering, 4th Edition, McGraw-Hill, New York.

Nijsing, R., 1964, Diffusional and kinetic phenomena associated with fouling. Euratom 1964, Eur. 543e. European Atomic Energy Community.

Nixon, A.C., 1962, Autoxidation and Antioxidants of Petroleum, in: Lundbery, W.O. ed. Autoxidation and Autoxidants, Vol. 2, Interscience, New York.

Onifer, L.G. and Knudsen, J.G., 1983, Modelling chemical reaction fouling under sub-cooled body conditions - Atlanta A.I.Ch.E. Symposium Series, Vol. 89, pages 308 - 313.

Scarborough, C.E., Cherrington, D.C., Diener, R. and Golan, L.P., 1979, Coking in crude oil at high heat flux levels. Chem. Eng. Prog. 75, No. 7, 41 - 46.

Shah, Y.T., Stuart, E.B. and Sheth, K.D., 1976, Ind. Eng. Chem. Proc. Des. Dev., 15, 518 - 524.

Shalhi, A.M., 1993, Use of HITRAN inserts to enhance heat transfer and control fouling from hydrocarbons. PhD Thesis, University of Bath.

Smith, G.T., 1969, Ind. Eng. Chem. Proc. Des. Dev., 8, 229 - 308.

Steele, G.L., Brinkman, D.W. and Whisman, M.L., 1981, Predictive test method for coking and fouling tendency of used lubricating oil, in: Somerscales, E.F.C. and Knudsen, J.G. eds. Fouling of Heat Transfer Equipment. Hemisphere Publ. Corp., Washington, 483 - 488.

Stephenson, W.K. and Rowe, C.T., 1993, Fouling control in ethylene recovery systems. Heat Transfer - Atlanta A.I.Ch.E. Symposium Series, Vol. 89, pages 335 - 340.

Takatsuka, T. *et al*, 1989, J. Chem. Eng. Japan, 22, No. 2, 149 - 154.

Taylor, W.F. *et al*, 1967, Kinetics of deposit formation for hydrocarbons - general features of Process. Ind. Eng. Chem. Prod. Res. Dev., 6, 258 - 62.

Taylor, W.F. *et al*, 1968, Kinetics of deposit formation for hydrocarbons - effect of trace sulphur compounds. Ind. Eng. Chem. Prod. Res. Dev., 7, 198 - 202.

Tseregounis, S.I., Spearot, J.A., Kite, D.J., 1987, Formation of deposits from thin films of mineral oil base stocks on cast iron. Ind. Eng. Chem. Res., 26, 886 - 894.

Vranos, A., Mortency, P.J. and Knight, B.A., 1981, Determination of coking rate in jet fuel, in: Somerscales, E.F.C. and Knudsen, J.G. eds. Fouling of Heat Transfer Equipment. Hemisphere Pub. Corp. Washington.

Watkinson, A.P., 1988, Critical review of organic fluid fouling: Final Report AWL/CNSV-TM-208, Argonne National Laboratory, Argonne.

Watkinson, A.P., Epstein, N., 1969, Gas oil fouling in a sensible heat exchanger. Chem. Eng. Prog. Sym. Ser. 65, No. 92, 84 - 90.

Yen, L.C., Wrench, R.E. and Ong, A.S., 1988, Oil and Gas J. Jan. 67 - 70.

Zhang, G., Wilson, D.I. and Watkinson, A.P., 1993, Fouling of a cyclic olefin on plain and enhanced surfaces - Atlanta A.I.Ch.E. Symposium Series, Vol. 89, pages 314 - 319.

# CHAPTER 12

## Biological Growth on Heat Exchanger Surfaces

### 12.1   INTRODUCTION

The accumulation of living matter, i.e. deposition and growth, on heat exchanger surfaces, may be divided into two groups depending on the size of the organism:
1.  Micro-organisms including bacteria, algae and fungi.

2.  Macro-organisms including mussels, barnacles, hydroids and serpulid worms and vegetation such as seaweed.

The problem of fouling from living matter is usually, although not exclusively, associated with aqueous systems where the temperature is not too different from that in the natural environment. Cooling water systems for instance, particularly "once through" or open recirculating systems, are prone to this form of fouling since the temperature is usually near to the optimum for biological activity and in general, nutrients are present in the water. Even bore hole water and towns water in recirculating systems quickly acquire nutrients (and micro-organisms) from exposure to the atmosphere in cooling towers, spray ponds and lagoons. The presence of micro-organisms in the fouling layer is often associated with "fresh" water drawn from rivers, canals, or lakes, whereas the larger organisms are usually present in sea or esturine water. Non-aqueous fluids may also be subject to biological activity. Metal working fluids usually emulsions, formulated with organic components can provide nutrients for sustaining growth of micro-organisms [Cook and Gaylarde 1993].

The presence of biological material on heat exchanger surfaces may promote other fouling mechanisms. It is not uncommon to find that scale formation, particulate deposition and corrosion accompany biological growth. In general this is associated with metabolic activity involving chemical reactions and changed conditions, particularly $pH$ beneath the biological mass. Micro-organisms produce extracellular material which may be sticky and hence facilitates the adhesion of particulate material. The appearance of the foulant can be very varied depending on the organism involved and the conditions under which growth is taking place. It is possible therefore to have long filaments of bacteria or algae, slime layers or extremely rough surfaces from the presence of crustaceans.

The structure and function of biofilms has been reviewed by Characklis and Wilderer [1989].

## 12.2   BASIC BIOLOGY

In a consideration of fouling mechanisms affecting the transfer of heat such as biofouling, it is not possible or necessary, to discuss all the aspects that contribute to the basic mechanism.  It is useful however, to give something of the background science so that the important factors that affect the fouling of surfaces, can be more readily understood.  It will be helpful to divide this discussion into two parts, i.e. fouling due to micro-organisms and to the presence of larger animals.

### 12.2.1  Microbial biology

All living organisms have a common basic structure containing microscopic subunits generally referred to as cells.  In general the chemical composition is similar with the occurrence of DNA and RNA and protein, with common chemical activities that constitute the metabolism processes.

The cell wall surrounding the organism, is the primary barrier that separates the cell from the environment in which it exists.  The selective permeability property of the cell wall maintains the appropriate levels of nutrients, metabolites, salts and the internal *pH* of the cell.  The cell membrane is the site of nutrient, ion and proton transfer processes.  The membrane is semi permeable which means that water is able to pass in or out of the cell through the influence of osmotic pressure.  In addition the cell wall must be able to control the passage of macro molecules.  The chemical formulation of the cell wall components determines the surface properties of the cell, such as electrical charge, binding of ions, immunogenicity, and the recognition of bacteriophages.

In many organisms the cell wall is rigid which determines the shape of the cell as a coccus, rod or filament.  In general a rigid cell wall means that the cell is limited to utilising nutrients that are soluble but at the same time a rigid wall allows the formation of structures.  Some organisms do not have a rigid cell wall and require a way of controlling osmotic effects to prevent damage.  Many freshwater organisms can pump out water via contractile vacuoles, while some algae control the concentration of soluble metabolites in response to changes in the extracellular osmotic pressure.

In addition to the cell wall, micro-organisms can display other extracellular layers such as slimes and mucilages which, for some organisms, are used to help attach to a substrate.

During metabolism the single cell imbibes nutrients from which it synthesises macromolecules and will therefore grow in size and eventually it must divide.  The process of cell division in bacteria is well regulated; the new (or daughter)'cells are very uniform both dimensionally and in genetic information.

Some yeast cells divide by the production of a bud on the cell surface.  The bud is much smaller than the cell itself and grows till it reaches adult dimensions.  The release of each bud leaves a "bud scar" on the parent cell.  It is not possible for this area to be used to produce further buds so that there is a maximum number of buds possible from each adult parent cell, dependent on the surface area.

Fungi grow almost exclusively by what is known as apical growth, with synthesis of new wall material predominantly at the tips of the growing cells at the periphery of the colony and at branching locations.

Micro-organisms in general are capable of movement either by the use of extracellular flagella as in some bacteria or gliding due to fibrils (e.g. blue green algae).

The availability of suitable nutrients in the particular microbial environment will determine whether or not a micro-organism will live and grow. In general micro-organisms are versatile in their nutrient utilisation and many micro-organisms, particularly bacteria, have or can acquire, the ability to use many different nutrients.

Bacteria through their metabolic processes convert the nutrients to waste products that include $CO_2$ and $H_2O$.

Lynch and Poole [1979] list four different ways to classify potential nutrients:

1.  Essential or useful but dispensable

2.  Building blocks for micro-molecules or as energy sources or both

3.  Macro-nutrients required in large quantities or as micro-nutrients

4.  Macro-molecules requiring breakdown before entry into the cell or as small molecules readily entering as soluble nutrients

In the biological reactions associated with metabolism systems of electron and hydrogen transport are associated with both respiration and photosynthesis. The electron transport or respiratory chains, involved with the oxidation of organic and inorganic substances by the micro-organisms include a series of components each of which can exist in two forms, i.e. oxidised or reduced forms. Each component of the system is characterised by a constant redox potential.

A detailed discussion of nutrient requirements and interactions are outside the scope of this book but it is useful to summarise some of the aspects of cell requirements since this has a direct bearing on the fouling of surfaces in heat exchangers. As might be expected the elements that are found in the make up of micro-organisms are those generally associated with organic chemistry and include carbon, hydrogen, oxygen, nitrogen, sulphur, phosphorous and other inorganic elements (minerals). Nutrition of micro-organisms involves a source or sources, of energy, carbon and reducing equivalents.

Although carbon dioxide is plentiful in the natural environment it is only used in algae by photosynthesis. The usual source of carbon is via carbohydrates. Monosaccharides such as glucose are not found in nature, probably because if they are present they are rapidly utilised by cells [Lynch and Poole 1979]. Laboratory experimentation on the other hand, often uses glucose as a carbon source for the study of bacterial growth because of its convenience.    Monosaccharides are incorporated directly into macromolecules and represent good sources of energy. Although there are disaccharides in nature, most of the carbohydrate available to micro-organisms is in the form of polysaccharides.  Organic acids are readily used as a carbon source by many micro-organisms.  Other sources of carbon that may be used by certain organisms include fats, hydrocarbons such as methane, and proteins.  Cellulose may be utilised by some fungi.

The abundance of nitrogen in the atmosphere does not generally make this a useful source of nitrogen for the majority of micro-organisms.  A more usual source of nitrogen is from nitrogen containing compounds such as amino acids, ammonia, nucleotides, uric acid, and urea.

Sulphur is found widely in nature.  Inorganic sulphates can be reduced via sulphide to be incorporated into amino acids.  Some micro-organisms can use hydrogen sulphide as a source of sulphur, organo-sulphur compounds may also provide microbial sulphur.

The origin of phosphorous for microbial metabolism is usually via phosphates usually as soluble inorganic compounds.  Micro-organisms require minerals containing potassium, magnesium and iron while some may require calcium, sodium or silicon.  Other trace elements required may include; zinc, copper, cobalt, manganese and molybdenum.

It is possible to classify micro-organisms in terms of their nutrition:

1.  Phototrophs obtain their energy directly from the sun.

2.  Chemotrophs acquire their energy from the oxidation of organic and inorganic substances.

3.  Autotrophs are capable of synthesising their cell carbon from simple compounds such as $CO_2$.

4.  Hetrotrophs require fixed organic sources of carbon.

5.  Lithotrophs produce the reducing equivalents required for cell synthesis from inorganic substances like $H_2S$ and ferrous iron.

6.  Orgotrophs obtain chemical reduction from the oxidation of organic molecules.

Table 12.1 gives data on the nutrient used for the laboratory growth of *Pseudomonas fluorescens* and illustrates the relative amounts of different constituents including trace elements to sustain healthy growth of the cells.

Of considerable importance to the fouling of heat exchangers is the behaviour of cells within the environment in which they are located, especially in relation to surfaces.  Bacteria and other organisms, respond to a wide range of physical and chemical stimuli, which serve to place the cell in a more favourable environment to co-ordinate concerted growth.

TABLE 12.1

Medium composition for *Pseudomonas fluorescens* culture (g/l)

| Mineral salts | | Trace elements | | | Glucose (carbon source) | |
|---|---|---|---|---|---|---|
| $NaH_2PO_4.2H_2O$ | 1.01 | $MnSO_4.4H_2O$ | 13.3 | x 10⁻³ | $C_6H_{12}O_6$ | 5.0 |
| $Na_2HPO_4$ | 5.50 | $H_3BO_3$ | 3.0 | x 10⁻³ | | |
| $K_2SO_4$ | 1.75 | $ZnSO_4.7H_2O$ | 2.0 | x 10⁻³ | | |
| $MgSO_4.7H_2O$ | 0.10 | $Na_2MoO_4.2H_2O$ | 0.24 | x 10⁻³ | | |
| $Na_2EDTA.2H_2O$ | 0.83 | $CuSO_4.5H_2O$ | 0.025 | x 10⁻³ | | |
| $NH_4Cl$ | 3.82 | $CoCl_2.6H_2O$ | 0.024 | x 10⁻³ | | |

In order to utilise these advantages many bacteria have the ability to move towards a favourable chemical environment, i.e. towards a nutrient source. For instance some bacteria demonstrate aerotactic response that is to say, aerobic bacteria move towards oxygen whereas anaerobic bacteria move away from the presence of oxygen. Predatory bacteria tend towards their hosts such as other bacteria, or towards fungi, algae and plants depending upon their particular characteristics. Although chemotaxis is apparent in bacteria it is not so developed in fungi and algae. The adsorption of nutrients on surfaces may influence the colonisation of the surface by chemotaxis.

Some micro-organisms move towards or away from light of particular intensities, depending on their need to utilise light in photokinetic reactions. Other responses to the environment include a reaction against for instance, the effects of gravity. Some blue green algae can counteract gravity by producing gas vacuoles as a flotation mechanism. Response to contact and pressure is also widespread.

Orientation to the presence of heat is also a characteristic of many micro-organisms so that they move along the temperature gradient to the most favourable temperature for optimum growth. A classification of micro-organisms may be made on the basis of the temperature ranges suitable for growth. Table 12.2 is an illustration of such a classification.

Cooling water systems generally operate in the range 20 - 50°C so that the micro-organisms likely to be encountered in these systems are mesophiles, but biofilms are likely to be encountered at any temperature below 90°C.

TABLE 12.2.

Classification of micro-organisms according to temperature

| Temperature range °C | Classification |
|---|---|
| -12 - 20 | psychrophile |
| 20 - 45 | mesophile |
| > 45 | thermophile |

### 12.2.2 Micro-organisms on surfaces

The importance of interfaces to microbial activity has been comprehensively received by Marshall [1976]. It would appear that many micro-organisms prefer to reside on a surface in colonies or biofilms and it is not difficult to appreciate the importance of surfaces to their well-being. Under natural conditions, e.g. in rivers, streams and lakes as Costerton *et al* [1978] point out in relation to bacteria, 90% of micro-organisms in the biosphere live in biofilms and that if cells were not adherent, the water would be virtually sterile since the cells would be swept away faster than they could swim against the prevailing current. These authors draw attention to the fact that a square centimetre of submerged surface may have as many as a million bacteria attached to it, whereas a cubic centimetre of water contained in a flowing stream over the surface, may only have as few as a thousand cells. The advantages of being attached to a surface include the continuous supply of nutrients, protection and the removal of metabolic waste products. The continuous movement of the aqueous environment over a surface also usually means that the water is aerated, which will be a further advantage for aerobic bacteria. The situation may be contrasted with planktonic (suspended) cells moving at the velocity of the stream that are only able to use nutrients in close vicinity; nutrient replacement being largely by Brownian motion. Heat exchanger surfaces would seem to represent an ideal location for micro-organisms, notably bacteria, since not only all the advantages of surfaces are available, but in general in cooling water applications, the surface temperature is such as to encourage growth.

Fletcher [1992] suggests that bacteria in biofilms appear to be protected from surrounding perturbations, and their survival may, in fact, be enhanced in biofilms, compared to suspension in the liquid phase. Flemming [1991] has given a more precise statement of the ecological advantages of the biofilm mode of growth. Some of these advantages are:

1. Scavenging and enrichment of nutrients in the gel matrix

2. Protection from
   - short term *pH*-fluctuations, salt- and biocide concentration shocks

   - shear forces (cells are kept in place and can scavenge nutrients from the flowing medium)
   - dehydration

3. Development of micro-consortia
   - symbiosis with other living matter
   - utilisation of less readily degradable substrates by specialised organisms, e.g. degradation of cellulose or xenobiotics; nitrification
   - creation of ecological niches, e.g. anaerobic zones under aerobic biofilms in aerobic environments; nutrient-rich compartments in oligotrophic systems
   - feasibility of gene transfer because of the long retention times of the micro-organisms

Hamilton [1987] on the other hand, concludes that the experimental evidence for the advantages of cell attachment to a surface and the associated higher metabolic activity is inconclusive. He observes that over a wide range of organisms and experimental systems, processes (e.g. growth) and activity assessment has variously been found to be increased, decreased or unaffected by adhesion to a solid surface.

Mozes and Rouxhet [1992] draw attention to the fact that when micro-organisms are attached to a surface three zones may be distinguished:
1. The primary layer of cells which are in direct contact with the solid surface (the substratum)

2. The bulk of the biofilm

3. The surface layer of cells that are in direct contact with the liquid phase.

The activity of the cells in each of these zones will be affected by the mass transfer conditions. Stone and Zottola [1985] for instance, reported that for *Pseudomonas fragi* attaching to a sterile 304 stainless steel, growth rates were enhanced under conditions of agitation. Marine biofilms initially developed more rapidly on mild steel where there was appreciably more action as opposed to calm sea water [Eashwar *et al* 1988]. The cells nearest the liquid layer will have full access to the nutrients available, limited only by the mass transfer resistance in the liquid phase. The biofilm itself restricts the penetration of nutrients towards the solid substrate; not only does the biofilm offer resistance to mass transfer but nutrients are consumed by the cells in the succeeding layers of biofilm. In all but very thin biofilms, i.e. during the early stages of growth, it is more than likely that the cells near the substrate are starved of nutrients. Where the biofilm is made up of a range of micro-organisms, the concentration gradients of oxygen, essential inorganic nutrients and dissolved organic carbon through the biofilm may give rise to cross feeding from exudates (waste products) and cell lysis (death). The range of conditions may be fortuitous and serve only to increase the metabolic versatility of the biofilm or, in nutrient-depleted conditions, it may be an obligatory requirement for the activity and growth which require a fixed carbon source [Haack and McFeters 1982].

Characklis [1981] and Hamilton and Characklis [1989] identify five steps in the formation of biofilms:

1.  Formation of an organic conditioning film of macro-molecules on the surface

2.  Transport of micro-organisms to the surface

3.  Adhesion of cells at the solid-fluid interface

4.  Nutrient transport and assimilation by the growing biofilm resulting in the replication of attached cells and the production of extracellular material

5.  Removal of biofilm

In general steps 1 - 4 favour the development of biofilms unless there are problems of inhibition by substrate or waste products. Step 5 is an adverse process as far as the biofilm is concerned.

The discussion so far has attempted to consider micro-organisms in general but because of the differences between bacteria, algae and fungi in terms of metabolism and behaviour it is necessary to consider separately each group. More attention will be paid to bacteria since these are the micro-organisms usually associated with heat transfer surfaces.

*Bacteria*

Probably the most important factor in the attachment of bacteria is the bacterial surface itself, although this is difficult to characterise [Fletcher *et al* 1983]. There is considerable variation between bacteria and this is due, at least in part to the differences in composition of cell surface polymers [Neufeld *et al* 1980].

Fletcher [1992] draws attention to the fact that bacteria attached to surfaces appear to differ metabolically from their free living counterparts. She suggests two possible reasons for these physiological differences.

1.  Physiochemical conditions at the solid surface-liquid interface are not identical to those in the bulk phase, i.e. the environment at the interface therefore will be different in terms of the concentrations of ions, small molecules or polymers. The chemical species concerned may be beneficial to the cells or toxic [Mozes and Rouxhet 1992]. Unlike natural surfaces such as rocks and stones the surfaces found in heat exchangers are usually metallic and may give rise to ions that are generally not present in natural aqueous systems. Vieira *et al* [1993] report that metallic ions such as $Zn^{2+}$ and $Cu^{2+}$ seem to interfere with the initial adhesion and development of biofilms formed from *Pseudomonas fluorescens*, whereas no such effect was observed with aluminium ions.

2. Bacteria in biofilms are usually in resident proximity to other bacteria and micro-organisms and sometimes macro-organisms. It would be expected that there would be interaction because the requirements of the different species will be different. The effects may be described by the definitions listed in Table 12.3.

TABLE 12.3

Definition of interactions between species of organisms (based on the glossary [Melo *et al* 1992])

| Term | Definition |
|------|-----------|
| Synergy | An interactive association between two populations of organisms which is not necessary for survival, but from which each population benefits. |
| Mutualism | Interactions between populations of two organisms by which both populations benefit. |
| Competition | Interactions between two population groupings of organisms that require the same resource (e.g. space, limiting nutrient) that results in growth restrictions below the optimum for the two species. |
| Autoganism | Interaction between populations of two organisms in which one or both populations suffer. |

It is probably as a result of the changes brought about by the micro-environment and the wide ranging interactions that have led to conflicting reports on the effects of immobilisation of microbial activity identified by Mozes and Rouxhet [1992].

Fig. 12.1 shows cells of the bacterium *Pseudomonas fluorescens* residing on a surface.

As the colonies of bacteria grow on a surface, extracellular products are formed and accumulate. The result is a highly complex matrix of cells linked by strands of extracellular polymers consisting largely of polysaccharide. Cooksey [1992] emphasises however, that the extracellular material is a complex agglomeration of different polymeric materials and he considers the definition proposed by Geesey [1982] of extracellular polymeric substances (EPS) is a justifiable description and could be applied to all types of cells not just bacteria. Fig. 12.2 shows part of such a matrix. The spaces between the components of the matrix will be water and it is for this reason that biofilms have been found to contain 90 - 95% water [Patel 1986]. In an industrial system the "glue-like" character of the polysaccharide will retain other material such as mineral particles, corrosion products and other solid debris suspended in the water. The structure and the density of the biofilm will be largely dependent on the hydrodynamic conditions at the biofilm/water interface.

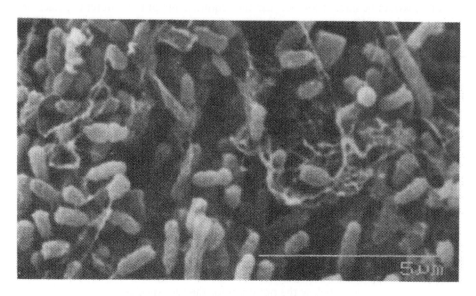

FIGURE 12.1. Cells of *Pseudomonas fluorescens* attached to a surface (courtesy of M. Pujo)

FIGURE 12.2. Part of a biofilm matrix showing the "open" structure (courtesy of M. Pujo)

The structure of the biofilm will affect the resistance to the mass transfer of nutrients through the matrix. Veira *et al* [1993] have concluded that mas transfer within a biofilm is strongly dependent on the hydrodynamic conditions imposed by the fluid flow across the biofilm, at least for turbulent flow. In part this dependence is due to the effect of velocity on the compactness of the biofilm. Under high velocity the biofilm is generally thin and dense with relatively low internal mass transfer. These authors suggest that there is not a unique value for diffusivity and furthermore they have reason to believe that the compactness of the biofilm will vary according to the depth. The mass transfer resistance of biofilms is probably the explanation why bacteria in biofilms, often appear to be protected from excursions in conditions in the bulk water phase away from the biofilm. Survival may be enhanced by the presence of the matrix with implications for the use of biocides (see Chapter 14), in comparison to the vulnerable position of corresponding planktonic bacteria. It is also possible that within the biofilm, the microbial activity is reduced so that changes in nutrient or toxic chemical levels do not exert the same influence as for free living counterparts.

In addition to the protection from environmental changes afforded by the extracellular polysaccharide substance and the capture of debris, Characklis and Cooksey [1983] speculated about other possible functions of the extracellular material. The possibilities include:

1. Provision of cohesive forces within the matrix

2. Adsorption of nutrients

3. Adsorption of heavy metals

4. Provision of intercellular communication within the biofilm

5. Enhancement of intercellular transfer of genetic material

6. Provision of short term energy storage

Although in industrial systems the presence of particles is common there has been little investigation of the interaction of particles with biofilms. Roe [1992] reports work that demonstrates that biofilms trap and hold particles. Bott and Melo [1992] made the observation that the effect of suspended particles is very dependent on the operating conditions, particularly the nature of the particles and bacteria present. The result can either be an enhancement of biofilm growth and stability, or the reverse. For instance particles of kaolin would appear to assist biofilm development, whereas sand grains appear to act as a scouring agent.

The discussion relating to the metabolism of bacteria in biofilms is of necessity short. Nevertheless it does draw attention to the complexity of bacterial biofilm formation and development, commonly encountered in heat exchangers associated with aqueous industrial systems such as cooling water. It is on account of these complexities that a general overall and somewhat empirical, approach to the problem of bacterial accumulation is often adopted.

*Algae*

Algae are much larger than bacteria in size and are structurally much more complex. They may exist as simple single cells, as long strands of colonies with a filamentous structure or residing on a solid surface. Algal films may develop on any surface that is adequately illuminated and submerged in water or where mosisture is readily available provided suitable nutrients are present. In natural conditions algal films are to be found on rocks and stones in rivers or along seashores. They can also reside on man-made structures such as off-shore oil rigs, bridge supports and on exposed structures in cooling water systems (e.g. within cooling towers on the packing structure or in the basin below the tower, or in spray ponds). Algae utilise sunlight, carbon dioxide and inorganic chemicals as their primary source of nutrients using the sunlight to photosynthesise sugars from the carbon dioxide. Fig. 12.3 shows an example of algae.

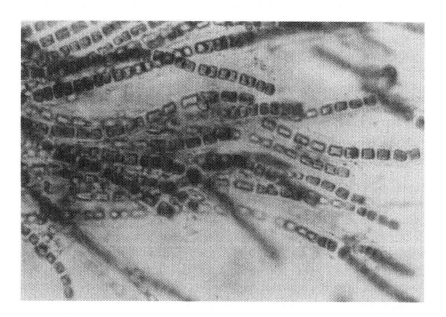

FIGURE 12.3. An example of green algae (courtesy of M. Callow)

Although biofilms are generally composed of unicellular or simple filamentous organisms, in certain circumstances the surrounding environment, e.g. at high water velocities or in the presence of toxic substances, causes stunted growth forms such as macro-algae [Leadbeater and Callow 1992]. Under these conditions the algae can become compact or become incorporated into a microfouling slime layer. The composition of the biofilm depends essentially on the nature of the substrate, nutrient supply and competitive conditions that will modify the colonisation and growth processes. Photosynthetic organisms may also, by the production of oxygen through the photolysis of water to oxygen, encourage the growth of aerobic bacteria. Other photosynthetic cells may release organic compounds that provide a source of nutrients for other micro-organisms.

Leadbeater and Callow [1992] outline the sequence of events that lead to the formation of algal biofilms. Dissolved organic material including glycoproteins [Baier 1980] and colonisation by bacteria [Characklis and Cooksey 1983, Marshall 1985] adsorbed onto the solid surface initiates the algal biofilm. Diatoms (unicells often characterised by a beautiful symmetrical silica outer surface, see Fig. 12.4) are generally the early algal colonisers. In certain conditions the biofilm is almost entirely composed of diatoms and a few entrapped miscellaneous unicells [Jackson and Jones 1988].

FIGURE 12.4. Examples of diatoms (courtesy of M. Callow)

Algae are not likely to be a problem in heat exchangers since in general the surfaces do not receive sunlight. Serpentine coolers over which water falls in contact with the atmosphere, are subject to algal growth and possibly glass heat exchangers through which industrial cooling water is passing. The principle difficulty involving the growth of algae in cooling towers and spray ponds, is the organic debris produced that might be a source of nutrients for other micro-organisms residing on heat transfer surfaces in the same circuit.

*Fungi*

All fungi require fixed organic sources of carbon and the rigid cell wall limits them to being saprophytic on organic substrates, or parasitic living on animals, plants or algae or other fungi [Lynch and Poole 1979]. Fungal growth may be found in stored foodstuffs, decaying paint films, leather, paper, adhesives and plastics; in fact anything that provides the organic substrate. Lacking chlorophyll they cannot carry out photosynthesis. When compared with bacteria fungi have simple nutritional requirements, and their metabolic and biosynthetic processes are not generally diverse. There are two major cell types, the lypha and the yeast [Barnett 1976] with very different properties being determinants for growth in different environments.

In the context of heat exchanger fouling the effects may be compared with algae. In wooden structures (e.g. cooling towers) fungi may be responsible for the degradation of the wood. Fungi produce enzymes that are able to open up the close packed cellulose fibres and so make the wood susceptible to attack from other organisms [Higgins and Burns 1975]. The breakdown products and debris may contribute to fouling on heat exchanger surfaces in a cooling water system and may provide nutrients for bacteria. The use of other materials of construction may obviate the problem of rotting wood, but other cooling tower infills made from PVC or other plastics may also sustain fungal growth. In some instances the sheer weight of saturated biomass on cooling tower internals, has resulted in a complete collapse of the internal structure. The adhesion of filamentous fungi to plastics and plasticisers and subsequent biodegradation has been extensively studied, Parkhurst *et al* [1972], Osmon and Klausmeier [1977], Pantke [1977], Mogilnitskii *et al* [1987], and Hyde *et al* [1989]. It would appear that little work has been undertaken in respect of fungal films on metal surfaces [Mota and Melo 1992].

## 12.2.3 Macrofouling

The problem of macrofouling is not usually associated with heat exchangers in cooling systems [MacNair 1979] but a brief discussion is included here because there are usually implications for the system as a whole and hence heat exchanger operation.

Macrofouling consists of macro-invertebrates such as clams, mussels, barnacles and oysters, together with fish and other living creatures, vegetable matter such as seaweeds and grasses, dead vegetation and manmade debris that could include plastic material. It is the living matter that is generally regarded as macrofouling. The problem is more likely to be manifest in cooling water circuits employing marine or esturine water rather than in fresh water applications, and it is likely to be apparent at intakes and open culverts. It is usual to employ elaborate screening and filtering techniques in order to reduce the accumulation of deposits on surfaces. Restrictions in the intake systems caused by accumulated water borne material (living or dead), may cause reductions in water flow that could affect the operation of the system, including the associated heat exchangers.

Whitehouse *et al* [1985] describe the conditions peculiar to the development of a macrofouling community in power plants. Included are the continuous flow of seawater supplying oxygen and food and reduced salt deposition, the lack of competition from algae in the culverts and a reduction in the presence of predators. The waste products from marine life activity together with decomposition products from dead material, is a potential source of nutrients for microfouling.

According to Whitehouse *et al* [1985] the most important fouling organisms in saline waters are:

Mussels and barnacles
Hydroids, serpulid worms and
Sea mats, sea squirts

Many different varieties exist and the spectrum of organisms will be different between different locations and with the seasons (e.g. Maruthamuthu *et al* 1990].

There is a well defined sequence of events in the development of marine fouling on surfaces recently introduced to the aqueous environment. Initially bacteria attach to the surface forming a slime layer, very much as described in Section 12.2.2. The colonisation of the surface by the bacteria may be regarded as a conditioning process, making it more suitable for the settlement of barnacles and mussels.

*Mussels*

The growth of mussels is highly seasonal, at low temperatures (winter) growth is restricted. Above 10°C the growth can be rapid. Reproduction is also temperature dependent. After fertilisation the eggs develop into free-swimming larvae and depending on water and nutrient availability, the larvae develop and through metamorphosis to the so-called mussel spat. Settlement follows an elaborate pattern to locate a permanent adult site. The size of the animals at this stage is of the order of 0.9 - 1.5 *mm* [Whitehouse *et al* 1985, Lange 1990]. Filtering to remove these potential macrofoulers during settlement is not feasible because of the flow restrictions imposed by such a small filter mesh size. A regulatory limitation of the mussel population is the availability of a solid surface, but the presence of cooling water intakes and culverts encourage growth. The growth of mussels can be rapid, most individuals reaching a length of 30 *mm* within a year of settlement [Holmes 1970].

*Barnacles*

Barnacles give birth to planktonic larvae developing in stages till metamorphosis into a small barnacle. Barnacles can settle in much higher water velocities than mussels and may settle on filter screens causing blockage problems. Barnacles are not such a serious problem in Europe as compared to tropical areas.

*Hydroids*

Hydroids are primitive animals that resemble plants in colonies of very numerous microscopic units called polyps, but there is a free-swimming stage. Hydroids are pioneer organisms and can enhance the settlement of mussels by providing a better substrate than a bare surface.

In hydroids the body wall is not calcified, so that their fouling potential is not so great as with barnacles, mussels and serpulids. Large accumulations of hydroids however, can give rise to severe water flow restriction.

*Serpulid worms*

Serpulid worms build calcaneous tubes and frequently they adopt a helical form. Problems arise if these tubes are allowed to develop and form encrustations on moving parts of equipment or valves [Whitehead *et al* 1985].

*Sea mats and sea squirts*

Serious fouling problems with soft bodied sea mats and sea squirts have not been reported in power stations [Whitehead *et al* 1985]. In large numbers however, it is clear that they could become a nuisance.

*Other organisms*

Some crustaceans such as crabs, although not requiring attachment to a surface for survival, can become occluded on screens and filters. Similar problems occur seasonally from a wide range of other living matter including kelps and seaweeds.

As far as heat exchanger fouling is concerned problems may occur as a result of the small larvae of mussels and barnacles being carried forward in the water stream with settlement occurring in the regions of heat exchangers where flow velocities are low. Unless controlled, the favourable conditions can give rise to accumulations of these creatures, i.e. true macrofouling. Eimer and Renfftlen [1987] suggest that low cooling water velocities within partially plugged tubes promote settling and subsequent growth of barnacle larvae. In turn this can cause pitting corrosion at the growth site and potential erosion - corrosion when the blockage has been cleared and normal cooling water flow is restored. It has also been known for small fish (fry) to pass into heat exchangers and remain say, in the header box of a shell and tube heat exchanger, where they can develop in size and affect the flow distribution through the tubes.

## 12.3   BIOFILM FORMATION

In order that a biofilm can develop on a heat exchanger surface through which a contaminated water, e.g. cooling water is flowing, micro-organisms must arrive at the surface and remain attached so that reproduction and growth can occur. Colonisation and subsequent growth are the result of a complex interaction between factors that are of importance to the metabolism of the cells and the physical conditions, e.g. velocity and temperature, under which the heat exchanger is operating. An early review of the influence of different factors on the formation of biofilms was published by Characklis [1973]. Much research in this area has been conducted by Bott and co-workers [Bott 1993].

A typical growth curve for constant velocity conditions is given in Fig. 12.5. It shows three distinct regions. An initial period where colonisation is taking place and there appears to be little growth. The delay in apparent growth may in fact, be due to the necessary conditioning of the surface with macromolecules before attachment is possible. A second stage involves a period of exponential growth till a third more or less steady state condition is reached. The thickness of the biofilm at the plateau may vary from a few micrometers to several millimeters depending on the prevailing conditions, particularly the velocity and nutrient availability. The effect of marine biofouling on heat transfer performance of titanium condenser tubes was clearly shown by Nosetani *et al* [1981]. A rapid increase in pressure drop over a relatively short time was observed.

Since the micro-organisms may be regarded as particles suspended in a flowing system it is reasonable to assume that they arrive at the heat transfer surface under the mechanisms described in Chapters 5 - 7. Undoubtedly deposition together with the motility that some micro-organisms possess, will be responsible for the initial colonisation of the surface once the adsorbed layer is in place, enhanced by turbulence in the bulk flow. It could be supposed that this is how the biofilm develops, but once on the surface cells will divide and biofilm growth will occur provided nutrients are available. The importance of growth rather than deposition of cells was well demonstrated by Bott and Miller [1983]. Figs. 12.6 and 12.7 show the experimental development of biofilms of *Pseudomonas fluorescens* under identical conditions. Fig. 12.6 shows that after 18 days when no bacteria were added to the system the biofilm continued to grow and there was no apparent change in the rate of development. On the other hand on Fig. 12.7 when the nutrient source was eliminated after about 17 days there was a corresponding drop in the accumulation of biofilm on the surface. The "saw tooth" appearance of the curves may be attributed to a random removal mechanism generally attributed to the prevailing hydrodynamic conditions in the region of the test surface.

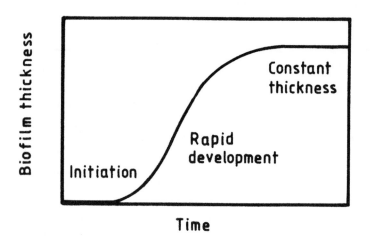

FIGURE 12.5.  An idealised growth curve for a biofilm on a surface

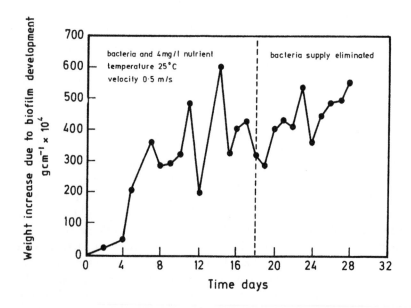

FIGURE 12.6.  The effect of elimination of bacteria from the flowing system

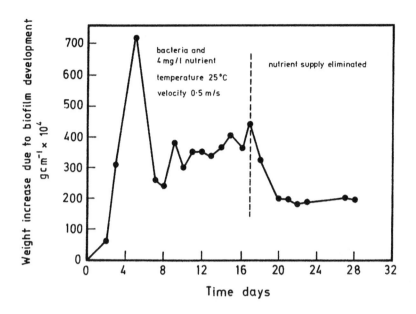

FIGURE 12.7. The effect of elimination of nutrient from the flowing system

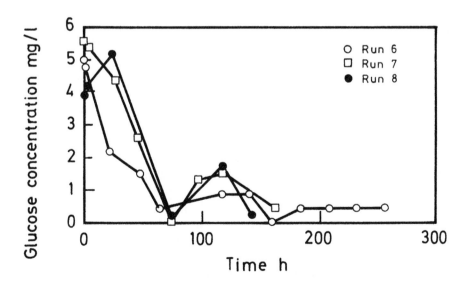

FIGURE 12.8. The reduction of nutrient availability due to biofilm growth

Fig. 12.8 shows how in an experimental system, once the biofilm has started to develop following the initiation period the available nutrients in terms of glucose, are depleted [Kaur 1990]. The quality of the nutrient in terms of $C:N$ ratio also influences the growth of biofilms [Bott and Gunatillaka 1983].

### 12.3.1 Velocity effects

Fig. 12.9 shows how velocity affects the fouling resistance of biofilms. These data were presented by Pinheiro *et al* [1980] for the growth of *Pseudomonas fluorescens* on aluminium surfaces across which water contaminated by the organism and containing nutrients was flowing. *Pseudomonas fluorescens* was chosen as a test organism since it is a slime forming bacterium representative of micro-organisms found in cooling water. It is seen that at the higher velocities the plateau thickness as represented by the fouling resistance, is less than at lower values of velocity. Other work [Nesaratnam 1984] using the same bacterium shows that at low velocities the biofilm thickness is relatively small. Fig. 12.10 shows that biofilm thickness reaches a peak at a water velocity in the region of 1 *m/s* across the surface. It is interesting to note that a "rule of thumb" often quoted, is that velocity to minimise biofouling in steam condenser design should be > 1 *m/s*.

The explanation for the effects of velocity on biofilm thickness usually put forward, is that at very low velocities the resistance to mass transfer of nutrients to the colonising bacteria is high. As the velocity is gradually increased the rate of transfer of nutrients to the biofilm increases so that a greater thickness of biofilm can be sustained on the surface. As velocity is further increased, although the greater turbulence will provide increased mass transfer of nutrients, the shear forces also increase and at the higher velocities it is difficult for a thick biofilm to remain on the surface. Even at relatively low velocity it is possible for areas of biofilm to become detached from the surface. The phenomenon is usually referred to as "sloughing".

There is some evidence [Santos *et al* 1991] that bacteria in contact with flowing water become orientated so that each cell offers the least resistance to flow thereby reducing the tendency to slough, i.e. for rod shape bacteria the major axis was in line with the flow velocity. It is not clear whether this is a "deliberate" attempt by the bacteria to remain on the surface, or whether it is due to the natural elimination of cells that are susceptible to shear forces due to their position in relation to flow.

Data reported in the literature are often obtained from tubes with similar diameters operated at different velocities, i.e. the Reynolds numbers (and hence the turbulence level) were different for each tube. Work by Pujo and Bott [1991] sought to examine the effect of velocity at fixed Reynolds number. An example of their work is shown on Fig. 12.11. For a given Reynolds number of 12,200 it may be seen that velocity does still affect the extent of the biofilm. Biofilm thickness passes through a peak value with increasing velocity till at a velocity of 2 *m/s* the biofilm thickness is severely limited and may be attributed to the shear effects.

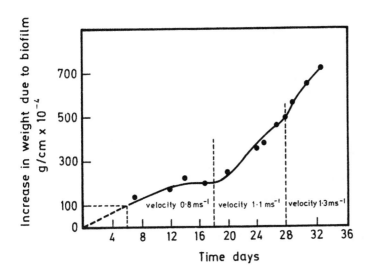

FIGURE 12.9. The change in biofilm thickness with velocity for *Pseudomonas fluorescens* growing on aluminium tubes

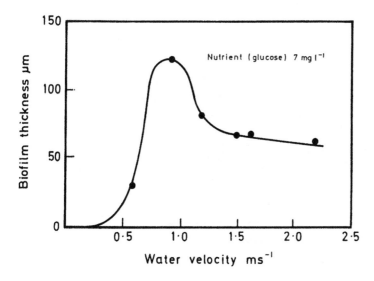

FIGURE 12.10. The effect of velocity on plateau values of biofilm thickness

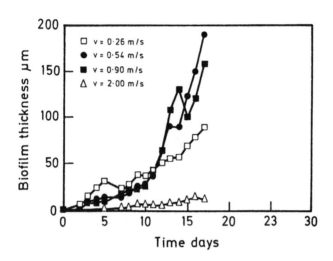

FIGURE 12.11. The development of biofilm with time at a Reynolds number of 12,200 and at different velocities

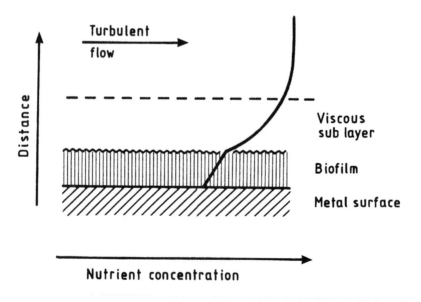

FIGURE 12.12. Idealised interfacial diffusion after a period of time

Kawabe *et al* [1985] reported the effects of velocity on the attachment of marine organisms on toxic copper and copper alloy sea water pipes.   They observed that a wall velocity of 0.4 *m/s* would be sufficient to prevent attachment of barnacles over a twelve month period (i.e. full range of seasonal effects).  On the other hand mussels were not similarly affected because the attachment to the pipe surface reduced the effects of the toxicity of copper ions.

Fig. 12.12 shows the situation that might prevail in the region of a biofilm comprising micro-organisms that are generally aerobic in character.  The laminar sub-layer offers resistance to the transport of nutrients to the biofilm and the removal of waste products back into the region of the bulk water flow. Furthermore the presence of the biofilm itself restricts the flow of nutrients and oxygen through the biofilm to the micro-organisms closest to the heat exchanger surface.  Utilisation of nutrients and oxygen by micro-organisms in the biofilm will also limit their availability at the interface.  It is possible that these micro-organisms will be starved of nutrients and oxygen and that their growth characteristics will be affected.  The restrictions may even produce anaerobic conditions near the metal surface.  It is possible that these changed conditions will be responsible or at least be a contributory factor, in the sloughing process.  Hoehn and Ray [1973] noted that nutrient utilisation by mixed culture bacterial films, decreases when some limiting biofilm thickness eliminates the oxygen availability at the base of the biofilm.    Patel and Bott [1991] published data to show how the oxygen concentration fell at the metal-biofilm interface as the biofilm developed (see Fig. 12.13). The water velocity across the surface was also shown to affect the oxygen available at the biofilm/metal interface.  The reduced resistance of the laminar sub-layer at the higher velocities is illustrated by these data, but eventually the resistance to oxygen transfer by the thicker biofilm and the consumption of oxygen by the cells in the biofilm, dominates and the oxygen concentration at the interface becomes zero.

The driving force for oxygen transfer is limited by the saturation concentration of oxygen in the flowing water and this provides the maximum possible concentration gradient down to zero at the biofilm-metal interface.   The concentration of nutrient is not so limited and high levels of nutrient will increase the nutrient mass transfer rate to the biofilm for a given bulk flow rate and more rapid growth of biofilm could be expected.   Fig. 12.14 shows the effect of increasing the nutrient level (expressed in terms of glucose) for a given velocity of 1.2 *m/s* on the development of a biofilm consisting of *Pseudomonas fluorescens* [Nesaratnam 1984].  The weight increase in a simulated heat exchanger tube gives a measure of the biofilm development.

Changes in nutrient concentration are also reflected in the number of cells per unit mass of biofilm as illustrated in Table 12.4.

The effect of velocity on the development of the biofilm affects its density [Bott and Pinheiro 1977, Vieira *et al* 1993].  At higher velocities the biofilm density is increased and this will in turn, affect the mass transfer through the biofilm as the higher density biofilm will offer a greater resistance to molecular diffusion.

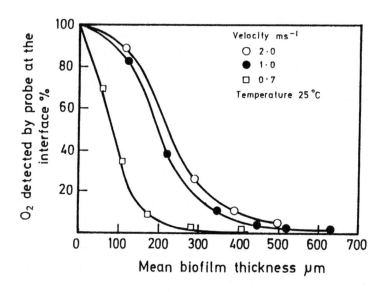

FIGURE 12.13.  The change in oxygen concentration through a growing biofilm

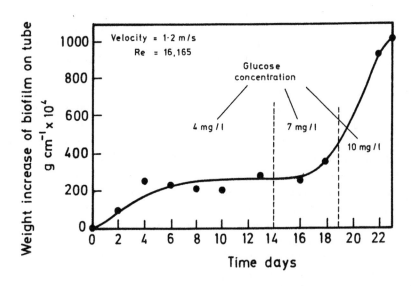

FIGURE 12.14.  The change in the mass of biofilm on a surface at different nutrient concentrations

TABLE 12.4

Cell numbers and nutrient concentration [Patel 1986]

| Nutrient concentration as *mg/l* glucose | Cell numbers *cells/gm* biofilm |
|---|---|
| 0 | $2.5 \times 10^5$* |
| 5 | $11.0 \times 10^9$ |
| 100 | $2.8 \times 10^{12}$ |

* This number of cells is considered to have arrived at the surface by transport from the bulk suspension

It is likely also that biofilms will become more dense with the passage of time since the voids within a biofilm, will gradually fill with cells as a result of reproduction. The effects of the age of a biofilm of *Pseudomonas fluorescens* on mass transfer through the biofilm were demonstrated by Vieira *et al* [1994]. Using *Li Cl* as the diffusing solute there was a fall of around 60% in mass transfer coefficient in a period of about 150 hours. The Reynolds numbers used in the experiments were 8250 and 14700 but there appeared to be little dependence as would be anticipated, of mass transfer through the biofilm due to changes in Reynolds number.

## 12.3.2 Temperature

The discussion so far has concentrated on the effects of velocity but another variable of considerable influence on the development of biofilms is temperature. In Section 12.2 it was stated that all micro-organisms have an optimum temperature for maximum growth.

Fig. 12.15 shows a remarkable increase in biofilm thickness for only 5°C temperature change. These data were obtained for *Escherichia coli* in an experimental apparatus with flow Reynolds number of $6.3 \times 10^3$, i.e. turbulent [Bott and Pinheiro 1977]. The optimum temperature for maximum growth for this species is around 35 - 40°C. Work by Harty [1980] using the same species (*Escherichia coli*) demonstrated similar effects. Fig. 12.16 also shows that as velocity is increased (i.e. increased shear at the surface) in the particular system studied, the effect of temperature becomes less. It would be expected that the effects of shear are likely to be more pronounced than the effects of temperature on growth.

Tests on an industrial cooling water system reported by Bott *et al* [1983] also showed that temperature (as measured by hours of sunshine) influenced the extent of biofilm formation for the mixed population involved. The data are reproduced on Fig. 12.17 and shows the increased biological activity during the summer months. Kaur *et al* [1994] demonstrated the remarkable difference in algal cell

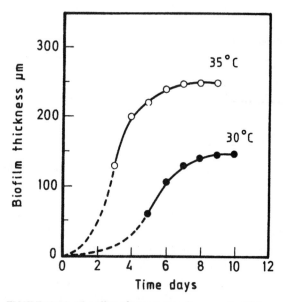

FIGURE 12.15. The effect of temperature change on the thickness of a biofilm containing *Escherichia coli*

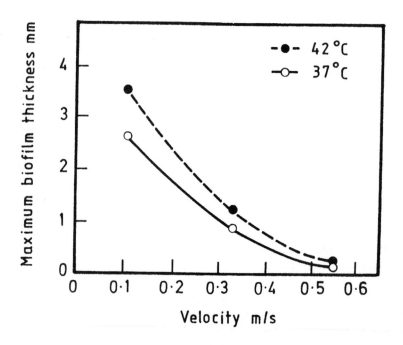

FIGURE 12.16. Maximum biofilm thickness versus velocity for temperatures 37 and 42°C

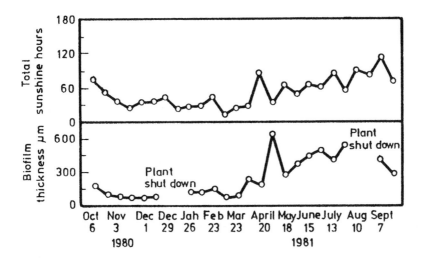

FIGURE 12.17. Comparison of hours sunshine with mean biofilm thickness

numbers in a reservoir water (originating in a river) between the months of August and October 1989. The ratio of average cell numbers/*ml* in August compared with October was 24:1.

12.3.3 The effect of surface

The material of construction of a heat exchanger and its surface roughness will influence the development of biofilms [Mott *et al* [1994]]. Vieira *et al* [1992] showed that the initial deposition of *Pseudomonas fluorescens* was more pronounced on aluminium plates than on brass or copper. Furthermore the number of bacteria deposited increases with fluid velocity with aluminium. It is possible the copper and the brass were producing toxic ions (copper) that inhibited growth. Vieira *et al* [1993] conclude that the presence of metal ions is likely to affect the metabolism of the micro-organisms as they colonise a surface that will modify the adhesion properties of the cells. On the other hand the effect of different materials of construction may be due to differences in surface roughness. In extensive trials with spirally indented and plain tubes fitted to power station condensers Rabbas *et al* [1993] observed fouling from river water on the indented tubes to be about double that on plain tubes. At least some of the fouling could be attributed to the presence of biofilm. It would be expected that rough surfaces would be more hospitable to micro-organisms than smooth surfaces and this was clearly demonstrated by Reid *et al* [1992] and Mott and Bott [1991]. Suspensions of *Pseudomonas fluorescens* together with appropriate nutrients and trace elements were passed through tubes of glass, fluorinated ethylene polypropylene, "as received" and electropolished 316 stainless steel. The water velocity through the tubes was 1 *m/s*. The accumulation of biofilm with time is recorded on Fig. 12.18. It can be seen that there is an appreciable additional biofilm on the "as received" 316 stainless steel compared with the other materials, glass, fluorinated ethylene polypropylene, and electropolished stainless steel 316 that may be considered smooth.

In contrast, Heukelekian and Crosby [1956] suggested that for sewage works, where high concentrations of micro-organisms are likely to be present and velocities are different from those generally experienced in heat exchangers, that surface roughness was not significant in the amount of slime formed.

Once the surface is covered with biofilm the characteristics of the interface between the solid and the flowing stream are altered. As the biofilm thickens it will become "compliant", this is it will deform under the effects of the flowing fluid [McKee 1991]. In addition the apparent roughness of the biofilm changes.

Characklis *et al* [1979] showed by experiment, how the equivalent sand grain roughness increased with time. Fig. 12.19 shows some of their data. Increased equivalent roughness will increase the pumping requirements to maintain a given velocity (see Chapter 3).

### 12.3.4 The effects of suspended solids

Particulate matter in the water stream will influence the retention of biofilms as mentioned in Section 12.3.2. Data over a 12 month period on the amount and size of suspended particulate matter in the River Tees is given on Fig. 12.20 [Mansfield 1983]. The high levels in the winter months may be attributed to the increased "wash off" due to rain and melting snow. It has been reported [Bour *et al* 1981] that power station condensers using cooling water containing particulate matter are remarkably free from biofilms. The cleanliness is attributed to the erosive effect of the suspended solids. The morphology of the particles which largely depends on their constituent material, influences their effect on biofilms. For instance research by Lowe [1988] shows that sand particles as opposed to kaolin (clay) particles are much more efficient at controlling biofilm development. There is evidence [Lowe 1988] that some of the particles contained in the flowing stream will be incorporated into the biofilm. The presence of these particles may have implications for the stability of the biofilm.

### 12.3.5 *pH* changes

Since natural biofilms on rocks and surfaces in contact with rivers and streams form at a *pH* near neutral, excursions from *pH* of around 7 are likely to affect the development of biofilms. Some tolerance of *pH* change is displayed by micro-organisms since through their own activity, there is a tendency to discharge

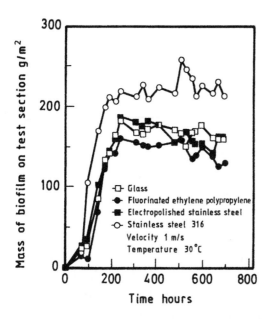

FIGURE 12.18. Biofilm (*Pseudomonas fluorescens*) development on various surfaces

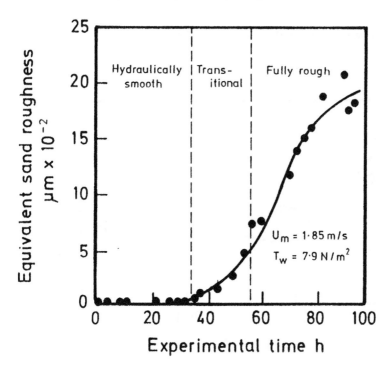

FIGURE 12.19. Change in equivalent sand grain roughness over a 100 *h* experiment

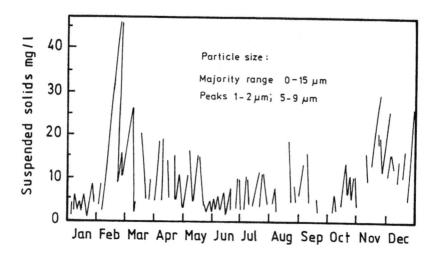

FIGURE 12.20.  The variation of suspended matter in the River Tees

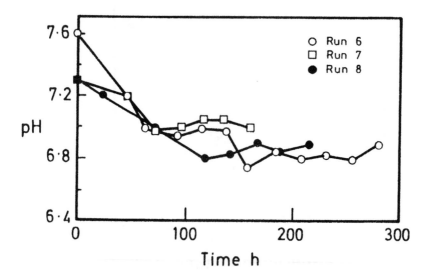

FIGURE 12.21.  *pH* change during biofilm (*Pseudomonas fluorescens*) growth

products that lower the *pH*. Kaur [1990] showed that as biofilm activity increased the *pH* gradually fell. The range of the *pH* change (7.5 - 6.6) however is relatively small as seen on Fig. 12.21.

### 12.3.6 Trace elements

In Section 12.2 the need for trace elements was discussed. Work by Santos [1993] reveals that unless trace elements are present biofilm growth on a surface exposed to a flowing system of bacteria (*Pseudomonas fluorescens*), is severely restricted. Fig. 12.22 shows that when trace elements were added (after approximately 800 hours) there was a remarkable sudden growth in biofilm.

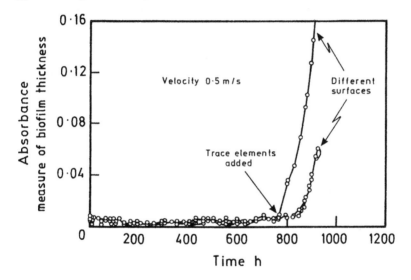

FIGURE 12.22. The effect of trace elements on biofilm (*Pseudomonas fluorescens*) development

## 12.4   MODELS OF BIOFOULING

The previous sections in this chapter have demonstrated the complex behaviour of micro-organisms in relation to surfaces and any attempt to describe the process mathematically will require some simplifying assumptions. An early attempt at modelling was made by Atkinson and Daoud [1968] who based their analysis on the analogy between micro-biological "reaction" and heterogeneous catalysis. The importance of diffusional limitations in the mass of micro-organisms at a support surface was demonstrated. More recently a number of workers, notably amongst them the late Professor Bill Characklis, has attempted with some success, to analyse the physical and biological processes that contribute to biofouling. The work of these authors will be used extensively in this section.

Characklis [1981] lists the steps that are involved in the biofouling process:

1. Transport of dissolved and particulate matter from the bulk fluid to the surface

2. Firm attachment of micro-organisms to the surface

3. Microbial transformations, i.e. growth reproduction and death

4. Detachment generally due to fluid shear forces

He also emphasises that confusion between process rates and rates of change must be avoided if an accurate mathematical assessment of the problem is to be achieved. The process rates are those that can be directly related to system variables such as temperature, pressure, composition, velocity and geometry (e.g. flow area). These fundamental quantities involve conservation of mass, energy, momentum and chemical species and may be generalised and simply correlated. Rates of change on the other hand, cannot be simply correlated or generalised, and involve the rate of accumulation of the biofilm and the net rate of input by virtue of the flowing system.

12.4.1  Transport to the surface

A general model based on the concepts of Equation 4.1 was suggested by Trulear and Characklis [1979]. The relationship had the form:

$$A\rho\frac{dx}{dt} = \phi_D - \phi_R \tag{12.1}$$

where $A$ is the area for deposition

$\rho$ is the biofilm volumetric density

$x$ is the biofilm thickness

In this equation the system variables will have a pronounced effect on the magnitude of $\phi_D$ and $\phi_R$.

Of primary interest in the analysis is the transport of micro-organisms and nutrients towards the surface. Pinheiro *et al* [1988] based an analysis of deposition flux in terms of physical processes involving transport of micro-organisms and adhesion, and biological processes including transport of nutrients and biological growth. The size range of the cells involved is likely to be in the range 0.5 - 10 $\mu m$ [Kent 1988].

In Chapter 5 some of the principles of mass transfer were discussed and these are now applied to microbial fouling. When a clean surface comes into contact

with a suspension of micro-organisms transport towards the surface occurs. Mass transfer involves several mechanisms [Characklis and Cooksey 1983] some of which have already been mentioned. They include:
Brownian diffusion
sedimentation (i.e. the effects of gravity)
thermophoresis (i.e. movement down a temperature gradient)
motility (taxis)

It is probably true to say that in a heat exchanger involving turbulent flow, these mechanisms are overshadowed in the bulk flow by the more prominent effects of the fluid flow forces and turbulence, i.e. eddy transport and inertia. Within the viscous sub-layer however, it is likely that all these mechanisms can have an influence on the deposition process together with the phenomena that may be associated with the viscous sub-layer, i.e. instabilities including turbulent bursts (lift forces and fluid downsweeps) and surface drag.

Equation 5.6 may be applied to the mass transfer of micro-organisms to the surface, i.e.

$$N_A = -(D_{AB} + E_D)\frac{dc_A}{dx} \tag{5.6}$$

where $D_{AB}$ is the diffusivity $E_D$ is the eddy diffusion

$N_A$ refers to the number of cells moving towards the surface under the concentration gradient $dc_A/dx$, i.e. the mass flux of cells.

It is possible to integrate Equation 5.6 [Beale 1970] to give the flux at the wall

i.e. $N_A = K_d c_{AV}$ $\tag{12.2}$

where $K_d$ is the deposition coefficient

$c_{AV}$ is the average cell concentration over the flow area

The deposition coefficient $K_d$ may be further defined by the following equation:

$$K_d = \frac{Kv_r P}{K + u_r P} \tag{12.3}$$

where $K$ is the transfer coefficient

$u_r$ is the radial velocity of the cell

and     $P$ is the sticking probability (defined in Chapter 7)

Characklis [1981] considered that the radial velocity $u_r$ is made up of a number of components dependent on the different forms of movement previously listed including Brownian motion, motility and sedimentation, together with the component due to the effects of fluid flow on mass transfer as summarised in Equation 5.6. The extent of thermophoresis is likely to be small since the temperature differences involved are generally small.

Simplifications of Equation 12.3 may be made based on the magnitude of $K$ the transfer coefficient. When $K \gg Pu_r$

$$N_A = Pu_r c_{Av} \qquad (12.4)$$

and when     $Pu_r \gg K$

$$N_A = K c_{AV} \qquad (12.5)$$

Equation 12.5 describes the general condition applicable to microbial fouling.

For the initial stages of biofouling when the growth on the surface is negligible the colonising development of the biofilm is directly related to the mass transfer of the cells and the macromolecules. Assuming that these two processes are occurring simultaneously, it is possible to write an equation that describes the total transport ($N_{TOTAL}$) to the surface as

$$N_{TOTAL} = N_A + N_M \qquad (12.6)$$

where $N_M$ refers to the number of macromolecules arriving at the surface, i.e. the mass flux of macromolecules

Assuming that Equation 12.4 applies then

$$N_A = K_A c_{A \, Av} \qquad (12.7)$$

and     $N_M = K_M c_{M \, Av} \qquad (12.8)$

where $K_A$ and $K_M$ and $c_{A \, Av}$ and $c_{M \, Av}$ are the transfer coefficients and average concentrations of the cells and macromolecules respectively.

The total flow (cells and macromolecules) to the surface will therefore be given by

$$N_{TOTAL} = K_A c_{A \, Av} + K_M c_{M \, Av} \qquad (12.9)$$

If Equation 12.9 is used to make calculations in respect of a particular system the units (or dimensions) of the various components need careful attention so that the equation is dimensionally consistent, i.e. cell numbers must not be added to mass flow.

In an analysis of the near initial rate of biofilm accumulation Bryers and Characklis [1981] related the development of biofilm to the concept of a continuous stirred tank reactor. They took into account the effects of shear forces and associated removal of biofilm and assumed that the subsequent effects were dependent on the biofilm already on the surface, i.e.

$$\frac{dN_A}{dt} \quad = \quad k_d(f_1 N_A) \quad + \quad k_g(f_2 N_A) \quad - \quad k_s(f_3 N_A)$$

i.e. | initial biofilm accumulation rate | = | micro-organism deposition and attachment rate | + | attached biofilm growth rate | - | attached biofilm removal rate |

$$(12.10)$$

$k_d$, $k_g$ and $k_s$ are the appropriate rate constants for attachment, growth and removal respectively, $(f_1 N_A)$, $(f_s N_A)$ and $(f_3 N_A)$ are functions of attached biofilm.

### 12.4.2 Adhesion and growth

An important aspect of biofilm formation is the attachment to the surface (see Section 12.2.2) and in some respects, this might be considered the biofilm accumulation rate determining step (excluding reproduction and growth within the film). Fletcher [1977] modelled the adsorption of micro-organisms to a surface with the following assumptions:

1.  Adsorption was limited to a monocellular layer. This is a necessary assumption since the mechanism of all adhesion to a surface will be completely different from cell incorporation into an existing biofilm.

2.  Adsorbed components are restricted to specific adsorption sites.

3.  Heat of adsorption is independent of surface coverage.

On this basis she concluded that:

$$N_A = kc_A(1 - \theta) \tag{12.11}$$

$N_A$ is the rate of attachment of cells and if the sticking probability is unity it is identical to the rate at which the cells approach the surface.

$k$ is the adsorption rate constant

$\theta$ is the fraction of the surface covered by the attached cells

The number of cells per unit area $X$ is related to the coverage of the surface so that

$$X = X_T\theta \qquad\qquad (12.12)$$

where $X_T$ is the number of cells required for complete surface coverage

i.e. $N_A = kc_A(1 - \dfrac{X}{X_T})$ \hspace{4cm} (12.13)

The surface coverage $X_T$ is likely to be in the range $2 - 3 \times 10^7$ cells/$cm^2$ [Characklis 1981]. Even for cultures of a single species of micro-organism, $K$ will vary with the organism and for mixed cultures, it is very likely that it will not be a constant but dependent on local conditions.

Some bacteria (e.g. *Cytophaga*) attach to surfaces in what might be termed a "loose" sort of way since they are capable of gliding across the surface. The adhesion is through a slime consisting of a linear polymer with the characteristics of a Stefan adhesive [Humphrey *et al* 1979] and may be thought of as "temporary". In contrast some bacteria remain attached to a specific point on a surface and do not migrate across the surface and adhesion is "permanent" in the one location. Despite the permanence of location and the relative closeness of other "permanent" cells, such bacteria are still capable of multiplication and progeny may remain at the location or be released to settle at some other point in the system. A number of authors, e.g. Marshall [1981] consider that the permanent attachment is due to so-called "polymer bridging" although even now, the exact mechanism is not fully understood. It is likely to depend on a physico-chemical process involving macromolecules on the surface and fibrils on the cells. Bryers and Characklis [1992] conclude that the extent of the adsorption of macromolecules will greatly affect the subsequent biofilm formation.

Using the original concept of ZoBell [1943], Characklis and Bryers [1992] consider the relation between cells that are "temporary" or reversibly attached and those that are irreversibly bound to the surface. Ignoring multiplication of cells at the surface an equilibrium will be established such that:

$$
\begin{array}{ccc}
X & \rightarrow & X_{rev} \\
\text{Suspended cells} & & \text{reversibly} \\
& & \text{attached cells}
\end{array}
\qquad (12.14)
$$

and $X_{rev}$ $\rightarrow$ $X_{irr}$ (12.15)
irreversibly
attached cells

The total cells attached to the surface at any time will be the sum of $X_{rev}$ and $X_{irr}$.

If the rate of cell attachment can be described as a first order reaction then

$$r_{x_{ads}} = k_{x_{ads}} c_x$$ (12.16)

where $r_{x_{ads}}$ is the rate of cell adsorption

$k_{x_{ads}}$ is the rate constant for adsorption

and $c_x$ is the concentration of cells in suspension

Cells attached to a surface will continue their metabolic transformations dependent on the transport of suitable nutrients to the biofilm and the removal of by products (usually soluble). Kent [1988] commented on the rate expressions that describe biofilm growth. They may be written in terms of biomass production rate or nutrient depletion rate. The corresponding rate constants will be dependent on the variables that generally affect chemical reactions. They include:
 *pH*
 temperature
 limiting nutrient concentration
 terminal electron acceptor
 organism concentration

The latter is not necessarily the usual effect of concentration but may be associated with the extent of by-products produced by the system. Depending on the prevailing conditions and the species present, the biofilm can assume different states, e.g. aerobic, or anaerobic.

The chemical reactions are required to account for the fundamental processes occurring within the biofilm including:
 cell growth and multiplication
 cell maintenance
 by-product formation (involving extra- and intracellular products)
 death by viability loss
and death by cell lysis

Capdeville *et al* [1992] have mathematically described biofilm growth in terms of active micro-organisms $M_A$ and deactivated micro-organisms $M_D$.

The total biomass $M_T$ may therefore be defined as

$$M_T = M_A + M_D \tag{12.17}$$

The rate of change in concentration of the biomass may be obtained from the expression

$$\frac{dc_{Bio}}{dt} = r_g c_{Bio} \tag{12.18}$$

or $\dfrac{dc_{Bio}}{dt} = r_g c_{Bio} - k c_{in} c_{Bio}$ \hfill (12.19)

where $r_g$ is the rate of growth

$k$ is the coefficient of arrested growth or mortality

$c_{Bio}$ is the concentration of biomass

$c_{in}$ is the concentration of inhibitory product

   These equations are based on the assumption that at any given moment the population of micro-organisms (bacteria) in a culture will multiply as long as either there is nutrient available or, the concentration of the inhibitory product is not limiting. The rate of multiplication within the biofilm will vary according to these criteria. Belkhadir *et al* [1988] described the fundamental growth phases in a biofilm. The growth of active biomass $r_{MA}$ is assumed to have an order of zero in relation to the nutrient and an order of unity in relation to the active bacteria so that:

$$r_{MA} = r_o M_A \tag{12.20}$$

where $r_o$ is the maximum growth rate

$M_A$ is the active mass per unit of surface area

Deactivation kinetics $r_{MD}$ depend on cell density and the accumulation of inhibitors, i.e.

$$r_{MD} = K_1 c_{in} M_A \tag{12.21}$$

where $K_1$ is the constant for deactivation

and $c_{in}$ the concentration of inhibitors

The rate of accumulation of micro-organisms on the surface is given by:

$$\left(\frac{dM_A}{dt}\right)_{acc} = r_{MA} - r_{MD} \tag{12.22}$$

The concentration of inhibitors will be a function of the active cells present in the system. At the end of the dynamic or exponential growth phase a plateau or maximum biofilm thickness will have been produced (i.e. $M_{A(max)}$) and (see Fig. 12.5)

$$\left(\frac{dM_A}{dt}\right)_{acc} = r_o M_a - K_2 M_A^2 \tag{12.23}$$

where $K_2 = r_o/M_{A(max)}$ (12.24)

Differential equations follow from this analysis, i.e.

$$\left(\frac{dM_A}{dt}\right)_{acc} = r_o M_A \left[1 - \frac{M_A}{(M_A)_{max}}\right] \tag{12.25}$$

and $r_{MD} = \left(\frac{dM_D}{dt}\right)_{acc}$ (12.26)

$$= r_o \frac{M_A^2}{(M_A)_{max}} \tag{12.27}$$

The experimental observations of Belkhadir [1986] show that the density of the microbial colonies depends on the surface area of the support. Using this fact Capdeville *et al* [1992] were able to relate the accumulation of biofilm to the area available for support.

$$\left(\frac{dM_A}{dt}\right)_{acc} = r_o M_A B \frac{(A_o - A_t)}{A_o} \tag{12.28}$$

where $a$ is the area covered by micro-organisms at time $t$

$A_o$ is the initial area available for colonisation
$B$ is a correction term for the difference between the surface area of a model and the true situation in reality

Equation (12.27) demonstrates that as long as there is sufficient area available the quantity of active biomass will increase.

For relatively short times Capdeville *et al* [1992] provide the expression

$$M_A = (M_A)_o e^{r_o' t} \qquad\qquad (12.29)$$

where $(M_A)_o$ is the initial active mass per unit of surface area

## 12.5 CONCLUDING REMARKS

Biofouling is a highly complex phenomenon involving mass transport, physico-chemical reactions, and biological metabolism. The interactions between these processes and mechanisms are extremely difficult to understand let alone incorporate into mathematical models. The brief résumé in section 12.4, of the approaches that have been made to the solution of the problem of mathematical modelling, are useful in quantitatively setting out the effects of the different variables. It will be many years before a reliable model will be developed that may be used to calculate the accumulation of biofilm on a heat exchanger surface and hence provide the basis for an estimation of heat transfer resistance for use in heat exchanger design.

## REFERENCES

Atkinson, B. and Daoud, I.S., 1968. The analogy between micro-biological "reaction" heterogeneous catalysts. Trans. Inst. Chem. Engrs. 46, T19-T24.

Baier, R.E., 1980, Substrata influences on the adhesion of micro-organisms and their resultant new surface properties, in: Britton, G., Marshall, K.C. eds. Adsorption of micro-organisms to surfaces, 59 - 104, Wiley, New York.

Barnett, J.H., 1976, Fundamentals of mycology. Edward Arnold, London.

Belkhadir, R., Capdeville, B. and Roques, H., 1988, Fundamental descriptive study and modelization of biological film growth. Nat. Res. 22, 1, 59 - 69.

Belkhadir, R., 1986, Etude fondamentale des biomasses fixées: description et modélisation des films biologiques anaérobies. PhD Thesis, No. 18 INSA, Toulouse.

Bott, T.R., 1993, Aspects of biofilm formation. Corrosion Rev. 11, 1 - 2, 1 - 24.

Bott, T.R. and Gunatillaka, M, 1983, Nutrient composition and biofilm thickness, in: Bryers, R.W. and Cole, S.S. eds, Fouling of heat exchanger surfaces. United Engineering Trustees Inc. 727 - 734.

Bott, T.R. and Miller, P.C., 1983, Mechanisms of biofilm formation on aluminium tubes. J. Chem. Tech. Biotechnol., 33B, 177 - 184.

Bott, T.R. and Melo, L.F., 1992, Particle-bacteria interactions in biofilms, in: Melo, L.F., Bott, T.R., Fletcher, M. and Capdeville, B. eds. Biofilms -

Science and Technology, 199 - 206, Kluwer Academic Publishers, Dordrecht.

Bott, T.R. and Pinheiro, M.M.V.P.S., 1977, Biological fouling - velocity and temperature effects. Can. J. Chem. Eng., 55, 473.

Bott, T.R., Miller, P.C. and Patel, T.D., 1983. Biofouling in an industrial cooling water system. Process Biochem. 18, 1, 10 - 13.

Bour, D.P., Battaglia, P.J. and Burd, K.M., 1981, Biofouling control practice and assessment, EPRI Report CS., 1796, Palo Alto.

Bryers, J.D., and Characklis, W.G., 1981, Kinetics of initial biofilm formation within a turbulent flow stream, in: Somerscales, E.F.C. and Knudsen, J.G. eds. Fouling of Heat Transfer Equipment. Hemisphere Pub. Corp. 313 - 333.

Bryers, J.D. and Characklis, W.G., 1992, Biofilm accumulation and activity: a process analysis approach, in: Melo, L.F., Bott, T.R., Fletcher, M. and Capdeville, B. eds. Biofilms - Science and Technology. Kluwer Academic Publishers, Dordrecht, 221 - 250.

Busscher, H.J. and Weerkamp, A.H., 1987. Specific and non-specific interactions in bacterial adhesion to solid substrata. FEMS Microbiol. Rev. 46, 165 - 173.

Capdeville, B., Nguyen, K.M. and Rols, J.L., 1992, Biofilm modelling: structural, reactional and diffusional aspects, in: Melo, L.F., Bott, T.R., Fletcher, M. and Capdeville, B. eds. Biofilms - Science and Technology. Kluwer Academic Publishers, Dordrecht, 251 - 276.

Characklis, W.G., 1973, Attached microbial growths - I. Attachment and growth. Water Res. 7, 1113 - 1127.

Characklis, W.G. and Wilderer, P.A., 1989, Structure and Function of Biofilms. John Wiley and Sons Ltd.

Characklis, W.G., Bryers, J.D., Trulear, M.G. and Zelver, N., 1979, Biofouling, film development and its effect on energy losses: A laboratory report. Condenser biofouling control symposium, EPRI, Atlanta.

Characklis, W.G. and Cooksey, K.E., 1983, Biofilms and microbial fouling. Adv. Applied Microbial, 24, 93 - 138.

Characklis, W.G., 1981, Microbial fouling: a process analysis, in: Somerscales, E.F.C. and Knudsen, J.G. eds. Fouling in Heat Transfer Equipment. Hemisphere Publishing Corp. Washington, 251 - 291.

Cook, P.E. and Gaylarde, C.C., 1993, Microbial films in the light engineering industry, in: Denyer, S.P. and Sussman, M. eds. Microbial Biofilms: Formation and Control. Blackwell Scientific Publications, London.

Cooksey, K.E., 1992, Extracellular polymers in biofilms, in: Melo, L.F., Bott, T.R., Fletcher, M. and Capdeville, B. eds. Biofilms - Science and Technology, 137 - 147, Kluwer Academic Publishers, Dordrecht.

Costerton, J.W., Geesey, G.G. and Cheng, K.J., 1978, How bacteria stick. Scientific American, 238, No. 1.

Eashwar, W., Chandrasekaran, P. and Subramanian, G., 1988, Marine microbial films and the corrosion of steel. Bul. Electrochem. 4, 2, 115 - 119.

Eimer, K. and Renfftlen, R., 1987, Evaluation of various condenser and cooling water system macro-biological fouling control methods. EPRI Symposium Providence RI.

Flemming, H.C., 1991, in: Flemming, H.C. and Gessey, G.G. eds. Biofouling and Biocorrosion in Industrial Water Systems. Springer-Verlag, Berlin.

Fletcher, M., 1977, The effects of culture concentration and age, time and temperature on bacterial attachment to polystyrene. Can. J. Microbiol. 23: 1 - 6.

Fletcher, M., 1992, Bacterial metabolism in biofilms, in: Melo, L.F., Bott, T.R., Fletcher, M. and Capdeville, B. eds. Biofilms - Science and Technology, 113 - 124, Kluwer Academic Publishers, Dordrecht.

Fletcher, M., Bright, J.J. and Pringle, J.H., 1983. Attachment of aquatic bacteria to solid surfaces, in: Bryers, R.W. ed. Fouling of Heat Exchanger Surfaces. Limited Engineering Trustees Inc., New York, 33 - 49.

Geesey, G.G., 1982, Microbial exopolymers: ecological and economic considerations. An. Soc. Microbiol. News, 48, 9 - 14.

Haack, T.K. and McFeters, G.A., 1982, Nutritional relationships among micro-organisms in an epilithic biofilm community. Microb. Ecol. 8: 115 - 126.

Hamilton, W.A., 1987, Biofilms: microbial interactions and metabolic activities. Symp. Soc. Gen. Microbiol. 41, 361 - 385.

Hamilton, W.A. and Characklis, W.G., 1989, Relative activities of cells in suspension and in biofilms, in: Characklis, W.G. and Wilderer, P.A. eds. Structure and Function of Biofilms. John Wiley and Sons Ltd.

Harty, D.W.S., 1980, Microbial fouling of a simulated heat transfer surface. PhD Thesis, University of Birmingham.

Heukelekian, H. and Crosby, E.S., 1956, Slime formation in polluted waters II. Factors affecting slime growth. J. Sewage Ind. Wastes, 28, 1, 78 - 92.

Higgins, I.J. and Burns, R.G., 1975, The Chemistry and Microbiology of Pollution. Academic Press, London.

Hoehn, C. and Ray, A.D., 1973. Effects of thickness of biofilm. J. Water Pol. Control Fed. 45, 11, 2302 - 2320.

Holmes, N., 1970, Mussel fouling in chlorinated cooling systems. Chem. and Ind. 1244 - 1247. 26th September.

Humphrey, B., Dickson, M.R. and Marshall, K.C., 1979, Physico-chemical and *in situ* observations on the adhesion of gliding bacteria to surfaces. Archives Microbiol. 120, 231 - 238.

Hyde, K.D., Moss, S.T. and Gareth Jones, E.B., 1989, Attachment studies in marine fungi, Biofouling, 1, 287 - 298.

Jackson, S.M. and Jones, E.B.G., 1988, Fouling film development on antifouling paints with special reference to film thickness. Int. Biodeterioration, 24, 277 - 287.

Kaur, K., Bott, T.R., Heathcote, G.R., Keay, G. and Leadbeater, B.S.C., 1994, Treatment of algal-laden water: pilot plant experiences. J. Inst. Water and Enviro. Man. 8, No. 1, 22 - 32.

Kaur, K., 1990, Ozone as a biocide in cooling water systems PhD Thesis, University of Birmingham.

Kawabe, A., Yasui, K., Ikeda, T., Nishijima, N., Yamamoto, H. and Kawabe, A., 1985, Effect of flow dynamics on attachment of marine organisms for coper and copper alloy sea water pipes. Denki Kagaku, 53, 6, 422 - 423.

Kent, C.A., 1988, Biological fouling: basic science and models, in: Melo, L.F., Bott, T.R. and Bernardo, C.A. eds. Fouling - Science and Technology, Kluwer Academic Publishers, Dordrecht.

Lange, A., 1990, Money saving by the improved operation of condenser tube cleaning systems, the installation of cooling water debris filters and newly developed condenser monitoring. VGB-Kraftwerkstechnik., 8, 1 - 8.

Leadbeater, B.S.C. and Callow, M.E., 1992, Formation, composition and physiology of algal biofilms, in: Melo, L.F., Bott, T.R., Fletcher, M. and Capdeville, B. eds. Biofilms - Science and Technology. Kluwer Academic Publishers, Dordrecht.

Lowe, M.J., 1988, The effect of inorganic particulate materials on the development of biological films. PhD Thesis, University of Birmingham.

Lynch, J.M. and Poole, N.J., 1979, Microbial Ecology - a Conceptional Approach. Blackwell Scientific Publications Oxford.

MacNair, E.J., 1979, Fouling in normal condensers and heat exchangers. Paper C74/79 39 - 44. I. Mech. E.

Mansfield, G.H., 1983, Private communication.

Marshall, K.C., 1976, Interfaces in Microbial Ecology. Havard University Press Cambridge.

Marshall, K.C., 1985, Mechanisms of bacterial adhesion at solid-water interfaces, in: Savage, D.C. and Fletcher M. eds. Bacterial Adhesion, 133 - 161, Plenum Press, New York.

Marshall, K.C., 1981, Bacterial behaviour at solid surfaces - prelude to microbial fouling, in: Somerscales, E.F.C. and Knudsen, J.G. eds. Fouling of Heat Transfer Equipment, Hemisphere Pub. Corp. 305 - 312.

Maruthamuthu, S., Eashwar, M., Manickam, S.T., Ambalavanan, S., Venkatachari, G. and Balakrishnam, K., 1990, Marine fouling on test panels and in-service structural steel. Tuticorin Harbour, Indian J. Marine Sci., 19, 68, 0, 70.

McKee, C.J., 1991, The interaction between turbulent bursts and fouled surfaces. PhD Thesis, University of Birmingham.

Melo, L.F., Bott, T.R., Fletcher, M. and Capdeville, B., 1992, Biofilms - Science and Technology. Kluwer Academic Publishers, Dordrecht, 687 - 688

Miller, P.C., 1982, Biological fouling film formation and destruction. PhD Thesis, University of Birmingham.

Mogilinitskii, G.M., Sagatelyan, R.T., Kutischeva, T.N., Zhukova, S.V., Kerimov, S.I. and Parfenova, T.B., 1987, Deterioration of protective properties of PVC coating by micro-organisms, Zasch. Met. 23, 173 - 175.

Mota, M. and Melo, L.F., 1992, Fungal biofilms on metallic surfaces, in: Melo, L.F., Bott, T.R., Fletcher, M. and Capdeville, B. eds. Biofilms - Science and Technology, 175 - 180, Kluwer Academic Publishers, Dordrecht.

Mott, I.E.C., Santos, R. and Bott, T.R., 1994, The effect of surface on the development of biofilms. Proc. 3ème Réunion Européenne Adhesion des Micro-Organismes aux Surfaces. Châtenay-Malabry, C.21.1 - 2.

Mott, I.E.C. and Bott, T.R., 1991, The adhesion of biofilms to selected materials of construction for heat exchangers. Proc. 9th Int. Heat Trans. Conf. Jerusalem, vol. 5, 21 - 26.

Mozes, N. and Rouxhet, P.G., 1992, Influences of surfaces on microbial activity, in: Melo, L.F., Bott, T.R., Fletcher, M. and Capdeville, B. eds. Biofilms - Science and Technology, 126 - 136, Kluwer Academic Publishers, Dordrecht.

Nesaratnam, R.N., 1984, Biofilm formation and destruction on simulated heat transfer surfaces. PhD Thesis, University of Birmingham.

Neufeld, R.J., Zajic, J.E. and Gerson, D.F., 1980, Cell surface measurements in hydrocarbon and carbohydrate fermentations. Appl. Environ. Microbiol. 39, 511 - 517.

Nosetani, T., Sato, S. and Onda, K., 1981, Effect of marine biofouling on the heat transfer performance of titanium condenser tubes, in: Somerscales, E.F.C. and Knudsen, J.C. eds. Fouling in heat transfer equipment. Hemisphere Publishing Corp. Washington, 345 - 353.

Osmon, J.L. and Klausmeier, R.E., 1977, Techniques for assessing biodeterioration of plastics and plasticizers, in: Walters, A.H. ed. Biodeterioration Investigation Techniques, 76 - 90, Applied Science, London.

Pantke, M., 1977, Test methods for evaluation of susceptibility of plasticized PVC and its components to microbial attack, in: Walters, A.H. ed. Biodeterioration Investigation Techniques, 51 - 76. Applied Science, London,

Parkhurst, E.S., Davies, M.J. and Blake, H.M., 1972, The ability of polymers or materials containing polymers to provide a source of carbon for selected micro-organisms, in: Walters, E.H. and Hueck van der Plas, E.H. eds. Biodeterioration Investigation Techniques, 51 - 76, Applied Science, London.

Patel, T.D. and Bott, T.R., 1991, Oxygen diffusion through a developing biofilm of *Pseudomonas fluorescens*. J. Chem. Tech. Biotechnol. 52, 187 - 199.

Patel, T.D., 1986, Film structure and composition in biofouling systems. PhD Thesis, University of Birmingham.

Pinheiro, M.M., Bott, L.F., Bott, T.R., Pinheiro, J.D. and Leitão, L., 1988, Surface phenomena and hydrodynamic effects on the deposition of *Pseudomonas fluorescens*. Can. J. Chem. Eng., 66, 63 - 67.

Pinheiro, M.M., Melo, L.F., Pinheiro, J.D. and Bott, T.R., 1988, A model for the interpretation of biofouling. Second UK Nat. Conference on Heat Transfer. Inst. Mech. Engrs. 187 - 197.

Pujo, M. and Bott, T.R., 1991, Effects of fluid velocities and Reynolds Numbers on biofilm development in water systems, in: Keffer, J.F., Shah, R.K. and Ganic, E.N. eds. Experimental Heat Transfer, Fluid Mechanics and Thermodynamics. Elsevier, New York, 1358 - 1362.

Rabas, T.J., Panchal, C.B., Sasscer, D.S. and Schaefer, R., 1993, Comparison of river water fouling rates for spirally indented and plain tubes. Heat Trans. Eng., 14, 4, 58 - 73.

Reid, D.C., Bott, T.R. and Miller, R., 1992, Biofouling in stirred tank reactors - effect of surface finish, in: Melo, L.F., Bott, T.R., Fletcher, M. and Capdeville, B. eds. Biofilms - Science and Technology, 521 - 526, Kluwer Academic Publishers, Dordrecht.

Rippon, J.E., 1979, UK Biofouling Control Practices. CERL, Leatherhead.

Roe, A., 1992, Biofilms observed to trap and hold particles, in: Process Fuel 1992, 4-5, Interfacial Microbial Process Engineering, Montana State University.

Santos, R.C., 1993, Polymer coatings in relation to single and mixed population biofilms. PhD Thesis, University of Birmingham.

Santos, R.C., Callow, M.E. and Bott, T.R., 1991, The structure of *Pseudomonas fluorescens* biofilms in contact with flowing systems. Biofouling 4, 319 - 336.

Stone, L.S. and Zottola, E.A., 1985, Relationship between the growth phase of *Pseudomonas fragi* and its attachment to stainless steel. J. Food Sci. 50, 957 - 960.

Trulear, M.G. and Characklis, W.G., 1979, Dynamics of biofilm processes. 34th Annual Purdue Industrial Waste Conf., West Lafayette.

Vieira, M.J., Melo, L.F. and Pinheiro, M.M., 1994, Private Comunication to be published.

Vieira, M.J., Oliveira, R., Melo, L.F., Pinheiro, M.M. and van der Mei, H.C., 1992, Biocolloids and biosurfaces. J. Dispersion Sci. and Tech. 13, 4, 437 - 445.

Vieira, M.J., Melo, L.F. and Pinheiro, M.M., 1993, Biofilm formation: hydrodynamic effects on internal diffusion and structure. Biofouling, 2, 63 - 80.

Vieira, M.J., Oliveira, R., Melo, L.F., Pinheiro, M.M. and Martins, V., 1993, Effect of metallic ions on the adhesion of biofilms formed by *Pseudomonas fluorescens*. Colloids and Surfaces B: Biointerfaces, 1, 119 - 124.

Whitehouse, J.W., Khalanski, M., Saroglia, M.G., Jenner, H.A., 1985, The control of biofouling in marine and esturine power stations. CEGB North Western Region.

ZoBell, C.E., 1943, The effect of solid surfaces upon bacterial activity. J. Bacteriology, 46, 39 - 56.

# CHAPTER 13

## The Design, Installation, Commissioning and Operation of Heat Exchangers to Minimise Fouling

### 13.1 INTRODUCTION

The first stage in the mitigation of heat exchanger fouling is in the concept of the unit and its thermohydraulic design, in relation to the potential fouling problem. Some appreciation, if not a complete understanding of the fouling likely to be encountered, is required so that the appropriate decisions may be made about the general design. The way the heat exchanger is installed and its commissioning can also have an effect, sometimes a profound effect, on its ultimate performance.

### 13.2 Basic considerations

Previous chapters have revealed the importance of the different variables that contribute to the rate and extent of deposit accumulation on heat transfer surfaces. It is these variables that have to be properly accounted for if the incidence of fouling is to be controlled or eliminated.

*Velocity*

The principle variable is the velocity of the fluid across the heat transfer surface. Velocity of all the variables, is the one over which the designer has the greatest control. In general the designer has freedom to set the velocity on both sides of the exchanger, it is usual to match its specification in terms of pressure drop, at minimum capital cost. Fig. 13.1 demonstrates the effect of velocity on costs and without considering the fouling potential the exchanger could be designed to give the minimum total cost provided the resulting pressure drop is acceptable. As the velocity is increased the overall heat transfer coefficient increases which means that for a given temperature difference, the heat transfer area (and hence the capital cost is reduced). On the other hand as the velocity increases the pressure drop rapidly increases (i.e. as the square of the velocity).

The difficulty that arises however, in regard to fouling, is that the optimum velocity for acceptable pressure drop may not be the same as that required to minimise the incidence of fouling. For relatively low fouling conditions (e.g. clean organic liquids at low temperature) this approach may be satisfactory but not necessarily so, where heavy fouling is anticipated. For these situations the philosophy would be in general, to design for the highest velocity commensurate

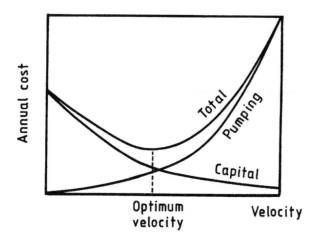

FIGURE 13.1. The dependence of cost on velocity. (The capital cost may be related to an annual figure through the depreciation, e.g. over 5 years)

with the allowable pressure drop. It also has to be borne in mind that if the velocity is too high erosion of surfaces, particularly if droplets or particles are present in the fluid stream, may become a problem. The final choice of velocity is still very much part of the skill and experience of the engineer.

*Temperature*

The temperature at the solid/fluid interface can exert an influence on the development of fouling layers. In Chapter 12 it was demonstrated how a 5°C change in temperature could make a substantial difference in the biofilm thickness. In general raising the temperature can increase the rate of corrosion and the rate of chemical reactions and for inverse solubility salts, raising the temperature may initiate deposition. Although the temperatures required in the operation of a particular heat exchanger will be specified in the proposal, the designer has some limited control of the temperature at the solid/fluid interface through the choice of velocity.

If steam is condensing at a temperature of 111°C on the outside of a tube and an organic liquid at a temperature of 52°C is flowing at 1 *m/s* through the tube it is possible to estimate the temperature distribution across the surface for clean conditions assuming knowledge of the heat transfer coefficients on either side of the tube and the thermal resistance of the tube wall. If the heat transfer coefficients on the steam and liquid side are 8460 and 1462 $W/m^2K$ respectively and the thermal resistance of the tube wall is 0.0000042 $m^2K/W$ (copper 1.65 *mm* thick) the total resistance to heat flow $R_T$ (using Equation 2.5) is:

$$R_T = 0.0000042 + \frac{1}{8460} + \frac{1}{1462}$$

assuming that the effect of the curvature is negligible

i.e. $R_T = 0.0000042 + 0.00019 + 0.000684$

$$= 0.000878 \; m^2 K / W$$

The total temperature difference is 111 - 52 = 59°C

The organic/metal interface temperature is

$$52 + \frac{0.000684}{0.000878} \times 59 = 98°C$$

The temperature fall across the copper is small as would be expected, since the thermal conductivity of copper is high.

If the velocity of the organic liquid is doubled to 2 *m/s* it can be assumed everything else remaining constant, that the heat transfer coefficient for the organic liquid is also doubled [Walker and Bott 1973].

i.e. the new heat transfer coefficient on the organic liquid side will be 2924 $W/m^2K$

The total thermal resistance under the new conditions will be

0.0000042 + 0.00019 + 0.000342

$= 0.000536 \; m^2K/W$

The new temperature of the organic liquid/metal interface will be

$$52 + \frac{0.00019}{0.000536} \times 59 = 72.9°C$$

i.e. a reduction of approximately 25°C of the interface temperature between operating at 1 and 2 *m/s*.

If fouling due to chemical reaction is anticipated on the organic liquid side of the tube, operating at the higher velocity would reduce the problem since chemical reactions are temperature sensitive. Furthermore the shear force will be increased by a factor of around 4 which is also likely to reduce the extent of the fouling. In addition, because of the increased overall heat transfer coefficient at the higher liquid velocity, a heat exchanger based on these data would require a smaller heat transfer area with lower capital cost. These advantages would have to be set against the increased pumping costs.

*Concentration of foulant or foulant precursor*

In general the greater the concentration of foulant or its precursor, the greater the incidence of fouling is likely to be. In design the fouling stream has to be accepted together with its deposit forming properties; there is little the designer can do to get around the problem. The heat exchanger has to be designed minimising as far as possible the potential fouling problem. In serious cases of likely fouling where the flow stream is laden with say particulate matter, it might be prudent to call for filtration prior to entry into the heat exchanger. The difficulty that arises however, is the back pressure likely to be generated by this procedure may not be acceptable.

In other examples the fouling may result from the presence of impurities in very low concentration, e.g. oxygen in deposition in crude oil preheaters. Here again the designer is forced to accept the problem and attempt to accommodate it within the limits of the variables under the designer's control such as velocity, and to some extent the temperature. Usually crude oil storage tanks are designed to exclude oxygen or may even be provided with a nitrogen blanket with the object of reducing or eliminating potential chemical reactions involving oxygen (see Chapter 11). The choice of material of construction may also affect the extent of fouling under those circumstances.

*Surface condition of the heat transfer surface and pH*

There are two principal categories of surface condition that affect deposit accumulation. The first is the result of corrosion of the material of construction (usually a metal) in contact with the fouling fluid even if the corrosion only takes the form of a protective layer. As mentioned in Chapter 10 the corrosion can produce a resistance to heat transfer, and it is possible that metallic surfaces may act as inhibitors or indeed catalysts in chemical reaction fouling.

The second effect of the material on the fouling process is the extent and nature of the surface roughness. In Chapter 12 for instance, data were presented that showed that there was a tendency for smooth surfaces to attract less biofouling than surfaces with more substantial roughness elements. Unless special materials of construction are used, for example electropolished stainless steel, few heat exchangers may be regarded as having smooth surfaces. Commercially available materials of construction have roughness that are related to the method of manufacture [Bott and Walker 1973]. Surface roughness can be visualised as having two effects [Bott 1990]:
1. The provision of "nucleation" sites which encourage the laying down of the initial deposit - to be followed by further layers that gradually renders the

original surface roughness less significant. Furthermore "rough" surfaces are likely to provide a more positive keying of the deposit to the surface.

2.  If the roughness elements are relatively substantial they may interfere with the stability of the viscous sub-layer near the heat transfer surface. In turn there may be an effect on the rate and extent of the deposit accumulation.

It also has to be remembered that once a surface has become covered with deposit its characteristic surface condition may have changed with consequent effects on deposit retention. Surfaces carry electrical charges as described in Chapter 6. The magnitude and sign of this charge depends on the material involved, and in turn this may affect the accumulation of deposits. Where the depositing species is of opposite sign to the surface, initial deposition may be facilitated whereas where the charges are similar, the reverse is likely to be the case. The presence of a deposit may modify these effects of attraction and repulsion. Some mitigation techniques depend on changing surface charges either of the surface itself or the "particles" that approach the surface. The *pH* of an aqueous solution may also influence the charge distribution that in turn affects the fouling potential.

*Additional considerations*

In assessing the potential fouling of a heat exchanger at the initial design stage, it is necessary to consider influences that might occur external to the heat exchanger.

Individual heat exchangers are generally part of a complex network of heat exchangers and other equipment such as distillation columns, evaporators and reactors. The performance of any one heat exchanger will be dependent on the performance of the other components in the process complex. Long residence times in a distillation column may give rise to chemical reactions that produce insoluble particulate matter. For instance in a distillation column used to recover a solvent from a solvent/water mixture, the solvent having previously been used to dissolve resin for coating purposes, insoluble particles of high molecular weight may be produced. The particles may eventually accumulate in the reboiler where they may generate fouling problems. Furthermore particles could be carried by the hot bottoms stream to the heat recovery system, e.g. a preheater on the feed to the distillation column. As a result particulate fouling and possibly further reaction fouling may occur on the "hot" side of the feed preheater.

Corrosion products released upstream of a heat exchanger may give rise to particulate fouling on the heat transfer surfaces downstream.

Attention to the *pH* of an aqueous solution may reduce the corrosion potential, but in general it is not feasible for the heat exchanger designer to call for changes in *pH*, since he or she has to work within the limits of the specification of the heat exchanger.

Although the suitable choice of material of construction for the heat exchanger is a possibility to alleviate the problem of fouling, that choice is unlikely to be invoked in all but exceptional examples on account of cost. In food processing to avoid problems of contamination, fabrication from special materials may be entirely justified. In general special treatment such as electropolishing is expensive, and corrosion resistant alloys that may retain a reasonably smooth surface (as compared to a corroded surface) usually carry a high cost. Table 13.1 gives an indication of the relative costs of different materials of construction used in heat exchangers. The data are based on figures given by Redman [1989] and show just how much costs depend on the choice of material of construction.

TABLE 13.1

The relative costs of three materials of construction in relation to mild steel

| Metal or alloy | Sheet | 50 *mm OD* tube |
|---|---|---|
| Stainless steels | 3.75 - 8.25 | 3.5 - 8.25 |
| Titanium | 43.75 - 87.5 | 35 - 70 |
| Tantalum | 625 | Not available |

Note: Tantalum sheet is over 600 times more expensive than mild steel sheet.

*Residence time*

Residence time may be linked to chemical reaction or "reaction like" mechanisms, but it is only likely to become an influence where the reaction rate is relatively high. In most industrial heat exchangers the residence time is generally low so that the effects on fouling are not substantial. It is possible however where velocities are low or the fluid is made to flow through substantially long pathways, that residence time may assume importance. Bott [1990] cites the hydrolysis of $SiO_2$ in solution in geothermal water as an example. As the pressure (which at the bottom of the bore hole may be quite high) is lowered and the temperature changes, the solubility of the $SiO_2$ is affected. Simultaneous hydrolysis reactions may account for substantial fouling if long residence times in the transfer lines to the heat exchanger and in the heat exchanger itself, are encountered. Chemical reaction in transfer lines may give rise to particulate matter that could deposit in heat exchangers downstream.

Residence time in the laminar sub-layer will be a consideration in the design of heat exchangers to handle heat sensitive materials such as food products or petroleum fractions. The effects of temperature may be to decompose the material and produce fouling. Steps will be necessary to reduce or eliminate the problem (see Section 13.3 or Chapter 15).

13.2.1  Summary of practical preliminary design concepts

The brief discussion concerning the effects of certain variables reveals some basic design concepts:

1.  The design should be as simple as possible, particularly so if heavy fouling is expected.

2.  Design for the highest velocity allowable under the constraints of pressure drop and erosion.

3.  Minimise surface temperature in conjunction with the choice of fluid velocity.

4.  If at all possible reduce the incidence of potential fouling by reducing the concentration of foulant precursors.

5.  Within the constraints of cost, choose a material of construction that discourages the accumulation of deposits.

6.  Ensure that the residence time is as low as possible.

7.  Attention to *pH* and corrosion not only within the heat exchanger, but also in transfer lines and equipment upstream of the heat exchanger.

8.  Potential fouling due to conditions external to the heat exchanger.

13.3   THE SELECTION OF HEAT EXCHANGERS

The assessment of the potential fouling problem associated with the heat exchanger under consideration and the scope for choosing the design features required, enables a selection of the type of heat exchanger to be made. A major consideration for fouling conditions must be the ease with which the heat exchanger may be cleaned in service (techniques for cleaning heat exchangers are described in Chapter 15), allowance for the method of cleaning should be a feature of the preliminary concept. It goes without saying that, unless there is some other overriding consideration, the fluid that is likely to create the greatest fouling should be made to flow through the "side" of the heat exchanger that is most easily cleaned; for shell and tube heat exchangers this means through the inside of the tubes. Table 13.2 [Bott 1990] lists some of the features of common heat exchangers and may be used as a preliminary guide in heat exchanger selection.
In some special situations it might be possible when fouling is likely to be a major problem, to eliminate the need for a heat exchanger. For instance where a dilute aqueous solution of an organic solvent containing complex molecules that can polymerise under the action of heat, the reboiler could be replaced by a steam

TABLE 13.2

Features of some typical exchanger types [Bott 1990]

| Type | Materials of construction | Ease of cleaning | Notes |
|------|---------------------------|------------------|-------|
| Shell and tube | Most materials | Tubes relatively easy to clean, shell more difficult | widely used |
| Gasketted plate | Stainless steel (usually) | Easy to clean | Compact |
| Double pipe | Commonly in carbon steel | Inner tube relatively easy, annular space more difficult or impossible | Only useful for small heat transfer areas |
| Immersed coils | Most materials | Inside tubes impossible except by chemicals, outside of tubes possible but may be difficult | Limited application |
| Spiral | Most materials | Easy access to whole channel length | Compact: useful for slurries and fouling conditions |
| Graphite block | Graphite | Impossible to clean mechanically, chemical cleaning possible | Useful for corrosive conditions |
| PTFE tube | PTFE | Only chemical cleaning of inside of tubes possible | Flexible nature of the small tubes may reduce incidence of fouling |
| Scraped surface | Most materials | Self-cleaning generally | Incorporates moving parts |
| Plate-fin | Aluminium, stainless steel, titanium | Only chemical cleaning possible | Highly compact |
| Air-cooled | Aluminium fins on carbon steel tubes common, other combinations possible | Inside tubes relatively easy, finned surface more difficult | Large plot area required |

sparge. It should also be borne in mind that although this eliminates the need to clean a reboiler the fouling problem may be only transferred to another heat exchanger downstream. Furthermore the cost of the steam would have to be considered against the saving in cleaning costs and downtime (see Chapter 3).

Another example could be the cooling or condensation of particulate laden gases or vapours. It is possible to cool the hot gas or condense the vapours by direct contact with a cold liquid, usually water. Although this might eliminate particulate fouling of a cooler and the associated cleaning costs, it could generate other costs for effluent treatment or the need to separate the condensate from the coolant phase. The latter may not be a problem if the two liquids are immiscible.

## 13.4    THE CHOICE OF FOULING RESISTANCE

Having selected the type of heat exchanger, the configuration, the material of construction and taking into account the expected fouling, it is necessary to make an estimate of the heat transfer area required to meet the specification. The physical size of the heat exchanger (and its cost) for a particular type (e.g. shell and tube or plate) will of course depend on the heat transfer required. In general terms the area is estimated from using Equation 2.7.

i.e. $Q = \dfrac{A \Delta T}{R_T}$  (13.1)

where  $R_T = \left(\dfrac{x_1}{\lambda_1}\right) + \left(\dfrac{x_2}{\lambda_2}\right) + \left(\dfrac{x_m}{\lambda_m}\right) + \dfrac{1}{\alpha_1} + \dfrac{1}{\alpha_2}$  (2.5)

where  $\Delta T$ is the overall temperature difference

and     $A$ is the required heat transfer area

It is usual to refer to $R_T$ the total thermal resistance, in terms of an overall heat transfer coefficient $U_D$ so that

$Q = U_D A_D \Delta T$  (13.2)

where  $U_D = \dfrac{1}{R_T}$  (13.3)

For clean conditions where there is no deposit on either side of the heat exchanger

$Q = U_c A_c \Delta T$  (13.4)

where $A_c$ = the required heat transfer area for clean conditions

Under these circumstances Equation 2.5 reduces to

$$\frac{1}{U_C} = \frac{1}{\alpha_1} + \frac{1}{\alpha_2} + \frac{x_m}{\lambda_m} \tag{13.5}$$

For given conditions (i.e. specified) $Q$ and $\Delta T$ are the same for fouled or clean conditions

so that $U_C A_C = U_D A_D$ (13.6)

or $\dfrac{A_D}{A_C} = \dfrac{U_C}{U_D}$ (13.7)

Equation 13.7 demonstrates that in order to cope with the expected fouling the heat exchanger has to have additional area over and above the area required to execute the same rate of heat transfer in the clean condition. In simple terms if the overall heat transfer coefficient for the fouled condition is half the coefficient for clean conditions the heat transfer area is doubled. It will be readily appreciated (as mentioned in Chapter 2) that the fouling resistances are likely to have a profound effect on the ultimate size of the heat exchanger. The choice of fouling resistances therefore is crucial not only to the efficient operation of the heat exchanger and therefore its operating cost, but also to its capital cost. Although fouling resistances are published in the literature or by bodies such as TEMA, as mentioned in Chapter 2, usually there is no supporting detail with regard to process conditions such as flow velocity, temperature or foulant precursor concentration or even what fouling mechanism is responsible for deposit accumulation. Chapters 7 - 12 have indicated that for all fouling mechanisms the extent of deposit is strongly dependent on these variables. The indiscriminate use of published fouling data that may have been obtained for conditions very different from those pertaining to the design in hand, could therefore lead to serious under or overdesign with attendant effects on capital and operating costs. It must be stated however, that published data do give an indication of the extent of fouling likely to be encountered, and may be used as a basis for assigning fouling resistances in design calculations. Previous experience or laboratory data with similar streams and conditions, may serve to reinforce or discount the published or recommended values.

Bott [1990] has given a case study which illustrates the difficulties faced by designers in terms of the fouling resistances, and involves the design of a plate heat exchanger.

The heat exchanger was required to cool 30 *kg/s* of demineralised water from 75 to 40°C prior to use in a chemical process using works cooling water at rate of

52.5 *kg/s* and a temperature rise of 25 - 45°C. No fouling was anticipated on the demineralised water side of the exchanger.

An engineer who had experience of shell and tube heat exchangers operating under similar conditions to those specified for the plate heat exchanger, considered that the fouling resistance on the cooling water side should be 0.0006 *m²K/W*. An engineer with experience of plate heat exchangers considered that a fouling resistance ten times smaller, i.e. 0.00006 *m²K/W* would be adequate.

Other relevant data were:

The "cold side" heat transfer coefficient = 13,000 *W/m²K*
The hot side heat transfer coefficient = 10,000 *W/m²K*
The thermal resistance of the exchanger plates = 0.00003 *m²K/W*

Table 13.3 gives a summary of the calculations using the fouling resistances suggested by the two engineers. Using the higher value of the fouling resistance increased the required area by:

$$\left(\frac{164 - 54}{54}\right)100\%$$

$$\approx 200\%$$

TABLE 13.3

Summary of plate heat exchanger calculations

| Calculated data | Fouling resistance *m²K/W* | |
|---|---|---|
| | 0.0006 | 0.00006 |
| $\Delta T_m$ *K* | 21.6 | 21.6 |
| $U_C$ *W/m²K* | 4831 | 4831 |
| $U_D$ *W/m²K* | 1239 | 3763 |
| Heat transfer area *m²* | 164 | 54 |

When operating the heat exchanger based on the higher fouling resistance it was found that the demineralised water temperature was reduced to 26.5°C rather than the required 40°C. In order to achieve the desired temperature the cooling water flow rate was drastically reduced to 22.2 *kg/s*. The resulting velocity reduction in the cooling water flow and the associated temperature rise, led to fouling rates greater than anticipated so that the final fouling resistance of 0.0006 *m²K/W* was obtained well below the allotted time of 12 months. After six months the pressure drop on the cooling water side was such that the design flow rate of 52.5 *kg/s* could not be achieved.

The specification for the heat exchanger required cleaning to be carried out on a 12 month cycle. The alternative design based on the lower fouling resistance of

0.00006 $m^2K/W$ operated satisfactorily within the limits set by the specification in terms of cleaning cycle and pressure drop.

It may be possible to predict the fouling resistance from the mathematical models given in connection with the various mechanisms. There are two severe limitations however to the use of these correlations:

1. There are usually unknowns in the expressions for which the designer has no information such as rates of reaction, sticking probability, mass transfer rates or thermal conductivity of the foulant layer. It is possible of course to hazard a guess at the magnitude of these unknowns, but this could lead to serious error.

2. The correlations themselves are generally subject to simplifying assumptions (so that the mathematical solution may be achieved). The errors so introduced could be substantial.

The resultant calculations based on these models may therefore provide values of foulant thermal resistance that are just as much in error as the TEMA type recommendations. The principle advantage of the models however that must not be underrated, is they do show the interaction of the variables and by an iterative process the designer is led towards an improved heat exchanger design in respect of fouling.

It may be possible to use the so-called "expert system" or "fuzzy logic" as an aid to the design of heat exchangers. The techniques lend themselves to making allowance for potential fouling when the full effect of the variables on a particular problem is not well understood. Mathematical models of fouling are helpful but as discussed elsewhere in this book, their application is limited through there being no information on vital terms, e.g. mass transfer coefficients or rates of reaction.

Bullock *et al* [1994] point out that no matter how many experiments are carried out, because of its multi-dimensional nature in terms of practical chemical engineering there is always a shortage of data. Pooling of knowledge items of different origin (experience, experiments, literature and so on) can partially eliminate difficulties caused by the sparseness of data through the application of fuzzy logic. Reppich *et al* [1993] have applied the technique to fouling factors and Dohnal and Bott [1995] have applied it to problems in biofouling. In general the principle is to make a qualitative assessment of the problem then to apply a logical reasoning to point towards a potential solution.

Afghan and Carvalho [1993 and 1994] have applied the expert system procedure to the assessment of fouling problems. For a specific heat exchanger the expert assessment procedure was performed for different situations. Diagnostic variables describing the state of the system are identified and converted into the confluence parameters which describes qualitatively the instantaneous state of the heat exchanger system. A retrieval system is used to find the corresponding state in the knowledge base with the set of the parameters describing malfunctions of the system. Each malfunction state has a recommendation for correction specified by expert advice.

## 13.5   PLANT LAYOUT

In all but a few heat exchangers such as those processing clean organic liquids at low temperature, it will be necessary to clean sooner or later.  In order to minimise the cleaning costs it is necessary to be sure that there is adequate space for cleaning operations to take place.  If for instance, the tubes of a shell and tube heat exchanger are expected to become badly fouled it may be necessary to provide for drilling out the tubes (see Chapter 15).  In order to allow easy access then at least the length of the heat exchanger must be allowed at one end, preferably at both ends, to allow insertion of the drill and drive.  It is also beneficial if the heat exchanger can be placed at ground level, to avoid difficulties and safety problems, that could occur with heat exchangers located on elevated platforms or at the top of tall structures, e.g. distillation columns.  Making heat exchanger cleaning straightforward and as easy as possible, ensures that the labour involved with the cleaning process may be used to best effect.

Where space is at a premium possibly in retrofit situations, it might be desirable to consider the use of a plate heat exchanger that in general requires less space for cleaning operations.

The location of equipment on site may also have an influence on the cleanliness of a heat exchanger from the point of view of fouling.  For instance an air cooled heat exchanger placed where it is susceptible to particulate matter and dust, could lose efficiency due to fouling on the finned tubes.  Corrosive fumes from nearby processing may also be responsible for air cooled heat exchanger fouling.

It is not possible to discuss all the potential problems associated with heat exchanger location, but logical examination and analysis of likely events will reduce the difficulties associated with fouling and provide the maximum opportunity for effective cleaning.

At the design stage it is useful to consider the kind of cleaning that may be necessary whether it is mechanical or chemical cleaning (techniques are discussed in Chapter 15).  If frequent mechanical cleaning is anticipated a "quick release" facility on the heat exchanger could prove to be an asset, provided that general integrity and safety are not impaired.  Where chemical cleaning may be desirable the additional problems that this may generate need to be considered, e.g. the provision of access points for the addition of chemical solutions, adequate venting possibly to the site flare stack may also be required, adequate disposal of effluent with drainage to the site effluent disposal plant may be necessary, in order to meet safety and environmental standards.

Where "on-line" cleaning is considered to be necessary (see Chapter 15) either at installation or at a later date when the extent of a particular fouling problem has been established, the heat exchanger will of course, need to be designed with this in mind.  Although the capital investment in such a system will be higher than the conventional method of changing velocity, its use could make a substantial reduction in operational and maintenance costs.

## 13.6    HEAT EXCHANGER CONTROL

The way in which the outlet temperatures of a heat exchanger are controlled could have a profound effect on the extent of the fouling.  In general one of the outlet temperatures, usually the principal process stream is required to be fixed within a narrow range.  The inlet temperature of this stream for some reason may be variable.  The usual way of controlling the principal process fluid temperature is to restrict or increase the flow of the other fluid.  The technique is illustrated in Fig. 13.2).

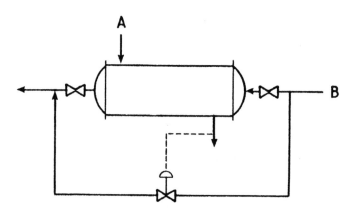

FIGURE 13.2. The conventional control of heat exchanger temperatures

If fluid $A$ is being heated by fluid $B$ and the outlet temperature of $A$ is fixed, the flow of $B$ will vary depending on the required input or removal of heat.  The discussion on flow velocity has shown that even relatively small changes in velocity can substantially affect the rate or extent of fouling.  In many respects the simple control illustrated in Fig. 13.2 is not generally satisfactory from a fouling point of view.  To overcome these potential difficulties it would be worth considering a "pump around" system to maintain the conditions in the exchanger.  The technique is illustrated in Fig. 13.3.

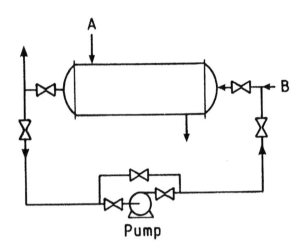

FIGURE 13.3  A "pump around" system to minimise heat exchanger fouling

## 13.7   ACCEPTANCE, COMMISSIONING AND START UP OF HEAT EXCHANGERS

In the earlier sections of this chapter stress was laid on good design to avoid or reduce the fouling problem.  Unless the concepts and intentions of the designer are maintained in the acceptance, commissioning and start up stages, it is entirely possible that heat exchanger fouling could be worse than anticipated.

It is imperative that time is taken to check the details of the heat exchanger delivered to site and to carry out testing procedures that do not lend themselves to increased fouling problems.  The way a heat exchanger is brought on stream may also have a profound effect on the incidence of fouling during subsequent operations.

Simple checks on the heat exchanger as delivered ensures that the design philosophy has been carried through to the finished product.  A general check list may be as follows:

1.  Material of construction

2.  Surface treatment (if any)

3.  Dimensions of the heat exchanger

4.  Correctness of connections

5.  Vents and drains

6.  Temperature control

7.  Auxiliary equipment such as pumps and their capacity, or the details of burners
    for combustion systems

    For a particular design of heat exchanger some additional checks will be
required.  For instance in shell and tube heat exchangers it is necessary to check
that the tube sizes are correct and the clearances between the components of the
bundle are as called for in the design (excessive leakage or "dead spot" creation
could accelerate fouling).  For plate heat exchangers the clearances between the
plates requires careful attention.  The use of incorrect gasket material may lead to
operational problems that assists foulant deposition.
    Although hydraulic testing of heat exchangers is carried out at the
manufacturers premises it is usual to carry out further independent tests on site as
part of the acceptance routine.  If testing is not carried out properly and in a
systematic manner, fouling of heat transfer surfaces may be encouraged and
deposition accelerated.  Attention to the following details will help to avoid the
problems.
1.  Removal of construction debris such as welding rods, or pieces of timber from
    the heat exchanger to reduce the possibility of flow restriction and the creation
    of dead spaces.

2.  Use of good quality water with corrosion inhibitors to prevent deposition and
    corrosion.

3.  Adequate flushing after testing to be followed by filling with a good quality
    water with corrosion inhibitor and biocide.

4.  If the heat exchanger is to remain idle for some time following the hydraulic
    tests it should be checked for satisfactory condition on a regular basis.

5.  The heat exchanger should be drained and purged before it is brought on
    stream so that it is in good condition at the start of operations.

    It goes without saying that during plant construction the heat exchanger, along
with all other equipment should be protected from damage from construction
operations or adverse weather.
    During commissioning that may only involve part of the total plant equipment
and conditions different from those expected during normal operation, it is
imperative not to subject heat exchangers to conditions that might promote
fouling, particularly the principal variables that affect fouling; velocity and
temperature should be kept within reasonable limits.  Excursions of these variables
are to be expected during the start up procedure but care should be exercised to
keep these to an absolute minimum.  The exposure of heat exchanger surfaces to

high temperatures and very low velocities even for a short time, can lead to accelerated fouling and "difficult to remove" deposits that were not anticipated at the design stage.

It is not only the heat exchanger that has to be carefully monitored; conditions upstream of the heat exchanger may promote fouling of the heat transfer surface. For instance long residence times in upstream equipment during "warm up" procedures may generate particulate matter, e.g. from polymerisation reactions, that subsequently deposit in a downstream heat exchanger.

Prior to start up and commissioning it is useful to go through possible sequences of effects and operations - similar to Hazop analysis, to determine situations and conditions that should be avoided to reduce the potential for fouling of heat exchangers.

## 13.8   PLANT OPERATION

The care with which the designer of a heat exchanger has taken to minimise the possibility of fouling is of no avail unless the operator of the equipment maintains the design conditions.   Changes in temperature and velocity, even for relatively short periods, may lead to troublesome fouling quite different from that anticipated when the equipment was designed.   The very nature of processing means however, that steady conditions cannot always be guaranteed, and often the heat exchanger is considered to be auxiliary to the principal pieces of equipment such as reactors, distillation columns or extractors.   Even allowing for these twin facts it is vital that the operator appreciates the potential difficulties and operates in such a way that the impact of changed conditions on the fouling is reduced to a minimum.   So often adjustments are made to satisfy production needs with no thought about the long term consequences in respect of fouling.   The effects may not be immediately apparent since the effects of the fouling may be "lost" in the generally changed conditions only to appear when attempts are made to return to the original flow rates and temperatures.

Poor maintenance and lax process management may also contribute to enhanced fouling of heat exchangers.   Contamination of one process stream by another through leaking equipment can have serious consequences for heat exchanger fouling.   Even if the extent of the contamination is small and the stream contaminated is otherwise clean, the design may be such that because little fouling was anticipated the contaminated stream may have been put through say the shell side of a shell and tube heat exchanger, the result may be a deposit that is difficult or even impossible to remove.   The consequences of lack of attention may be serious and expensive, with adverse effects of profitability.

During operation it is vital that the equipment operator is aware of the potential fouling problems and accordingly devise strategies that minimise the risk of increased fouling.

*Fouling of Heat Exchangers*

REFERENCES

Afgan, N.H. and Carvalho, M.G., 1993, Heat exchanger expert system. ICHMT Symp. New developments in heat exchangers. Lisbon.

Afgan, N.H. and Carvalho, M.G., 1994, Heat exchanger fouling assessment by confluence based expert system. Proc. 10th Int. Heat Trans. Conf. 4, 339 - 334.

Bott, T.R., 1990, Fouling Notebook. Inst. Chem. Engnrs. Rugby.

Bullock, P., Dohnal, M., Exall, D.I., Lund, G.D. and Zebert, A.C., 1994, A fuzzy policy of chemical engineering knowledge. 7th Nat. Conf. SAIChE, Johannesburg.

Dohnal, M. and Bott, T.R., 1995, Application of fuzzy logic to heat exchanger fouling with special reference to biofouling. To be published.

Redman, 1989, Non metallic heat exchangers. Chem. Engr. No. 459, 17 - 26 April.

Reppich, M., Babinec, F., and Dohnal, M., 1993 Bestimmung des Foulingfaktors an Wärmeübertragungsflächen mit Hilfe einer Fuzzy-Wissensbasis. Forschung im Ingenieurwesen. Engineering Research Bd. 59, 5, 97 - 101.

Walker, R.A. and Bott, T.R., 1973, Effect of roughness on heat transfer in exchanger tubes. Chem. Engr. No. 271, March, 151 - 156.

# CHAPTER 14

## The Use of Additives to Mitigate Fouling

### 14.1 INTRODUCTION

Despite precautions taken at the design stage of a heat exchanger and its subsequent commissioning and operation, there will be situations where fouling still occurs. One way of overcoming, or at least reducing the problem, is in some way to interfere with the fouling process by the addition of chemicals to the flowing fluid. The additive may act in one or more of a number of ways.

1. It may react chemically with the fouling species to modify its fouling potential.

2. It may physically affect the fouling process by changing the physical interaction between the foulant and the heat exchanger surface.

3. The additive may modify the nature of the deposit residing on the surface so that it is more susceptible to the removal forces.

4. It may destroy the activity of the foulant, e.g. living matter, so that it can no longer attach to a surface.

In general the concentration of the additive in the stream to be treated is quite low, often only a few *mg/l*. The need to restrict the application is two-fold:

1. The cost of treatment chemicals or substances may be high (in Chapter 3 for instance the cost of additives to reduce the fouling in crude preheat stream of 100,000 barrels per day oil refinery, was of the order of $1.6 x $10^5$ in 1981). There is clearly an incentive to keep these costs low, otherwise they may approach the cost of the problem they are intended to alleviate.

2. The ultimate destination of the additive. In a process stream it will represent a contaminant that may or may not be tolerated in the final product. Large quantities of added chemicals may require removal at a later stage in the process. In aqueous systems the final discharge of the water may be of concern if the added chemical is noxious in any way. High concentrations of residual additives may require treatment before discharge to drains or to the environment. Similarly additives in gas streams, e.g. combustion and incineration processes, must not represent a pollution hazard when discharged to the atmosphere. Expensive treatment operations to meet discharge requirements may not be acceptable.

There are other aspects of the use of additives that have to be considered. A suitable and reliable dosing system has to be engineered to meet all aspects of cost control, safety and environmental protection. The choice of treatment method has to be an economic balance between some or all, of the following constraints:

1.  Additive cost.

2.  Labour requirements in relation to the additive dosing.

3.  Demands of the system in respect of the additive (i.e. possible chemical reactions other than the preventative action).

4.  Likely fluctuations in the foulant concentration in the system.

5.  Materials of construction.

6.  Environmental and toxicity standards.

Methods of applying the additive can vary from something simple like a single shot or dose of additive manually applied at a convenient point in the system, or a sophisticated method using proportional intermittent mechanical dosing. Table 14.1 compares the advantages and disadvantages of the alternative dosing techniques that are available [ESDU 1988].

The choice of dosing technique runs parallel to the choice of additive. The principal constraints that need to be considered are:

1.  Capital cost of equipment.

2.  System fluctuations likely to be encountered.

3.  Labour quality and cost.

A simple flow sheet of a continuous dosing plant for a liquid system (e.g. cooling water) is shown on Fig. 14.1.

Although there may be a temptation to use the method with the cheapest installation cost, where heat exchanger fouling is critical to throughput and product quality, serious consideration should be given to a totally integrated system.

If an added chemical is required at low concentration say 1 - 2 *mg/l* the dosing must be accurate and rapid mixing achieved, so that the full potential is realised. Poor additive metering may be expensive in terms of chemical cost, and may not provide the protection from fouling that is required. Many of the chemicals used for deposition control are noxious or toxic, and therefore protection of the operatives and the environment will be required for normal operating conditions. Adequate protection against accidental spillage will also be necessary. Additional equipment may be required to handle the results of the chemical addition, e.g. settling tanks where flocculation of suspended particles is undertaken.

TABLE 14.1

Dosing techniques for liquid systems - some advantages and disadvantages

| Technique | Advantage | Disadvantage |
|---|---|---|
| Shot or slug | No moving parts. Suitable for systems where the fouling develops over a relatively long period of time. | High labour cost. Difficult to control the additive concentration. Personnel safety problems. Potential to cause increased fouling due to unintentional high levels of additive. High additive costs. |
| Drip feed (liquid additives) | No moving parts as entry is effected by a header tank. Low labour costs. | The rate of dosing depends on the level in the header tank (as the level falls the dose rate falls). Insensitive to changes in the system. Failure to check the system may lead to gross inaccuracies in dosing. Overdosing (e.g. of acid) could lead to other problems (e.g. corrosion). |
| Set rate continuous | Constant injection rate. Cheap pumping equipment could be employed. Effective for constant fouling conditions. Low labour costs. | Insensitive to changes in foulant concentration or changes in operation. Difficult to control if the dosing requirements are low. |
| Set rate intermittent | Reliable dosing from relatively cheap pumping equipment. Low labour costs. Economical in terms of additive requirements. | Insensitive to changes in foulant concentration and system changes. |
| Proportional | A near continuous level of additive related to flow rates is maintained. Low labour costs. | Not necessary for systems that operate at near steady conditions. |
| Proportional intermittent | The dosing is related to some characteristic of the flowing stream, e.g. electrical conductivity. Low labour costs. | Relatively high installation cost. |

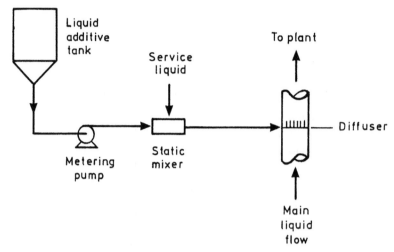

FIGURE 14.1. Simplified flow diagram of a continuous dosing for a liquid system

If the action of the additive is to be at the heat transfer surface rather than in the bulk, suitable flow conditions (turbulence) must be provided so that the mass transfer does not restrict the effectiveness of the treatment (see Chapter 5).

Whatever dosing technique is used the following are relevant to reliable chemical treatment:

1. Frequent checks on the functioning of any mechanical equipment involved (e.g. pumps).

2. Checks on levels in tanks on a routine basis so that replenishment schedules may be related directly to additive use.

3. Proper allowance should be made for lack of supervision when a full complement of personnel is not present (i.e. night time, weekends, etc.)

The choice of additive will depend upon the fouling process known to exist in the system and this presupposes that a proper assessment of the problem has been made. It is not possible to identify a single additive that will fulfil the requirements of any fouling problem and the choice has to be made from a whole range of potential mitigants. In some instances the additive will be a mixture of chemicals that provide a range of remedial processes. In any event many fouling problems are combinations of the processes described in Chapters 7 - 12 so that a single additive may not reduce or prevent deposition. It is not possible without careful consideration to blend different additives, because there may be detrimental interactions between the various chemicals that effectively nullifies the desired properties. For the same reason, i.e. the variety of effects brought about by the range of additives, it is not possible to discuss their function in a general way. Additives will therefore be described in terms of the fouling problem they are intended to combat. The discussion of additives will be divided into two major groups namely liquid and gaseous systems.

## 14.2  LIQUID SYSTEMS

In general the discussion will concern aqueous systems, largely because it is within these systems that much work has been carried out to develop suitable additives, principally for cooling water application. Because aqueous systems are common to many industries, and the potential need for suitable fouling mitigation is widespread, the financial incentive has been present to develop suitable and effective additives. The principles may be applied to other systems but it is unlikely that if the process stream is unique, a suitable additive will have been developed and tested. The development of a chemical that will restrict or prevent, a particular fouling problem may be expensive and the development cost unjustified. Under these circumstances either the deposition problem has to be tolerated or a different technique applied (see Chapters 15 and 16).

Common additives used in the control of fouling in aqueous systems are given in Table 14.2.

TABLE 14.2

Some common additives in water treatment

| Agent | Activity |
|---|---|
| Sequestering | Forms a chemical complex with the foulant to keep it in solution (or suspension). |
| Threshold | Retards precipitation of a scale-forming species while they remain in the system. |
| Dispersant | Imparts like charges to both the heat exchanger surface and suspended particulate matter to ensure that individual particles remain in suspension. |
| Chelating | Forms a strong chemical complex with the scale forming species. |
| Crystal modifier | Changes the crystal habit so that the formation of large crystalline structures is prevented (either in suspension or on the heat exchanger surface). |
| Biocide | Kills or disables micro-organisms so that they do not stick to surfaces. |
| Biostat | Reduces micro-organism activity. |

14.2.1 Particle deposition

In many instances, particularly where the velocity is high enough, particle deposition is not a problem in respect of heat exchanger fouling (see Chapter 7). In some situations however, the deposition of particulate matter does represent an operating difficulty, but even here the inclusion of particles in the deposit is often associated with some other fouling process, such as biofouling or chemical reaction. There are opportunities however, by suitable treatment, to reduce or eliminate the problem of particulate accumulation on heat exchanger surfaces.

With deteriorating raw water supplies particle deposition in cooling water is becoming an increasing problem. The particulate matter may be introduced into the cooling system either in the raw water, or from the atmosphere through the associated cooling tower acting as an air scrubber. The particles may deposit anywhere in the system where the flow velocity is reduced, for instance in the header box of a standard shell and tube heat exchanger. The techniques that are employed to reduce the difficulties encountered by the deposition of particulate matter in aqueous systems fall into two categories, namely to preclarify the feed water e.g. flocculation possibly followed by settling and filtration, or to restrict the opportunity for particles to migrate to heat exchanger surfaces. In very large installations for instance a once through cooling system at a large power station, preclarification may not be feasible because the size of the clarification plant required would be so large. It would be appropriate however, to filter the incoming water (e.g. from a river) using filter screens to remove the larger pieces of debris. Fine filters could not be employed due to problems of flow restriction. In the production of town's water the clarification stage of the process is often followed by an elaborate system of filters; this is often unnecessary in cooling water systems provided good control of the operation is practised. Side stream filtering (say 10% of the total throughput) may be a compromise to reduce capital and operating costs, where the excessive quantities of flocculated particles are not acceptable.

In Chapter 7 the importance of differences in the electrical charge between particles and the surface was established. It was demonstrated that in "pure" particulate systems *pH* had a dominant effect on the deposition process. In general with specific particles and surfaces the extent of the deposit may be controlled by the prevailing *pH*. Modification of *pH* in the system may be a means of reducing the fouling problem, but the process conditions may not permit the change to be made for one or more of the following reasons:

1. The process stream chemistry may not be compatible with the proposed *pH* change.

2. Changing the *pH* will affect the corrosion potential of the system (see Chapter 10). As a result unless the material of construction is suitable, particulate deposition in the heat exchanger may be replaced by corrosion fouling.

3. Relatively large volumes of acid (or alkali) may be required, representing excessive costs.

4. Treatment (neutralisation) may be required to render the stream to an acceptable *pH* level before discharge (say for cooling water).

The development of additives to deal with the problem of particulate deposition will essentially involve the modification of charges either of the surface or the suspended matter, and consequently they are flocculants, surfactants or dispersants.

*Flocculation and coagulation*

The action of coagulants or flocculating agents is to produce aggregation of particles (often colloidal) to larger size so that they are capable of being settled, filtered or made less susceptible to surface charges (e.g. on heat transfer surfaces). The action of the additive is to remove or reduce, the forces that stabilise the colloidal suspension. The destabilisation is assisted by interparticle collisions facilitated by gentle agitation and good mixing. The choice of coagulant will depend on the charge on the particles in suspensions. Excessive agitation will tend to break up flocs as they form. There are four mechanisms by means of which particle destabilisation can occur:

1. The action of an electrolyte on a colloidal dispersion can be explained as a compression of the diffuse layer of ions, to such an extent that the van der Waals attractive forces can achieve particle aggregation. The mechanism is sometimes referred to as "double layer compression".

2. Colloidal particles can adsorb chemical ions and radicals at their surface. If the adsorbed species carries a charge opposite to that of the colloid, the adsorption will lead to a reduction of the surface potential, causing the particle to become destabilised.

3. Destabilisation may be achieved by the enmeshment of the colloid in a precipitate. In this process a metal salt such as aluminium sulphate (alum) or ferric chloride, is added to the water forming positively charged species in the typical *pH* range of 6 - 7 for clarification. The hydrolysis reaction produces an insoluble gelatinous hydroxide according to the following equations:

$$Al_2(SO_4)_3 + 6H_2O \rightarrow 2Al(OH_3)\downarrow + 3H_2SO_4 \tag{14.1}$$

$$FeCl_3 + 3H_2O \rightarrow Fe(OH)_3\downarrow + 3HCl \tag{14.2}$$

Colloids present in the water to be clarified, may serve as condensation nuclei for the precipitation process or may become trapped in the precipitate as it settles. For this reason the process is also known as "sweep floc" coagulation.

4. It has been found that both positive (cationic) and negative (anionic) polymers are capable of destabilising negatively charged colloidal suspensions so that the double layer compression theory or the charge neutralisation model cannot be successfully applied. An alternative is the concept of interparticulate bridging. A number of naturally occurring, as well as synthetic coagulants, can form such bridging bonds. In general these additives have a large molecular size and most have multiple electric charges or ion exchange sites, along a molecular chain of carbon atoms. These are often referred to as polyelectrolytes and are used in many applications.

Polyelectrolytes are adaptable and may be suitably modified to interact effectively with the colloidal matter present in the system. The modifications include variations in molecular weight and in the ion exchange capacity. If the molecular variation does not have an ionic charge they are usually referred to as nonionic polymers, although they are not in the strictest sense polyelectrolytes.

Cationic polyelectrolytes are often either polyamines or quaternary amines. The polyamines hydrolyse in water to yield $OH^-$ ions so that these electrolytes are sensitive to changes in *pH*. Quaternary polymers are only slightly affected by *pH* retaining their positive charge over a wide *pH* range.

The anionic polymers incorporate a carboxyl group so that they give rise to hydrogen ions in solution. As a consequence anionic polymers become non-ionic at high solution *pH*.

Kemmer [1988] makes the following points regarding the use of organic polymers. Coagulants are positively charged molecules with a relatively low molecular weight. On the other hand flocculant polymers have a higher molecular weight providing bridges between small flocs to enhance particle growth. Flocculants may be cationic, anionic or nonionic. In general flocculants have a molecular weight of over $10^6$ whereas the coagulants have molecular weights below $10^5$. Unlike the inorganic salts used for flocculation, the polymers do not produce large volumes of gelatinous material which is a distinct advantage. The choice of flocculant generally cannot be made without laboratory screening in respect of the particular application, to be followed by plant trials. It is also necessary to recognise that the quality of the water may change throughout the year and this needs to be taken into account in the selection process. The following are some of the organic polymers used in water treatment:

Cationic coagulants:  Polyamines
Polyquaternaries
Polydiallyl-dimethyl ammonium chloride

Epichlorhydrin dimethylamine

Cationic flocculants:    Co-polymers of
                         Acrylamide and dimethyl-aminoethyl-methacrylate
            or           Acrylamide and diallyl-dimethyl ammonium chloride
                         Mannich amines

Anionic flocculants:    Polyacrylates
                        Co-polymers of acrylamide and acrylate

Nonionic flocculants:   Polyacrylamides

*Dispersion*

An alternative to coagulation and flocculation is to disperse the fine particles so that they remain in suspension and pass through the equipment without settlement and deposition onto surfaces. The action of added dispersants is to impose like charges to particles (and surfaces) so that the tendency to settle on the surface is reduced. Chemicals frequently used include surfactants and low molecular weight polymers. Synthetic dispersing agents have demonstrated considerable improvement over the more commonly used naturally occurring substances such as lignins and their derivatives like sulphonated lignins, for the control of calcium phosphate scale and magnetite in boilers. The synthetic dispersants have been designed for specific dispersion problems with specific molecules for the dispersion of particles of magnesium silicate, calcium phosphate and iron and copper oxide.

Dispersants may also be used in connection with fouling problems other than simple particulate deposition. For instance in corrosion prevention they may be used to prevent solids deposition and subsequent formation of oxygen concentration cells (see Chapter 10). In the contamination of surfaces by micro-organisms (see Chapter 12) dispersants (often called biodispersants) are used to prevent or restrict, the approach of micro-organisms towards heat exchanger surfaces.

14.2.2 The prevention of scale formation

The formation of scale on heat exchanger surfaces has long been a problem in boiler and evaporative systems. In the early days of the industrial revolution it often led to disastrous boiler explosions due to the effects of the precipitated scale on temperature distribution. The associated high metal temperatures resulted in metal failure aggravated by the steam pressure. The consequence is that anti scaling chemicals have their origins in boiler feed water treatment, although some developments have been specific to cooling water requirements and to evaporation systems (such as multistage flash evaporation for water desalination).

Jamialahmadi and Müller Steinhagen [1993] conclude that the theory and art of scale prevention in boilers has developed in a rather amorphous way. They cite the use of potatoes in 1821 for scale prevention in boilers. The origin of the technique was apparently that some labourers, wishing to cook their potatoes in the boiler, were distracted so that the potatoes were not retrieved. Some days later when the boiler was due to be cleaned, the operators were astonished to see that the heat transfer surfaces were unusually clean. The cleanliness was explained in terms of the starch particles from the potatoes that provided bulk nucleation sites and the creation of weaker deposits.

Crystallisation or scale formation on heat exchanger surfaces is influenced by a number of system variables (see Chapter 8) including temperature, *pH* supersaturation conditions and the flow velocity. In general these are fixed parameters and may not be changed without affecting the requirements of the process.

In aqueous systems the dilemma often facing the water chemist is whether or not a particular industrial water is likely to be scale forming or corrosive. The reason for this difficulty may be attributed to the presence of $CaCO_3$ (see Chapter 8). A simple test that may be applied to a water to provide a qualitative assessment of the problem is to add powdered $CaCO_3$ to the water. If the water is supersaturated in respect of $CaCO_3$ the addition of the solid particles will cause precipitation from the solution. Under these conditions there is a reduction in the *pH* of the water. It may be said that the water has a positive saturation index. If however the water is not saturated in respect of $CaCO_3$ (such waters are corrosive), some of the added solid $CaCO_3$ will enter the solution thereby increasing the hardness, alkalinity and *pH*. Water displaying these properties may be said to display a negative saturation index.

Such information is qualitative and for the technologist faced with the problem of dealing with scale formation in industrial equipment something more quantitative is required so that there is some measure of the magnitude of the problem. As a result of this need quantitative water indices were developed.

If the *pH* of a saturated solution of a particular water in contact with solid $CaCO_3$ is $pH_s$ and the actual *pH* of the water is *pH* then the so-called Langelier Index [Langelier 1936] is

$$pH - pH_s \tag{14.3}$$

If the actual *pH* is below $pH_s$ the result is a negative index and $CaCO_3$ will dissolve in the water. It is also generally assumed that this will indicate the water to be corrosive towards steel in the presence of oxygen. On the other hand if the Langelier index is positive (i.e. $pH > pH_s$) and the water is saturated with $CaCO_3$ scale formation is likely to occur.

An improved index is the Ryznar Index [Ryznar 1944] defined as

$$2pH_s - pH \qquad (14.4)$$

If the index is above 7 the likely problem is one of corrosion. An index below 6 suggests a tendency towards scale formation. In the range 6 - 7 there is uncertainty concerning the potential problem that is likely to be encountered.

Kemmer [1988] states that the Langelier index is most useful in predicting likely events for low flow situations, e.g. in storage tanks, and the empirical Ryznar index is applicable only to flowing systems, where conditions at the wall are quite different from the conditions in the quiescent bulk.

These indices have been used with success for many years. Recently Müller-Steinhagen and Branch [1988] have shown that the Ryznar Index may indicate scaling for conditions where the water is not saturated with $CaCO_3$. Such observations underline the difficulties that may attend the use of these indices.

Hawthorn [1992] points out that the Saturation Index (Langelier Index) was not derived from systems where there is heating, evaporative concentration, recirculation and *pH* changes. Langelier [1946] also suggested that the index is an indication of directional tendency and driving force, but is not related to capacity. Furthermore as seen from Equation 14.3 the index is related to equilibrium solubility conditions. Hawthorn [1992] concludes that in modern recirculation cooling water systems for instance, these concentrations are not reached, and he suggests that predictions of potential fouling should be based on what he calls scaling equations based on chemistry, kinetics and hydrodynamics. The basic relationship describing the deposition process is a simple mass balance, i.e.

$$\text{mass deposited} = \text{mass in - mass out} \qquad (14.5)$$

$$= \text{volume in x } c_{in} \text{ - volume out x } c_{out} \qquad (14.6)$$

where $c_{in}$ and $c_{out}$ are the concentrations of the scaling species in solution (or suspension) entering or leaving the system respectively.

To utilise Equation 14.5 in a predictive sense, requires a thorough knowledge of the chemical system and the plant operating conditions. Table 14.3 lists these variables and conditions.

Hawthorn [1992] describes an elaborate experimental programme to determine an acceptable equation for phosphate precipitation. Recognising the scientific merit of this approach, it has to be said that all the required data may not be readily available and despite their shortcomings, the indices may at least provide a guide to likely trends.

Since the deposition process is *pH* sensitive (see Equations 14.3 and 14.4 and Chapter 8) adjustment of *pH* is a possible way to control scale formation. Sulphuric acid has been used for this purpose and it is relatively cheap. In general

TABLE 14.3

System and plant variables affecting scale formation

| System variables | Plant factors |
|---|---|
| Total dissolved solids concentration. Concentration of particular species. Alkalinity/acidity defined as free alkali (e.g. carbonate) free acid (e.g. sulphuric) *mg/l* respectively. *pH*. Temperature. | Residence time. Evaporation (i.e. concentration effects). Changes in heat load that will affect the temperature distribution across the heat exchanger surface thereby changing the interface temperature. Fluid flow velocity. The general design of the equipment may also affect the problem of deposition. |

sufficient acid is added to reduce the alkalinity, but not necessarily to remove it altogether. The reduction in alkalinity is controlled sufficiently so that saturation is achieved and the saturation index qualitatively suggests non scaling conditions. In a consideration of calcium carbonate scale formation the following equilbria are possible (see also Chapter 8):

$$CaCO_3 \rightarrow Ca^{2+} + CO_3^{2-} \qquad\qquad (14.7)$$

also because in general, the water will be exposed to the air

$$H_2O + CO_2 \rightarrow H_2CO_3 \qquad\qquad (14.8)$$

$$H_2CO_3 \rightarrow H^+ + HCO_3^- \qquad\qquad (14.9)$$

$$HCO_3 \rightarrow H^+ + CO_3^{2-} \qquad\qquad (14.10)$$

It is possible therefore, for $Ca^{2+}$ ions to exist as $Ca(HCO_3)^+$ or $Ca(OH)^+$ by combination with the negative ions on the RHS of the above equations. There may be further association to give $Ca(HCO_3)_2$. In combination with sulphuric acid the following overall reaction is possible:

$$Ca(HCO_3)_2 + H_2SO_4 \rightarrow CaSO_4 + 2CO_2 + 2H_2O \qquad\qquad (14.11)$$

The acid treatment converts the bicarbonate to the more stable and more soluble $CaSO_4$. If there is already a substantial sulphate concentration in the water to be treated, then an alternative mineral acid (e.g. *HCl*) may be used to reduce alkalinity without increasing the risk of calcium sulphate deposition. One of the major problems with the use of acid to control scale formation, is the risk of corrosion. If the technique is employed careful control is imperative and may involve sophisticated dosing methods (see Table 14.1).

Since $CO_2$ may be involved in the process of carbonate precipitation Hollerbach and Krauss [1987] have suggested the use of $CO_2$ to treat water and process fluids. The hydrogen ions released by the interaction with water (Equations 14.8 - 14.10) are available to convert the $CaCO_3$ to $Ca(HCO_3)_2$. The advantage advanced by Hollerback and Krauss [1987] for the use of $CO_2$ is that an equilibrium is created by the chemical reactions because the bicarbonate acts as a buffer in the solution. Under these circumstances it is virtually impossible to create a highly acidic and corrosive mixture as illustrated by Fig. 14.2. It is also claimed that dosing with liquid $CO_2$ is safe and requires no special dosing equipment and *pH* adjustment is rapid. As a general rule however, treatment programmes usually employ more complex chemicals and the mechanisms involved are more complicated.

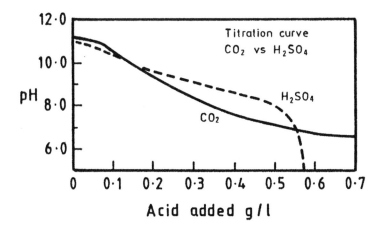

FIGURE 14.2. The change in *pH* with added acid ($CO_2$ and $H_2SO_4$)

More conventional chemical treatments (sometimes referred to as inhibition) use one or more of four groups of additives that includes; threshold agents, crystal modifiers, dispersants and surfactants.

*Threshold treatment*

Threshold agents have the ability, when added in very small concentrations, to inhibit the precipitation of alkaline salts and other scaling compounds on a non-stoichiometric basis. Reductions in the potential for precipitation of calcium salts, iron and manganese can be achieved. In certain cases the action of the additive includes the formation of a sludge rather than a scale. Although still a potential foulant, the sludge is more easily removed from the system. Sequestering additives include polyphosphates, organophosphorous compounds and polymers such as polyacrylates.

Sodium hexametaphosphate (sometimes called Graham's salt) was first used in water treatment to introduce phosphate into boilers without producing precipitation in the feed lines. It was used in the preparation with 14 parts to 1 part of calcium and was essentially a softening process. Because of the mode of action and the extremely small amounts of the additive required, the technique was called "threshold treatment". An alternative name is sequestration.

Sodium hexametaphosphate is one of a series of chain polymers termed polyphosphates, and threshold treatment using various members of the series has found extensive use in controlling calcium carbonate scale formation. Veale [1984] states that typically 5 *mg/l* of polyphosphate will inhibit the precipitation of 500 *mg/l* of $CaCO_3$. To control scale in sea water evaporation an additive mixture could include sodium tripolyphosphate, lignin sulphonates and an antifoaming agent.

Polyphosphonates represent a series of condensed phosphates derived from dehydrated phosphoric acid, i.e:

$$2H_3PO_4 - H_2O \rightarrow H_4P_2O_7 \tag{14.12}$$

$$\text{pyrosphosphoric acid}$$

$$3H_3PO_4 - H_2O \rightarrow H_3P_3O_{10} \tag{14.13}$$

$$\text{tripolyphosphoric acid}$$

$$H_3PO_4 - H_2O \rightarrow HPO_3 \tag{14.14}$$

$$\text{metaphosphoric acid}$$

Inorganic polyphosphates are used primarily in potable water systems whereas organophosphates, in particular phosphonates, are the preferred threshold treatment for cooling waters [Betz 1976], although Kemmer [1988] concludes that the threshold technique is not particularly suitable for such application because of the high dosage of polyphosphate required to form soluble complexes. Some organophosphorous compounds used for threshold applications are given in Table 14.4 together with some of their advantages and disadvantages.

The low molecular weight sodium salts of ethylenediaminetetracetic acid (EDTA) and nitrilotriacetic acid (NTA) soluble in water, are effective in the control of scaling in steam raising equipment. The action of these additives is to form complex ions with calcium and magnesium hardness. The action of EDTA is similar to ion exchange, and may be described by the simplified equation:

$$Ca^{2+} + Na_4\, EDTA \rightarrow CaNa_2\, EDTA + 2\,Na^+ \tag{14.15}$$

Jamialahmadi and Müller-Steinhagen [1991] have demonstrated experimentally the effectiveness of the use of EDTA in fouling reduction. Their work reflects the high heat fluxes encountered in modern boiler practice where the wall superheat is high with the high potential risk of scale formation. Fig. 14.3 presents some of the data provided by Jamialahmadi and Müller-Steinhagen [1991], and shows how EDTA addition can maintain high heat transfer rates in boiling calcium sulphate solutions.

TABLE 14.4

Some organophosphorous compounds used in threshold treatment

| Compound type | Advantages | Disadvantages |
|---|---|---|
| Polyphosphates | Hydrolysis to orthophosphate which reacts with calcium to form a sludge. | Nutrient for micro-organisms. |
| Phosphonates | Effective against both alkaline and non-alkaline scaling conditions. | If supersaturation levels are allowed to rise uncontrolled precipitation may occur. |
| Phosphate esters | Wide range available. Biodegradable. | Less effective than phosphonates. |
| Phosphoncarboxylic acid | Similar to phosphonates. | Some rise in supersaturation may be tolerated without precipitation. |

The generally higher cost of EDTA compared to phosphate usually limits the use of these chelating agents to feed waters with low hardness [Kemmer 1988]. There is the risk that at the temperatures currently encountered in modern steam raising equipment, breakdown of the organic additives is a distinct possibility. The result may be increased potential for corrosion. Excessive dosing can also lead to severe corrosion problems.

*Crystal modification*

As described in Chapter 8 robust crystal structures can be produced under certain conditions, but that crystal structure and growth may be affected by the presence of impurities. The effect is utilised in scale control by the use of additives. Unlike threshold treatment the use of crystal modifiers does not prevent

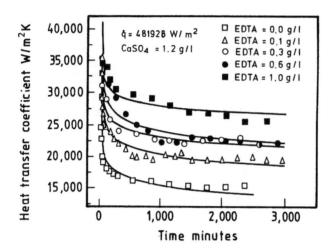

FIGURE 14.3. Heat transfer coefficient reduction with time at different EDTA concentrations

precipitation but reduces the opportunity for crystals to form a coherent structure on heat transfer surfaces. The resulting deposit is generally weak and therefore more susceptible to hydrodynamic and other removal mechanisms. Crystal modifying agents are widely used in cooling water systems. Naturally occurring organic material such as lignin or tannin, were the first additives of this kind to be used but they have been superceded by synthetic polymers.

Polycarboxylic acids are used for crystal modification and they can also provide threshold effects. These compounds have molecular weights generally in the range 1000 - 20000 and are effective against both alkaline and non-alkaline scales. Polymaleic acid gives a very soft deposit with precipitating $CaCO_3$ which is easy to remove.

Harris and Marshall [1981] have presented data on the effectiveness of various additives to control $CaCO_3$ precipitation. These data are included in Table 14.5.

In their paper Harris and Marshall [1981] also give some interesting comparisons between various additives under different conditions including the flow velocity of the water, the additive dose concentration and the water quality (concentration factor is defined in Chapter 16).

Using a water with the following characteristics:

| | |
|---|---|
| Concentration factor | 3 |
| Total alkalinity | 247 *mg/l* |
| Total hardness | 290 *mg/l* |
| Chloride concentration (*Cl*) | 110 *mg/l* |
| *pH* | 8.2 |

TABLE 14.5

Ability of inhibitors to retain $CaCO_3$ in solution

| Additive | Inhibition % at the stated dose level | | | |
|---|---|---|---|---|
| | *2.5 mg/l* | *5.0 mg/l* | *7.5 mg/l* | *10 mg/l* |
| Polyphosphate | 98 | 98 | 99 | 100 |
| Aminophosphonic acid | 97 | 96 | 95 | 94 |
| Acetodiphosphonic acid | 83 | 82 | 83 | 90 |
| Polyacrylate | 30 | 65 | 84 | 93 |
| Polymaleic acid | 26 | 35 | 44 | 56 |
| EDTA | 15 | 20 | 20 | 20 |

These authors demonstrate the effectiveness of a threshold agent (aceto diphosphonic acid) and a crystal modifying additive (polymaleic acid) Fig. 14.4 shows the plot of fouling resistance with time and it will be evident that the asymptotic fouling resistance has been reached in each example. The fouling resistance in the presence of both these additives is significantly less than that obtained with no inhibitor present.

The effect of different concentrations of the crystal distorting agent (polymaleic acid) on the rate of scale formation is included in Fig. 14.5. The scaling rate decreases rapidly for dose levels up to 3 *mg/l* but there is no effective change at higher concentrations. The scaling rate is defined as the *mg* scale per litre of water fed to the test apparatus.

Fig. 14.6 shows the effect of flow velocity on the effectiveness of polymaleic acid crystal distorting agent on reducing scale formation. Without the presence of the additive, the rate of scale deposition decreases roughly linearly with increased velocity, but Harris and Marshall [1981] observe that as the velocity increases the scale becomes harder, compact and more difficult to remove. In the presence of the crystal distorting agent, the rate of scale formation also decreases with velocity but instead of becoming harder the scale remains soft and easy to remove at all velocities. As a result of their studies Harris and Marshall [1981] were able to state that flow velocities greater than 1 *m/s* should be used for the best effects from crystal modifiers.

It is of interest to gauge performance of additives against the Ryznar Index for the water. Fig. 14.7 shows the effect of the threshold additive acetodiphosphonic acid and the crystal modifier polymaleic acid at different Ryznar indices. As would be expected from the definition of the Ryznar Index, the scaling is increased as the index is reduced when there is no additive present. In the presence of the threshold agent the rate of scaling is also reduced but below a Ryznar Index of about 4.5, the threshold effect is exceeded and the scaling rate rapidly rises.

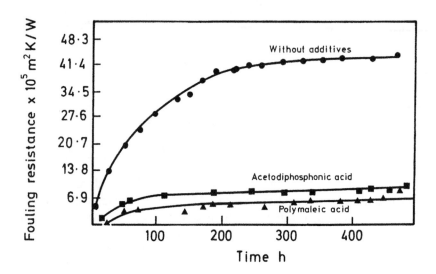

FIGURE 14.4.  Fouling resistance versus time with and without additives

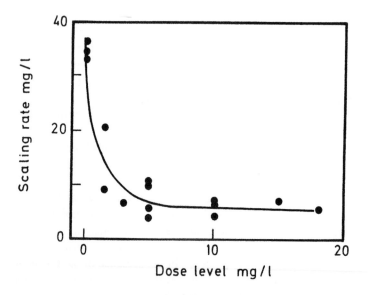

FIGURE 14.5.  The effect of dose level of polymaleic acid on the rate of scaling

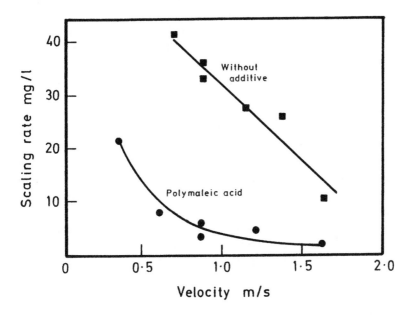

FIGURE 14.6. The effect of velocity on scaling rate with and without the addition of polymaleic acid

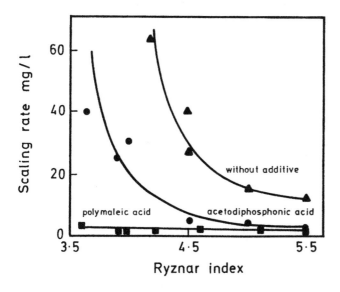

FIGURE 14.7. Scaling rate as a function of Ryznar Index in the presence of a threshold and crystal modifying agents

The beneficial effects of threshold treatment tends to breakdown when the hardness level is of the order of 350 - 400 *mg/l* and above these levels uncontrolled and rapid precipitation may occur [Harris and Marshall 1981]. To bring back effectiveness it might be necessary to add a mineral acid to reduce partially, the alkalinity of the water. In cooling water systems, reducing the number of cycles the water makes through the system, may restore the effectiveness of the threshold agent.

Fig. 14.7 also shows that the crystal modifying agent (polymaleic acid) maintains the scaling rate at a low value down to a Ryznar Index of 3.5. Harris and Marshall [1981] state that there is no limit on the total alkalinity or hardness levels in a system provided the crystal distorting additive is in proportion to the scaling potential entering the system. They concede however, that with very high hardness concentrations, it may be necessary to include a dispersant in the dosing programme. An alternative could be to use a side stream filter.

Crystal modifiers may be added to liquids subject to freezing fouling, e.g. crude oils to reduce the problem of wax deposition. The principal difficulty here is generally the cost of the treatment as large quantities of additive are required to be effective.

*Dispersion and surfactants*

The effect of an organic dispersant as described for particulate deposition control, is to add a charge to particles (e.g. crystallites) to hold them apart and this prevents stable crystal growth. Organic dispersants include organophosphorous compounds and polyelectrolytes. Surface active agents may be added to the system as effective wetting agents or to emulsify contaminants, to maintain them in suspension rather than allowing precipitation on a heat transfer surface.

Treatment programmes to deal with scale formation and crystallisation generally include mixtures of additives so that a wide spectrum of activity is accomplished and may additionally include corrosion control chemicals and biocides. Each system will have unique characteristics and will involve the need "to tailor" the programme to meet the specific needs of that system.

Atkinson, Gerard and Harris [1979] have provided a useful chart that sets out the opportunities for scale (and corrosion) control based on the Ryznar Index. Fig. 14.8 shows how the various additives may be used to advantage depending on the scaling/corrosion potential.

14.2.3 Prevention of corrosion fouling using additives

Corrosion is generally present in aqueous systems as described in Chapter 10. For corrosion to proceed all the elements of the corrosion circuit must be complete.

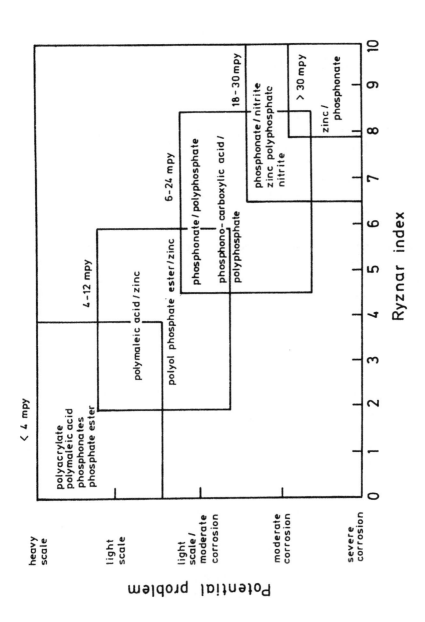

FIGURE 14.8.   Additive selection based on Ryznar Index

The water used in industrial operations (usually for cooling purposes) comes from some natural source, for example bore holes, rivers, lakes or from the sea. The total dissolved solids (TDS) in natural water ranges from 40 *mg/l* for ground water - 35,000 *mg/l* for sea water. Crittenden [1984] states that most land derived water has a TDS < 550 *mg/l* . He also observes that the hardness alone does not govern the corrosivity of water, but the following observations may be made:
1.  Soft, so-called upland waters are aggressive to most metals.

2.  Very hard waters are not usually aggressive provided there is supersaturation.

3.  Underground waters with a high $CO_3^{2-}$ content with a low *pH* are highly aggressive.

4.  The behaviour of water with an intermediate hardness is often determined by the presence of other constituents of the water. If loose scale is formed there is a distinct possibility that corrosion may be initiated beneath the scale.

5.  Salts in solution that are corrosive are primarily chlorides and sulphates. The predominating effect is with $Cl^-$ generally in the range 5 - 1000 *mg/l* but typically < 25 *mg/l*.

6.  Oxygen is the most significant dissolved gas since it is the usual reactant at the cathode in near neutral solutions (the rate of corrosion of active metals in near neutral solutions is proportional to the oxygen concentration). The removal of active oxygen however, may not be helpful since some cathodic current will be necessary if the formation of a passive film is desired.

7.  Dissolved $CO_2$ is usually < 10 *mg/l* and will be involved in the carbonate/bicarbonate equilibria (see Section 14.2.2).

The effect of *pH* is specific to the particular metal or alloy involved as described in chapter 10. Steel is a common material of construction and Fig. 10.4 describes the corrosion rate of steel in relation to *pH*. Minimum corrosion occurs at a *pH* of about 12. At higher *pH* the corrosion rate increases and soluble reaction products are formed. In the *pH* range 6.5 - 9.5 the corrosion rate is almost constant. The rate increases for low *pH* values (< 4.0).

There are two principal ways in which materials of construction may be protected from aggressive conditions in liquids and include corrosion inhibitors and controlled scaling. In general the protection arises from the imposition of an electrical barrier on the metal likely to suffer corrosion, so that the electrical circuit is broken or the flow of ions and electrons restricted. Some systems have "natural" corrosion control. A thin oxide of the metal or the constituents of an alloy, in fact the products of corrosion, act as an effective electrical barrier provided that a continuous film is maintained on the surface. Epstein [1983] mentions that the oxide layer may be removed under the influence of high velocities (i.e. > 30 *m/s*).

Corrosion involves among other things, anodic or cathodic reactions (see Chapter 10). Any chemical applied to the water to arrest the reaction at the anode, will halt the corrosion and conversely, any substance added to reduce or eliminate cathodic reactions, will also limit corrosion.

The terminology anodic and cathodic inhibitors is based on these functions. Anodic protection prevents or limits electron flow to the cathode area. Cathodic inhibitors generally reduce the corrosion rate by forming a barrier at the cathode thereby restricting the hydrogen ion or oxygen transport to the cathode surface. Tables 14.6 and 14.7 provide some information on common corrosion inhibitors. Specific corrosion control requirements are usually based on blends of two or more of the listed chemicals perhaps, in addition to chemicals to control scale formation and biological activity.

TABLE 14.6

Some common anodic inhibitors

| Additive | Notes |
| --- | --- |
| Chromates | Generally not now used due to unacceptable toxicity |
| Nitrites | Nutrient for aquatic plan life |
| Silicates | Slow to act |
| Tannins | Nutrient for biological activity |
| Benzoate | Not reliable except possibly for aluminium |
| Orthophosphates | Nutrient for aquatic plant life |

TABLE 14.7

Some common cathodic inhibitors

| Additive | Notes |
| --- | --- |
| Zinc salts | Toxic compounds and are likely not to be environmentally acceptable |
| Polyphosphates | Tends to form a sludge that is a nutrient for aquatic plant life |
| Calcium calmate | Scaling tendency |
| Polyphosphonates/ molybdates | Requires the presence of $O_2$ for good protection. Generally expensive |

Filming amines which cover the total surface with a monomolecular layer are also used to provide corrosion protection, particularly in the presence of steam condensates.

*Anodic and cathodic protection*

Chromates were commonly used for corrosion control because of their effectiveness and cheapness to apply. Their use now however, is severely restricted due to their toxic effects in the environment, so that they are likely only to be used where they can be retained in the particular system or where removal can be effected before disposal, and there is adequate provision against accidental discharge. In chromate corrosion control of ferrous materials the chromate reacts with the metal surface, forming a relatively hard film of reduced chromate. The resulting protective layer is generally long lasting. Knudsen *et al* [1984] have shown that the presence of zinc-chromate inhibitors alters the *pH* giving rise to $CaCO_3$ scale formation on hot surfaces. The work suggests that care is required in choosing operating conditions for a particular system in order that effective control may be achieved.

The toxicity of chromate treatment has led to the development of alternative anodic inhibitors. Orthophosphate forms an iron phosphate film that protects the surface from corrosion attack, but the layer is less adherent and therefore is not so long lasting, as chromate derived protection. At the same time orthophosphate corrosion control can be made effective provided the system is properly managed. Because of the biological nutrient properties of orthophosphates it is likely that their use will additionally involve the application of a biocide.

The use of tannin is not commonly practised and nitrite and silicate treatments are less popular than orthophosphates for anodic protection because of their technical limitations, and the need for constant supervision to maintain effectiveness. Among the shortcomings are the potential for the formation of deposits, and the encouragement of microbial activity.

The critical concentration of nitrite is proportional to the amount of aggressive species in the water and steel would be attacked if the weight ratio of inhibitor to total aggressive ion were less than unity [Bahadur 1993]. A concentration of nitrite of 300 - 500 *mg/l* is generally required and at this level, it is an uneconomic proposition. Furthermore the use of these high concentrations of nitrite makes the system susceptible to aerobic bacteria, with the oxidation of nitrite to nitrate in cooling water circuits exposed to the atmosphere. Bahadur [1993] also draws attention to the incompatibility of nitrite treatment in conjunction with chlorine, as a biocide (see Section 14.2.5). He also states that nitrite cannot be used in the presence of brass because ammonia, a decomposition product of nitrite treatment, can give rise to stress corrosion cracking.

The economics of using nitrite for corrosion protection may be improved by suitably blending with other additives, e.g. organic and inorganic phosphates. In his review Bahadur quotes a multicomponent additive consisting of nitrite, molybdate, phosphate and orthophosphate in the ratio of 3:2:1:1 for use in the *pH* range 6.5 - 8.5.

Silicates may be used as an anodic inhibitor in low salinity water in the presence of oxygen. The protection mechanism involves the formation of a silica layer in the presence of some corrosion products of the metal. Bahadur [1993] states that the silica film is self limiting in thickness and self healing when damaged. Continuous treatment is required. It is noteworthy that silica inhibits the further corrosion of steel already having an oxidised surface. Sodium silicate in liquid or dry form, may be used as the additive. A useful property of this treatment is that there are no toxicological problems since silicates are naturally occurring.

Benzoates used on their own are not considered to be reliable corrosion inhibitors since, if insufficient is added, corrosion is likely to remain ubiquitous. Benzoate has been used successfully to control corrosion of aluminium. The treatment however also included the simultaneous neutralisation of a high *pH* water, and it is possible, that the *pH* level could have been the controlling factor without which the benzoate may actually enhance corrosion [Neufeld *et al* 1987].

Orthophosphate acts as an anodic inhibitor for steel at concentrations of 12 - 20 *mg/l* provided sufficient of the additive is maintained in soluble form. The process of inhibition involves two stages; formation of phosphate followed by conversion to the oxide. At lower concentrations (i.e. 3 - 7 *mg/l*) orthophosphate acts as a cathodic inhibitor.

In general Kemmer [1988] states that if anodic inhibitors are not applied in sufficient quantities they do not neutralise all the available anodic sites. Under these circumstances the remaining anodic sites become points of intense corrosion activity, and deep pitting may be the result. Sufficient quantities of the inhibitor and good distribution throughout the system, are required to prevent such unwanted effects.

Zinc salts may be used to inhibit corrosion of iron and steel due to the precipitation of $Zn(OH)_2$ on the cathodic areas of the system. The barrier to oxygen transfer is not very durable and zinc is not generally used on its own. It is preferable to use zinc salts in conjunction with other additives, such as polyphosphates and phosphonate. Because of the toxicity problems the presence of $Zn^{2+}$ in plant effluents is generally not acceptable and if it is used, some form of effluent treatment is likely to be necessary before discharge.

Polyphosphonates may be used without blending with other compounds when the corrosion is not severe. The action of polyphosphonates is to form positively charged particles with divalent metals such as calcium that migrate to the cathodic areas of steel with which they are in contact. At the cathode a film is developed thereby restricting oxygen access and the cathodic potential. The advantages of a polyphosphate include relatively low cost and low toxicity. As seen in Section 14.2.2 it is also a scale inhibitor. Disadvantages of this additive include its slow action that may call for some kind of pretreatment, and its breakdown to orthophosphate that is less efficient and is a biological nutrient. Chemical reactions of polyphosphate also include hydrolysis to sodium dihydrogen phosphate at high temperatures. In the presence of $Ca^{2+}$ this produces a precipitate of calcium phosphate which may promote serious corrosion and in any case, is a foulant in its own right.

Blends of anodic and cathodic inhibitors are often employed to improve overall protection. The principle is called synergism.

Copper and brass may be included in the plant material of construction subject to corrosive aqueous environments. In general these metals have good corrosion resistance but under certain conditions, e.g. high *pH* and in the presence of high concentrations of chlorides and sulphates, corrosion of copper and dezincification of brass can occur. Under these conditions it is possible to use various heterocyclic compounds for example the sodium salt of 2-mercaptobenzothiazoleor benzotrizole, to limit corrosion by the formation of an insoluble chemical complex at the anodic areas.

*Film forming additives*

An alternative to anodic and cathodic protection is the use of organic compounds (e.g. amines and other long chain nitrogen compounds) that form a protective film on the metal surface. The technique is widely used in aggressive condensation systems where there is a significant reduction in the oxygen environment with a consequent reduction in the oxide layer on the metal. The long chain molecules have ends that are hydrophilic and hydrophobic. The hydrophilic group attaches to the metal and the hydrophobic end repels water. In principle the surface becomes nonwettable and provides a barrier against metal attack by water containing, $CO_2$, $O_2$ or $NH_3$. Since the molecules tend to repel each other, the film development on the surface is generally only one molecule thick. In steam condensers there is an additional benefit since the non-wetting surface leads to dropwise condensation and improvements in heat transfer as a result.

The film forming compounds are generally surfactants so that during initial application they clean the surface, and corrosion products are removed. Unless the addition of the agent is applied slowly to a system that has been in use for some time, the cleaning effect may produce quantities of particles of corrosion products that may give rise to a particulate fouling problem downstream.

There is a tendency for filming amines to react with contaminants in the aqueous stream such as hardness salts and ions of iron. The result particularly if excessive quantities of the amines have been applied, may be a sludge that represents another potential fouling problem. For these reasons the application of filming amines has to be undertaken with care and with an apreciation of the problems that may arise from high rates of addition and high concentrations. Rapid distribution throughout the system is essential for the prevention of secondary problems. A knowledge of the surface area to be protected is essential in determining the required dose.

Octadecylamine was one of the first film forming amines to be used industrialy for steam condensate system protection. It is only successful in corrosion protection within the relatively narrow *pH* range of 6.5 - 8.0, and good distribution is often difficult to achieve. Improvements to performance may be obtained by blending with emulsifying agents and dispersants. An alternative additive is octadecylamine acetate.

An alternative to the use of filming amines is to employ neutralising amines that raise the *pH* by reacting with the hydrogen ions released by the presence of dissolved $CO_2$ in the water (see Equations 14.8 - 14.10). The origin of the $CO_2$ is generally the breakdown of alkalinity under the operating conditions. The reaction involved may be described:

$$R - NH_2 + H^+ \rightarrow RNH_3^+ \qquad (14.16)$$
amine    hydrogen    neutralised
ion    radical

Amines commonly used for neutralisation include:
    morpholine
    diethylaminoethanol
    dimethylpropylamine
    cyclohexylamine

*Oxygen limitation and pH control*

Oxygen may be present in condensing steam systems due to air leakage brought about by the reduced pressure in the condenser (see Chapter 16). The oxygen may represent a corrosion potential. At high oxygen concentrations neutralisation is inadequate to control corrosion of steel and it will be necessary to use filming amines as a supplement.

Hydrazine ($H_2NH_2$) and other volatile oxygen scavengers may be used for *pH* correction as well as for oxygen control. Its use is expensive and will be uneconomical for high concentrations of $O_2$ and $CO_2$ in the equipment.

*Controlled scaling*

A fouling problem that may be used to advantage is scale or precipitation fouling. The presence of a continuous scale layer on a heat transfer surface could act as a barrier to corrosion reactions in a similar way to chemicals artificially added to a system. The technique relies on a knowledge of the scaling potential as might be provided by the Langelier or Ryznar indices. The method may be questioned however because the scaling index is likely to be different in different parts of the system, even with the possibility of scaling in one place and corrosion in another location. Changing residence times and throughputs will also affect the distribution of indices throughout the system.

In general each potential corrosion problem that may result in deposit formation on heat transfer surfaces, has to be considered in its own right. The brief description of some of the additives available to combat corrosion shows that there are numerous ways in which the problems may be handled. The final choice will depend on effectiveness required, the cost of the additive and the associated costs.

### 14.2.4 Additives to prevent chemical reaction

The discussion in this section is largely concerned with the processing of crude oil and petroleum products although the principles may be applied to any liquid system that may respond to the treatment. Apart from fouling due to particulate deposition, corrosion and scaling, petroleum processing is prone to chemical reaction fouling. The mechanisms include polymerisation of chemical components due to the presence of oxygen, polymerisation as a result of the metal ions, and coking reactions. The latter is usually due to the gradual increase in heat exchanger wall temperature as the organic deposit thickness increases. Control of chemical reactions leading to fouling will generally reduce the incidence of coking. Chemical aditives to reduce the incidence of chemical reaction, may be classified into four groups:

1. *Antioxidants (or inhibitors).* The action of these additives is to scavenge free radicals that may be formed in the process stream (see Section 11.2.4) in order to reduce or eliminate chain reactors. Equations 11.4 to 11.6 give the mechanism of free radical propagation, i.e.

$$R - H + X \rightarrow R + XH \tag{11.4}$$

$$R + O_2 \rightarrow ROO \tag{11.5}$$

$$\text{and } ROO + RH \rightarrow ROOH + R \tag{11.6}$$

With an inhibitor present, the peroxy radical reacts but the inhibitor does not form other peroxy radicals with oxygen. If $I\cdot$ represents an inhibitor group, chain termination reactions could be as follows:

$$ROO + HI \rightarrow ROOH + I\cdot \tag{14.17}$$

$$ROO + I\cdot \rightarrow \underset{INERT}{ROOI} \tag{14.18}$$

$$I\cdot + I\cdot \rightarrow \underset{INERT}{I_2} \tag{14.19}$$

The inerts produced by these reactions cannot react further and are therefore effectively removed from the potential fouling process.

Watkinson [1988] suggests that suitable additives include phenols and organic compounds containing phosphorous and sulphur. He lists the following chemicals as examples of antioxidant chain terminators:

2, 4 dimethyl 6 tert butyl phenol
2, 6 ditertbutyl 4 methyl phenol
cresols
catechols
N, N¹ disecbutyl p-phenylene-diamine
amine borates
amine guanidine derivatives
esters of dithiophosphoric acid

2. *Metal ion deactivators.* Certain chemicals prevent the initiation reactions catalysed by transition metal ions, e.g. $Ca^{2+}$, $Fe^{2+}$, $Zn^{2+}$ and $Mn^{2+}$. Unless the activity of these ions is restricted they can provide a source of free radicals for the oxidation chain reactions described by Equations 11.4 to 11.6. The activity of these metal ions in generating free radicals may be generalised by the following equations [Crawford and Miller 1963]

$$Me^{2+} + RH \rightarrow Me^+ + R\cdot + H^+ \tag{14.20}$$

$$Me^{2+} + ROOH \rightarrow Me^+ + ROO\cdot + H^+ \tag{14.21}$$

or the catalyst may be regenerated by a reaction such as

$$Me^+ + ROOH \rightarrow RO\cdot + Me^{2+} + OH^- \tag{14.22}$$

where $Me^+$ and $Me^{2+}$ represent appropriate metal ions.

Chelating the metal ions is a recognised way of preventing their entering into these undesirable chemical reactions. Some examples of metal ion deactivators are:

Ethylene diamine tetra acetic acid and its salts
Benzimidazole
Oxamine
Zinc diethyl dithiocarbamate
N, N¹ disalicylidene 1, 2 propylene diamine

3. *Dispersants/detergents.* The action of dispersants has been described elsewhere (Section 14.2.1 and 14.2.2). Their purpose is to impart an electrical charge to particulate matter (formed as a result of chemical reactions or already present in the process stream), so that it becomes a stable colloidal suspension thereby preventing agglomeration and potential deposition. Stephenson and Rowe [1993]

suggest that for a dispersant to be effective, it must have part of its molecule that anchors to the potential foulant material and a part that makes the resultant material soluble in the process stream. The anchoring sites are typically polar sites. Mayo *et al* [1988] claim that it is the solubilising properties of detergents that are the most important function in reducing deposit formation.

Watkinson [1988] lists some surfactants used for their dispersing action in organic liquids. He includes amongst them:

> organic and metal sulphonates
> metal phenolates
> metal dialkyl dithiophosphates
> sodium dialkyl sulphosuccinates
> polyoxyethylene alkyl and alicyclic amines
> monethanol ammonium phosphate salts
> co-polymers of N-substituted formamide
> fatty acid phosphates

Table 14.8, prepared by Professor Paul Watkinson [Watkinson 1988] illustrates something of the wide range of chemicals that are available for the treatment of chemical reaction fouling.

In an antifoulant programme to combat problems with chemical reactions it is usual to employ a blend of additives to provide wide ranging protection. The concentration of additive may be from a few *mg/l* up to several hundred *mg/l* depending on the problem and its magnitude. Many of the additives are proprietory and it is necessary to seek advice from suppliers as to the solution of particular problems, but specific information is difficult to obtain. It is also unwise for operators to attempt to select and devise treatment without advice. Chemical treatment companies can call on much experience and laboratory services in the search for an adequate solution to problems of chemical reaction fouling. Figs. 14.9 to 14.11 give some published data on un-named commercially confidential chemical additives that illustrate the effectiveness that may be achieved [Weinland *et al* 1967, Haluska and Shaw 1975], Fig. 14.9 shows how the increase in wall temperature may be controlled by the use of suitable antifoulants. Figs. 14.10 and 14.11 show how the rate of fall in heat transfer efficiency due to chemical reaction fouling may be reduced.

14.2.5 The control of biofouling with additives

The problem of biofouling resulting from the activity of micro-organisms and larger living creatures was outlined in Chapter 12. The problem is usually confined to aqueous systems such as cooling water circuits and water pipe lines although as indicated in Chapter 12, other liquid systems may be at risk. An effective way to control the incidence of deposits of living matter is to use an additive, generally referred to as a biocide, to kill the organism or at least to reduce its activity.

TABLE 14.8

Some antifoulants described in the patent literature [Watkinson 1988]

| Antifoulant | Application | Reference |
|---|---|---|
| Dimethyl sulfoxide | Naphtha hydrotreater | Ferm, R.L. Chevron Research USP 4, 469, 586 |
| *N, N*-diethyl-hydroxylamine | Heat exchangers to 810°C | USP 4, 551, 226 |
| Glycolic acid + polybuteneamine (product of) | Heat exchangers to 810°C | USP 4, 431, 514 |
| P-amyl phenyl acid thiophosphate | Crude oil to 540°C | Shell, D. Nalco CP 1, 071, 131 |
| Na di-2-ethylhexyl sulfosuccinate | Crude oil | Schener, C. UK Pat Appl'n 2, 017, 747 |
| Dodecylbenzine sulphonic acid + iso-octyl hydrogen phosphate | Sour crude to 315°C | Miller, R. Atlantic Richfield USP 4, 425, 223 |
| 10,10'-diphensthiazine + *N, N*-dibenzyl hydroxylamine | Hydrocarbon heating to 315°C | USP 4, 556, 476 |
| Diethylene triamine bis (oleic acid amide) sorbitan monoleate 2,6-ditertbutyl phenol Morpholine | Crude oil visbreaking | Pipe, A. Brit. Petrol. Eur. Pat. Appl'n EP 227 |

Additives that control bacterial growth may therefore exert both biostatic and biocidal effects. The mechanism of action responsible for each effect is not generally the same. Biostatic action is considered to represent metabolic inhibition which is released when the additive is removed. Biocidal activity causes irreversible damage to a vital structure or function of the cell. Dispersants may also be used to prevent attachment to a surface.

The following discussion generally applies to microbial fouling associated with cooling water since, in general, the principal problems for aqueous systems lie in the deposition and growth of micro-organisms on surfaces. Wherever appropriate reference will be made to macro-organism fouling.

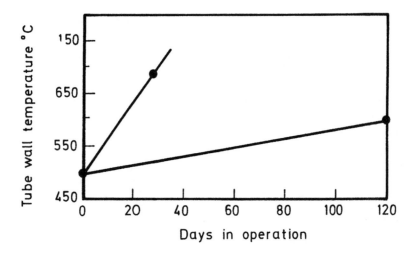

FIGURE 14.9. Effect of an antifouling agent on the tube wall temperature against operating time in a coker at a refinery [Weinland *et al* 1967]

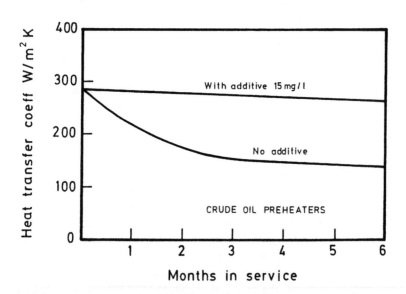

FIGURE 14.10. Effect of an antifouling agent on the heat transfer coefficient against operating time on a crude preheater [Haluska and Shaw 1975]

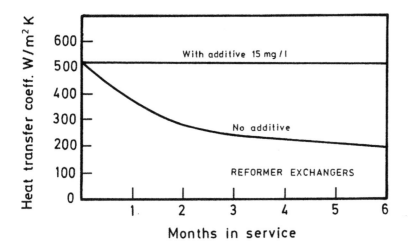

FIGURE 14.11. Effect of an antifouling agent on the heat transfer coefficient against operating time on a reformer unit [Haluska and Shaw 1975]

14.2.5.1 The ideal biocide and factors affecting choice

In any cooling water system a wide range of micro-organisms is likely to be encountered making up its own eco system and very dependent on the particular location, the climate and the season. At some stage in its use, the water is discharged (blowdown) either to the drainage system or to a natural water system (e.g. a river or lake), since it will have become saturated with potential scale forming salts, through evaporative cooling (see Chapter 16). The use of a chemically persistent biocide therefore, is unlikely to be acceptable because of the risk to the environment. It may be necessary to treat with attendant increase in operating cost, the water before discharge to reduce the potential pollution from certain toxic additives. Furthermore, as described elsewhere in this chapter, in addition to biocides, industrial water will generally require other treatment to prevent corrosion and scale formation, chemical interactions between these additives is clearly undesirable as this would reduce the efficiency of the overall treatment programme and represent unnecessary cost.

Bott [1988] has summarised these and other requirements for an ideal biocide:

1. Active against a wide range of micro-organisms.

2. Relatively low toxicity to other life forms.

3. Biodegradable, i.e. the necessary residual biocide in the water is rendered harmless by further biological activity.

4. Non-corrosive.

5. Effectiveness not impaired by the presence of inorganic or organic materials other than micro-organisms present in the system.

6. Does not deactivate other additives, e.g. corrosion inhibitors or scale preventatives used in the general treatment programme.

It is very unlikely that a particular biocide will meet all these criteria so that it will be necessary to make further assessments before making a final choice; not least in the consideration will be the likely cost involved. The resultant decision to use a certain biocide or combination of biocides, will be a compromise.

In addition to the cost of treatment other factors are of importance in making the final choice. They include:

1. Quality of the water used in the system.

2. Whether the system is recirculating or "once through".

3. Contact with the local atmospheric environment (i.e. the likely contamination of the water by airborne substances).

4. Potential process leaks which may:
    a) provide nutrients for biological growth
    b) act as a biocide
    c) react with chemical aditives

Although not strictly part of the consideration for treatment assessment the process fluid may represent an environmental pollutant when the water is discharged.

5. Residence time in the system.

Of these factors the most important is likely to be the quality of the water, involving the extent of the biological contamination dissolved solids and suspended matter. The chemical and particulate make up of the water and its *pH* may influence the efficacy of particular biocides, e.g. some of the constituents of the water may react chemically with the biocide. Because the quality of water is unlikely to be constant (see Chapters 12 and 16) it is necessary to make regular checks on the water quality.

14.2.5.2 System factors that affect efficiency of biocides

The activity of a biocide will also be affected by other factors in the system; in particular temperature and flow velocity will influence the biocidal efficiency in a heat exchanger (say a power station condenser).

Temperature will affect chemical reactions in which the biocide takes part either with other constituents in the water, or through simple temperature related decomposition. Fig. 14.12 and Fig.14.13 show the stability of glutaraldehyde as a function of temperature at two different *pH* values [Henry 1993]. The change of *pH* from 7 to 8 makes a considerable difference to the stability of the biocide at 82°C. In general of course, for cooling water applications, temperatures much above 55°C are not likely to be encountered so it has to be stated that the effects of temperature on the stability of glutaraldehyde may not be important in biocide applications.

The effect of flow velocity on biofilm removal is illustrated on Fig. 14.14 [Nesaratnam and Bott 1984]. The curves show that as the velocity is increased the rate of removal of established biofilm increases, and the apparent final biofilm thickness is less. There are two effects related to velocity and include:

1. As the velocity is increased the shear force on the biofilm is increased (see Chapter 12).

2. Mass transfer is related to velocity (i.e. turbulence) so that as the velocity increases the availability of biocide at the water/biofilm interface is facilitated. Furthermore as the biocide concentration at the interface is increased there is increased concentration driving force for mass transfer through the biofilm, thereby ensuring deeper penetration of the biofilm structure in a given time.

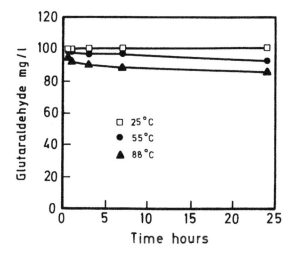

FIGURE 14.12. Glutaraldehyde stability at *pH* 7 and three temperatures

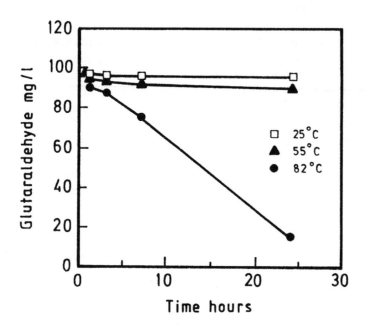

FIGURE 14.13.  Glutaraldehyde stability at *pH* 8 and three temperatures

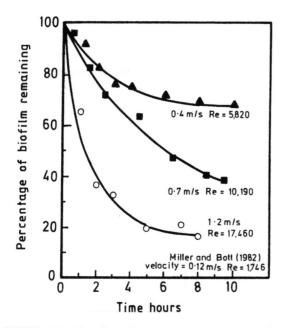

FIGURE 14.14.  The effect of flow velocity on the removal of biofilms of *Pseudomonas fluorescens* using sodium hypochlorite.

The concentration of bioicide will also affect the rate of removal of biofilm of an existing biocide. Fig. 14.15 [Nesaratnam and Bott 1984] shows how the initial rate of biofilm (*Pseudomonas fluorescens*) removal is dependent on the concentration of free chlorine in the water flowing at a velocity of 1.2 *m/s*. Again this is a mass transfer effect combined with fluid shear, the higher chlorine concentration in the bulk water provides a greater driving force for mass transfer and hence a greater biocide effectiveness.

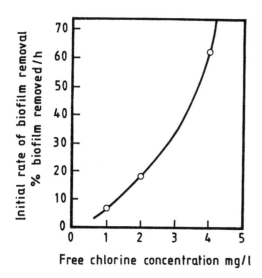

FIGURE 14.15. The dependence of initial rate of biofilm removal on free chlorine concentration

Although these data show the trends associated with changes in system variables the age and quality of the biofilm will also be a factor, i.e. whether the biofilm is open and "fluffy" or dense and compact. Diffusion of biocide into the biofilm will be facilitated by the existence of "pores" containing water within the biofilm matrix. The quality of the biofilm will be very much a function of the conditions under which it was laid down, particularly in terms of nutrient availability and flow rate (see Chapter 12). Apparent variations in biocidal efficiency reported in the literature may in fact be due to the different morphologies of biofilms of the same strain, but grown under different conditions.

Nichols [1994] concludes that in medically important biofilms, the glycocalyx (exopolysaccharide) material in bacterial biofilms up to 100 *μm* thick, is not a significant barrier to the diffusion of antibacterial agents. Nicholls makes some tentative calculations on the time for penetration for different biofilm thicknesses. The assessment is based on a molecule with a diffusivity of 6.8 x 10⁻⁶ *cm/s* presumably through water. It would appear also that no account has been taken of the depletion of biocide as it passes through the biofilm through chemical reaction. It is also presumably assumed that the concentration of agent on the outside of the biofilm is 100%. The theoretical calculations however, are interesting and show that for a biofilm of 10 $\mu m$ thickness the penetration time is 15 $s$ and for 1000 $\mu m$ biofilm the time is 1500 $s$.

Blenkinsopp *et al* [1992] have shown that the efficacy of some biocides against *Pseudomonas aeruginosa* biofilms grown on stainless steel studs, is increased by the application of a low strength electric field. Biocide concentrations lower than those necessary to kill planktonic cells were bactericidal within 24 $h$ when applied within the electric field. It is possible that this discovery could have implications for industrial biofilm control in the future.

### 14.2.5.3 Biocidal activity and industrial biocides

The action of biocides that leads to the mortality of microbial biofilms may be considered to be made up of three phases:
1. Adsorption onto the cell.

2. Penetration of the germicide into the cell.

3. Chemical combination of the biocide with essential cell structures or components, which is directly responsible for the death of the organism.

In macro-biological fouling the effects will be similar but related to the cells making up the larger structure.

It is possible to identify three target areas for biocidal attack. The outer cell wall, the membrane through which the metabolites pass into or out of the cell, and the cell interior (cytoplasm) [Hugo 1987]. In general these divisions do not represent mutually exclusive areas for biocide interaction and many biocides will have more than one target in the cell.

There is a wide range of bacteriocides available for the control of biofilm formation. Parkiss [1977] classified biocides according to their action:
1. Oxidation of cell material.

2. Breakdown of cell contents by hydrolysis.

3. Denaturisation of cell chemicals via insoluble salt formation.

4. Reactions with cell protein.

5. Cell permeation affecting metabolism.

The difficulty of such a classification is that some biocides act in more than one way. A more usual classification involves listing as either oxidising or nonoxidising action. Tables 14.9 and 14.10 give some examples of the wide range of common biocides but there are other chemicals available as proprietory formulations. It is not feasible in a book of this kind, to describe all biocides and their action in detail but it is useful to discuss some of the more common chemicals as this provides a background to their use.

*Chlorine*

Chlorine has been a preferred biocide for many years on account of its effectiveness and its relatively low cost. It has been used particularly in once through cooling water systems where large volumes of water are used, in order to minimise the cost of treatment. It is effective against micro- and macro-organisms.

Free chlorine reacts with water to form two acids.
The reactions are:

$$Cl_2 + H_2O \rightarrow HOCl + HCl \qquad (14.23)$$

hypochlorous hydrochloric
acid acid

TABLE 14.9

Oxidising biocides

| Biocide | Application |
| --- | --- |
| Chlorine | Liquid (compressed gas) |
| | Sodium hypochlorite solutions |
| | Calcium hypochlorite powder |
| | Organic compounds - chlorine donors |
| | *In situ* generation, e.g. electrolysis of brine |
| Chlorine dioxide | Usually generated on site by reaction between sodium chlorite and chlorine |
| Bromine | Liquid |
| | Hydrolysis of compounds like halogenated hydantoins |
| | Displacement from a bromine salt water by chlorine |
| Iodine | Solution |
| Hydrogen peroxide | Solution |
| Ozone | Generated on site by electrical discharge in dry air or oxygen followed by absorption in water |

TABLE 14.10

Some examples of non-oxidisng biocides

| Group | Example |
|---|---|
| Inorganic copper salts | Copper sulphate |
| Chlorinated organic salts | Sodium pentachlorophenate |
| Organo zinc compounds | Zinc dimethyldithiocarbamate |
| Isothiazolones and related chlorinated compounds | 3,5-Dimethyl tetrahydro-1, 3,5,-2H-thiadiazine-2-thione, 5-Chloro-2-methyl-4-isothiazolin-3-one |
| Bromine compounds | β-Bromo-β-nitrostyrene |
| Organo dibrominated compounds | 2,2-Dibromo-3-nitrilopropionamide 2-Bromo-4-hydroxyacetophenone |
| Chlorinated organic compounds | Poly[hydroxyethylene (dimethyliminio)-ethylene (dimethyliminio)] methylene dichloride |
| Thiocyanates | Methylene bis-thiocyanate |
| Carbamates | Sodium dimethyldithiocarbamate |
| Amines and nitrogen compounds | Tetrahydro-3,5-dimethyl-2H-1,3,5-thiadiazine-2-thione Tris-hydroxymethyl nitromethane |
| Aldehydes | 1,5 pentanedial cis or trans forms or polymers |

Notes:

1. The active additives listed above are generally added as solutions or emulsions. The solvents and suspending media may be water or organic liquids. In some instances the organic solvent may be a biocide in its own right.

2. The active ingredient may vary between a few percent up to 80 or 90% in the solution or suspension.

3. In many applications the additive consists of blends of different biocides to widen the biocidal spectrum. Biodispersants may be included in the blend.

4. The choice of biocides from this list will depend on the application, e.g. cooling water or paper making or other industries using water, to combat the particular microbiological system.

5. Blends of biocides (available from treatment companies) are usually supplied under a trade name.

6. Some biocides and blends may be the subject to patent protection.

7. The above list is by no means exhaustive.

The hypochlorous acid ionises in water according to the equilibrium

$$HOCl \; \rightleftarrows \; H^+ + OCl^-$$ (14.24)

The presence of hypochlorous acid as opposed to the hypochlorite ion, is responsible for the biocidal activity. Hypochlorous acid is estimated to be twenty times more effective as a micro-biocide than the hypochlorite ion. Unfortunately however, in a cooling tower there is a tendency to strip chlorine from the water by the countercurrent flow of the air, so that Equation 14.23 moves to the left and it is likely that by the time the water reaches the basin of the tower all the chlorine will have been lost to the air in addition to reactions with organic matter within the tower. It will be necessary therefore for maximum utilisation of the available chlorine to ensure that the equilibrium of Equation 14.24, is over to the LHS. Since the ionisation process produces hydrogen ions the equilibrium will be affected by the *pH* of the solution. The effect of *pH* on *HOCl* stability is shown on Fig. 14.16 [Marshall and Bott 1988]. In the figure 100% *HOCl* indicates that the equilibrium is completely to the left in Equation 14.24 and there is no ionisation of the *HOCl*. From Fig. 14.16 this condition occurs at *pH* of around 5. At a *pH* 7.5 hypochlorite ion and acid co-exist in approximately equal amounts. It will be seen that as the *pH* approaches 9 the hypochlorous acid content drops to a very low value and the biocidal effect is negligible. Values of *pH* lower than around 6.5 are not practical because of the potential risk of corrosion.

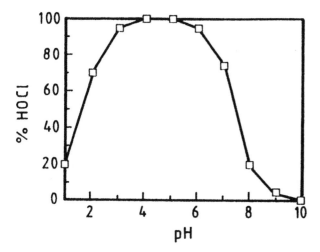

FIGURE 14.16. The effect of *pH* on the ionisation of hypochlorous acid

Sodium bromide is a constituent of sea water and may be as high as 70 *mg/l* in concentration. When chlorine is added to sea water, it displaces the bromine with the formation of hypobromous acid and hypobromide ions that are also biocides with a similar action to their chlorine counterparts. At a *pH* of 8 the effect of the addition of chlorine to sea water is first to produce hypochlorous acid according to Equation 14.23. The hypochlorous acid reacts rapidly with the bromine ion in solution.

$$Br^- + HOCl \rightarrow HOBr + Cl^- \qquad\qquad (14.25)$$
$$\text{hypochlorous}$$
$$\text{acid}$$

The *HOBr* ionises in water according to

$$HOBr = OBr^- + H^+ \qquad\qquad (14.26)$$

The biocidal activity of hypochlorous acid arises from the fact that it is able to diffuse through the cell wall and membrane to attack the cytoplasm within the cell to produce chemically stable nitrogen-chlorine bonds with cell proteins. In addition active sites on certain co-enzyme sulphydryl groups are affected, so that the intermediate chemical transformations that lead to the production of adenosine triphosphate (ATP), are blocked thereby affecting respiration.

Because the chlorine can react with many organic substances found in water systems, there is a "chlorine demand" that will affect the biocidal activity of a particular dose. It is usually necessary to add excess chlorine in order to leave a free residual of chlorine in the system to ensure efficient biocidal action. Some comments on chlorine treatment programmes are contained in Table 14.11 [Betz Laboratories Inc. 1976].

Low levels of residual chlorine may be regarded as a biostat, i.e. preventing colonisation or restricting biofilm development. In a particular laboratory experiment [Kaur *et al* 1988] chlorine levels as low as 0.2 - 0.6 *mg/l* inhibited biofilm attachment and growth. The effect of flow velocity and chlorine concentration are shown on Fig. 14.14.

Excessive use of chlorine for biofouling control can lead to corrosion of metal surfaces because of the acid conditions that are produced by the reactions of chlorine with water. If sodium hypochlorite is used as a source of chlorine this is not the situation.

Rippon [1979] suggests that it is necessary to distinguish between fresh water and esturine or sea water cooling systems in the use of chlorine as a biocide. She points out that a 2000 *MW* power station using inland water will require 200 - 300 tonnes per year of chlorine to control biofouling in the condensers. A similar power station using sea water for cooling will probably use 500 - 1000 tonnes per

TABLE 14.11

Effectiveness of chlorine addition

| Programme | Remarks |
| --- | --- |
| Continuous chlorination - free residual | Most effective |
| | Most costly |
| | Not always technically or economically feasible due to high chlorine demand |
| Continuous chlorination - combined residual | Less effective |
| | Less costly |
| | Inadequate for severe problems |
| Intermittent chlorination - free residual | Often effective |
| | Less costly than continuous chlorination |
| Intermittent chlorination - combined residual | Less effective |
| | Least costly |

year of chlorine because it will be necessary to prevent the growth of macro-fouling, e.g. mussels, barnacles and hydroids in the water intake and culverts. For these conditions it may be necessary to chlorinate the water continuously to maintain a low level of chlorine residual (< 0.5 *mg/l*) at the condenser through the mussel spawning season. The large quantities of chlorine used by just one power station indicate the huge amount of chlorine likely to be used world wide to combat biofouling. Although some of the added chlorine will be stripped from the water in the cooling tower or spray pond, the combined chlorine will pass through the system and be discharged with the blowdown. As mentioned earlier the chemical products of chlorination represent toxic and carcinogenic pollutants and are persistent in the environment, furthermore these products can enter the food chain. As a result of these concerns the use of chlorine is being restricted either on a voluntary basis or through legislation.

Because of the concern for the environment it may be necessary to remove the chlorine (dechlorination) from the water before discharge. Dechlorination methods may be listed as follows [EPRI 1984].

1. The use of reducing chemicals, e.g. sulphur dioxide, sodium sulphite and sodium thiosulphate. The technique is likely to involve elaborate dosing equipment, e.g. automatic control, in order to maximise effectiveness at minimum cost.

2. Adsorption onto granular activated carbon (i.e. similar to that used for water treatment).

3.  Physical methods, e.g. the use of air stripping.

Additional costs include capital equipment, operation including labour, and the cost of the reagents in (1). The increased expense involved in dechlorination, may make chlorine less attractive as a biocide than some of the so-called environmentally friendly biocides currently being developed.

Apart from the problems of discharge of chlorine there are hazards associated with the transport and storage of the required large quantities of chlorine. Apart from accidental spillage there could be a potential threat from terrorists who might threaten to release the chlorine to the detriment of the local population, unless their particular demands were met. An alternative to the transport and storage of chlorine is to generate chlorine on site against demand, i.e. using hypochlorite solutions or electrolysis of brine. The resulting chlorine cost however will be higher than bulk purchase from a chemical supplier.

The technique of targeted chlorine addition represents an alternative to the practice of maintaining a chlorine concentration throughout the whole system [EPRI 1985]. The method allows for the use of relatively high chlorine doses for short periods that are more effective than low concentrations for longer periods. In targeted chlorine dosing a predetermined fraction of the tubes in say a power plant condenser, is exposed to chlorine at any one time, thereby allowing higher chlorine concentrations to reach these tubes within the allotted period. At the water exit from the condenser, mixing of the water from all the tubes reduces the overall chlorine concentration circulating around the rest of the system and finally, in the discharge. For example if 10% of the tubes are targeted, it is possible to use 10 times the permissible allowable chlorine concentration since on recombination the chlorine concentration will be at the required level. Fig. 14.17 is a simplified diagram of equipment that could be employed for targeted chlorine addition. For a large condenser a series of these injectors would be required together with suitable control equipment.

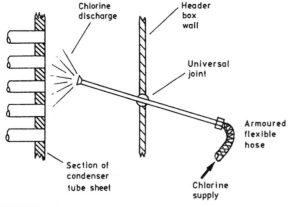

FIGURE 14.17. Concept of targeted chlorine dosing

The complex equipment required to achieve reliable targeted dosing represents a higher cost than that required for the more traditional blanket chlorination of all the circulating water. The targeted dosing technology however, lends itself to automatic control with little operator involvement. There is also the opportunity to optimise the dosing cycle to give the maximum protection for the minimum chlorine use. It is more than likely that the use of this advanced technology in a particular system, will represent lower chlorine use compared with the traditional method of biofilm control. Future developments of the technique may be even more sophisticated by injection into the viscous sub-layer adjacent to the surface subject to biofouling.

Whitehouse *et al* [1985] have discussed the use of chlorine to combat the macrofouling in cooling water. The growth of macro-organisms on the inlet culverts and tunnels can restrict the flow of water so that the quantity of water delivered by the pumps is reduced below that required to maintain optimum conditions in power station condensers. Although there is not a great risk that macrofouling organisms can themselves create a fouling problem on heat exchanger surfaces, sudden detachment of large numbers of mussels can exceed the capacity of the debris filters. The manual cleaning out of fouling from the cooling water intake tunnels and culverts can be expensive. Whitehouse *et al* [1985] report that 4000 man hours of effort were required to clean the culverts at the Dunkerque power station in 1971 with the removal of no less than 360 $m^3$ of mussels!

Chlorination has been used to prevent mussel proliferation. The optimisation of the chlorination process depends principally on the mussel growth rate [Whitehouse *et al* 1985]. At times of low growth rate intermittent chlorination is sufficient, but when mussel growth is rapid the application of chlorine must be continuous or at least semi continuous throughout the growth period.

According to work carried out by Whitehouse *et al* [1985] marine mussels (*Mytilus edulis* and *Mytilus galloprovincialis*) can survive intermittent chlorination at high concentrations between 3 and 5 *mg/l* for several hours per day without being killed. The initial concentration of chlorine which the mussel senses will be low due to the effects of distribution within the flowing water, the maximum chlorine concentration will be achieved some few seconds later. The time interval however, will provide an opportunity for the mussels to close their shells before the maximum chlorine dose reaches them. Even with continuous injection in the range of 0.2 - 1.0 *mg/l* residual chlorine, the time required for a complete mortality of the mussel population is likely to be many days and may be as high as six months. Chlorination does not appear to arrest the settlement of mussel larvae, death is generally attributed to restricted food supply and some pathological effects [Khalanski and Bordet [1980] and Lewis [1985]].

Holmes [1970] showed that water velocity had an effect on mussel removal. Chlorine residuals of 0.5 *mg/l* at water velocities in excess of 1.5 *m/s* accomplished effective mussel removal. With chlorine levels below 0.5 *mg/l* the critical velocity requirements is in excess of 1.5 *m/s*. Holmes [1970] suggested that the relationship between chlorine residual and water velocity, is probably due to some action of chlorine and chlorine compounds in weakening the attachment of mussels. As with other assessments of foulant removal discussed in this book, the effects will be a combination of mass transfer and shear forces.

*Chlorine dioxide*

Chlorine dioxide is generally manufactured against demand and is applied as a solution (see Table 14.8). It cannot be transported and stored because it is unstable and liable to decompose with explosive force. It is generated *in situ* from sodium chlorite and chlorine according to the equation:

$$Cl_2 + 2NaClO_2 \rightarrow 2NaCl + 2ClO_2 \qquad (14.27)$$
$$\text{chlorine}$$
$$\text{dioxide}$$

Chlorine dioxide is a strong oxidising agent and according to McCoy [1980] it disrupts protein synethesis in the microbial cell.

Stabilised chlorine dioxide has some advantages over chlorine including:
1.  Its effectiveness is not reduced by reaction with aromatics, ammonia or unsaturates.
2.  It is more effective than chlorine in the higher *pH* range.

3.  It is not markedly affected by *pH* changes although its toxicity is reduced at low *pH*.

Chlorine dioxide is generally more expensive than chlorine.

*Bromine*

Equations 14.23 and 14.24 give the reactions of chlorine in solution in water. If bromine is added to water it behaves in a similar way to chlorine. The reactions are:

$$Br_2 + H_2O \rightarrow HOBr + HBr \qquad (14.28)$$
$$\text{hypobromous} \quad \text{hydrobromic}$$
$$\text{acid} \qquad \text{acid}$$

The hypobromous acid ionises in water according to the equilibrium.

$$HOBr \rightleftharpoons H^+ + OBr^- \qquad (14.29)$$

Fig. 14.18 compares the hypochlorous and hypobromous acid concentration at different *pH* [Marshall and Bott 1988]. It will be seen that the ionisation of hypobromous acid occurs at a higher *pH* than hypochlorous acid so that bromine may be used as a biocide at higher *pH* values than chlorine. If ammonia is present in the water, bromine reacts with the ammonia to form bromamines that are also very effective biocides compared to the corresponding chloramines.

A convenient way of introducing bromine into a cooling water system is in the form of halogenated hydantoins. the most common compound used for this purpose is 1 bromo 3 chloro 5, 5 dimethyl hydantoin [Marshall and Bott 1988]. The compound hydrolyses in water to produce hypochlorous and hypobromous acids together with dimethyl hydantoin.

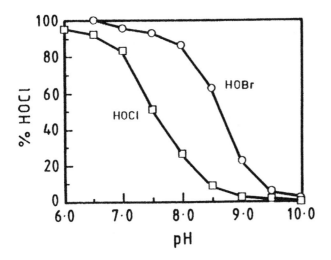

FIGURE 14.18. Hypochlorous and hypobromous acid concentration versus *pH*

*Ozone*

Ozone is a strong oxidising agent and it has been used for many years as a disinfectant in drinking water production. Bott [1990] has compared chlorine and ozone in cooling water treatment. Edwards [1983] compares the oxidising power of some common water treatment chemicals and he concludes that ozone is the most powerful. Table 14.12 gives the comparison. Bott and Kaur [1994] have reviewed the use of ozone in water treatment.

TABLE 14.12

Relative oxidation power of oxidising agents

| Agent | Oxidation potential volts | Relative oxidation power based on chlorine as 1 |
|-------|---------------------------|------------------------------------------------|
| Fluorine | 3.06 | 2.25 |
| Hydroxyl radical | 2.80 | 2.05 |
| Atomic oxygen | 2.42 | 1.78 |
| Ozone | 2.07 | 1.52 |
| Hydrogen peroxide | 1.77 | 1.30 |
| Hypochlorous acid | 1.49 | 1.10 |
| Chlorine | 1.36 | 1.00 |

Ozone is gaining importance in the water industry since it does not produce the organic decomposition products associated with chlorine and therefore ozone is considered to be the more acceptable. The breakdown products resulting from oxidation by ozone are not considered to be an environmental problem, but more work needs to be done in this area. Ozone itself decomposes to oxygen in water so that it is not persistent. Many water companies in the UK have introduced ozone treatment for primary disinfection prior to flocculation and filtration, but because of the unstable character of ozone, water is still treated with chlorine prior to distribution. Interest in ozone is growing in its use to control biofouling in cooling water systems. Donoghue [1993] reports that in the Boston area of the United States alone, twelve successful ozonation systems are in operation.

Ozone is an unstable allotrope of oxygen. In nature it is formed photochemically from oxygen by ultra violet light from the sun in the earth's stratosphere. It can sometimes be detected after a thunderstorm produced by lightning passing through the air. Commercially, ozone is produced by means of a silent electric discharge resulting from a high voltage alternating field between two electrodes separated by a dielectric (e.g. glass) and a narrow gap (1 - 2 *mm*). The feed gas (air or oxygen) passes through the gap.

Because ozone cannot be stored due to its general instability it has to be generated *in situ* as required. An ozonation system generally includes a number of steps. The ambient air used to generate the ozone is first compressed and dried to a very low dew point (a satisfactory corona discharge would not be achieved with wet gas). In the ozone generator the air is ionised with the consequent formation of ozone. Following generation, the ozonised air is injected into the cooling water through suitable contactors or venturi injectors. In general it is not economic to use oxygen for the generation of ozone since some form of oxygen recovery would be required to conserve the oxygen. There would be costs associated with drying the oxygen to an acceptable dew point before it could be returned to the ozoniser for recycling.

Edwards and Sellers [1991] comment on the production of ozone:

1. Concentrations of 1 - 3% $O_3$ and 2 - 8% $O_3$ are possible when air and oxygen respectively are used as the feed gas.

2. Dielectric materials used in ozone production include borosilicate glass, titanium dioxide and aluminium dioxide. Glass is the most common dielectric material in use.

3. The thinner the dielectric material the greater the efficiency of ozone production.

4. The smaller the gap between the electrodes, the higher the concentration of ozone produced but the lower the output. The wider the gap the greater the ozone production at lower concentration. A compromise will be necessary to achieve optimum conditions.

5. The higher the voltage the greater the production and concentration of ozone at a constant feed rate.

6. The higher the alternating cycles the greater the production and concentration at constant flow rate.

7. The lower the dew point of the feed gas (below -50°C) the higher the concentration and production at constant flow rate.

8. The higher the electrode temperature the lower the production.

Ozone is highly toxic, but according to Bell [1988] in 2000 installations world wide no person has been known to die as a result of exposure to ozone.

In water the following reactions are considered to occur:

$$O_3 \rightarrow O_2 + O \cdot \tag{14.30}$$

and $O \cdot + H_2O \rightarrow 2HO \cdot$ (14.31)
     hydroxyl
     radicals

Rice and Wilkes [1991] have summarised some aspects of ozone chemistry as they apply to biocide activity:

1. Ozone decomposes rapidly in water in comparison with other biocides used in water systems. In industrial waters the half life of ozone is in the range 1 - 10 minutes. In some dirty waters the half life of ozone may be even shorter.

2. The rate of ozone decomposition depends on *pH*. At low *pH* (< 7) the predominant species is molecular ozone. As *pH* is raised however, the rate of

ozone decomposition increases to produce hydroxyl free radicals ($HO\cdot$) that have an extremely short life. Above about *pH* 9 - 10 very little molecular ozone remains in solution.

3.  When the alkalinity of the water is high (> 200 *mg/l*) and the *pH* is high (> 9), the ozone decomposes rapidly to give hydroxyl free radicals that react rapidly with any bicarbonate and carbonate ions present. Equations 14.30 and 14.31 rapidly move to the RHS so that the addition of ozone has little benefit.

4.  If bromide ion is present in the water being treated with ozone, it is quickly oxidised to hypobromite ion and hypobromous acid (see earlier in this section). The hypobromous acid probably contributes to the biocide effect under these circumstances. The unique chemistry of ozone and bromide is discussed later.

5.  Some inorganic constituents of the water produce insoluble precipitates in the presence of ozone (Rice and Wilkes [1991] do not state what these reactions are likely to be).

6.  Although ozone reacts rapidly with organic material (including micro-organisms) in the water, the products of the reaction will be intermediate in character. Organic materials are rarely converted completely to carbon dioxide and water. The breakdown products are generally low molecular weight aldehydes, ketones and carboxylic acids. The most stable organic end product of ozone oxidation is oxalic acid [Rice and Gomez-Taylor 1986] which is unaffected by molecular ozone but can be reduced to $CO_2$ and water by the hydroxyl free radicals.

Microscopic examination of cells subject to attack from ozone has been made by Kaur *et al* [1991] leading to a speculative theory on the action of ozone on bacteria. The study revealed the presence of many membrane bounded vesicles that could be seen budding from the membrane of the cell wall after ozone treatment. The resulting lysis of the cells with the associated cytoplasmic degradation, suggests that the ozone first attacks the cell wall and the vesicles appear as the damage is repaired.

Sugam *et al* [1980] studied the effects of ozone treatment on a pilot scale condenser designed and constructed to duplicate a power station condenser system. They observed that although ozone was not as effective as chlorine on a weight basis, it could successfully be employed at low level. Based on some large scale tests Baldwin *et al* [1985], concluded that ozone at concentrations as low as 0.05 to 0.2 *mg/l*, are effective in controlling microbiological growth in the cooling tower and water recirculating system

Experimental work on a laboratory recirculating apparatus conducted by Kaur *et al* [1991], demonstrated the effectiveness of ozone as a biocide for biofouling control. Using ozonated water with a residual ozone concentration of 0.1 *mg/l* as a biocide solution, it was found that 80 - 99% of a biofilm consisting of *Pseudomonas fluorescens* was removed by a single application. As Table 14.13 indicates, the effectiveness of the ozone however, was dependent on morphology, thickness and age of the biofilms. These results indicate that as the age of the biofilm increases the amount of biofilm removed by a similar concentration of ozone decreases. It would appear that an older biofilm is more resistant to the effects of ozone and may be due to a greater degree of compactness discussed earlier. As might be expected from the resistance to ozone penetration associated with a thicker biofilm, the lower the rate of removal using similar ozone concentrations.

TABLE 14.13

Effectiveness of ozone application in removing established biofilms

| Biofilm | Age days | Ozone concentration mg/l | Removal rate % *mass/min* |
|---------|----------|--------------------------|---------------------------|
| a | 19 | 0.08 | 0.58 |
| b | 45 | 0.08 | 0.22 |
|   | initial mass mg/cm² | | |
| c | 51.5 | 0.07 | 0.58 |
| d | 5.8 | 0.08 | 1.22 |

Intermittent applications of ozone using concentrations < 0.1 *mg/l* were found to be capable of weakening the biofilm but a minimum period of 3 hours was required for the control to be effective. Table 14.14 gives some data obtained by Kaur *et al* [1991].

TABLE 14.14

Intermittent dosing data

| Dose sequence | Ozone concentration mg/l | Duration of application h | Initial removal rate % mass/min | Biofilm removed % |
|---------------|--------------------------|---------------------------|--------------------------------|-------------------|
| 1 | 0.06 | 3 | 1.56 | 93 |
| 2 | 0.04 | 2 | 1.35 | 93 |
| 3 | 0.03 | 3 | 0.93 | 73 |
| 4 | 0.02 | 3 | 1 30 | 88 |

These data show that low ozone concentrations for up to 3 hours, are sufficient to remove a large proportion of the biofilm. When the addition of ozone was stopped however, the remaining biofilm on the substrate rapidly regrew, suggesting the cells that were not removed were not damaged and because of the removal of the outer layers of biofilm, had full access to the available nutrients.

As with other biocides the effectiveness for a given concentration is very dependent on velocity, due to the effects of higher mass transfer effects at higher velocities. Kaur *et al* [1991] published data on the effect of velocity and are shown on Fig. 14.19.

The effect of ozone on planktonic bacteria was also studied by these authors. Fig. 14.20 gives some data and shows that if ozone residuals are maintained at 0.4 *mg/l* for 2 - 3 minutes, 100% kill of the cells is achieved.

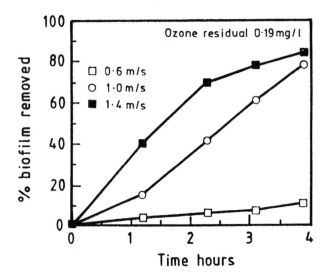

FIGURE 14.19. The effect of velocity on the biocidal activity of ozone at a concentration of 0.19 *mg/l*

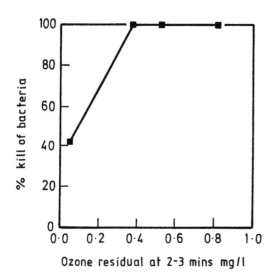

FIGURE 14.20. Viability of *Pseudomonas fluorescens* in suspension in the presence of ozone

A number of papers have been published on the merits and application of ozone in industrial cooling water systems. Claims made in a patent [Humphrey *et al* 1979] are that ozone in concentrations of 2 - 20 *mg/l* prevents scale formation and removes existing scale, passivates metal surfaces to inhibit further corrosion and disinfects the water to prevent algae growth. Edwards [1983, 1987] and Edwards and Sellers [1991] have published valuable reviews of ozone technology and cooling water. As well as biocidal activity the benefits of using ozone include:

1. Reduced blowdown reducing overall costs of make up water and effluent treatment or disposal charges.

2. Calcium carbonate scale forming part of the biofilm is released as the biofilm is destroyed.

3. Bacterial counts are lower in ozone treated water in comparison with multi-chemical treated systems.

4. Ozone is effective against *Legionella*.

5. The reaction of organic and some inorganic cooling water constituents with ozone modifying the fouling potential.

6. Because ozone is manufactured on site there are no storage or transport problems.

Rice and Wilkes [1991] have discussed the unique chemistry of ozone and bromide ion. Their observations are of interest where bromide ions exist in cooling water for instance, in sea water. Ozone reacts rapidly with bromide ion to produce hypobromite ion, i.e.

$$Br^- + O_3 \rightarrow OBr^- \qquad\qquad (14.32)$$

Hydrolysis of the $OBr^-$ ion follows to give hypobromous acid

$$OBr^- + H^+ \rightarrow HOBr \qquad\qquad (14.33)$$

Hypobromite ion can be further oxidised with additional ozone to bromate ion

$$OBr^- + O_3 \rightarrow BrO_3^- \qquad\qquad (14.34)$$

If ammonia is present it is possible for the hypobromous acid to react

$$HOBr + NH_3 \rightarrow NH_2Br + H_2O \qquad\qquad (14.35)$$

The monobromamine will be destroyed by reaction with the ozone present

$$NH_2Br + O_3 \rightarrow NO_3^- + Br^- + 2H^+ \qquad\qquad (14.36)$$

With organic compounds it is possible to produce brominated compounds such as bromoform.

In considering bromide ion chemistry in relation to ozone treatment of cooling waters it is apparent that if ozone is applied to water containing the bromide ion, hypobromous acid a biocide in its own right, is formed that does not decompose in the same way as ozone and is retained in the system. If the water contains ammonia in addition to bromide then the result of ozone addition will be to produce bromamine also a biocide, but less efficient than hypobromous acid [Mitchell 1985]. There have been suggestions that some of the biocidal activity attributed to ozone is in fact due to the bromine containing compounds resulting from the oxidising properties of ozone.

Since ozone is a strong oxidising agent it will be destructive to many organic materials of construction including packing materials and some polymers. Baldwin *et al* [1985] observed that nylon had only short term success against ozone. It is apparent that PVC is capable of withstanding the effects of ozone. There appears to be some controversy over corrosion of metals induced by the use of ozone. Bird [1988] concludes that the corrosion of mild steel in contact with ozonised cooling tower water is unacceptable. On the other hand Baldwin *et al* [1985] observe that their data suggest corrosion of steel by ozone is within acceptable industry standards.

One of the arguments against the use of ozone is the cost penalty and certainly there are indications that in comparison with chlorine, ozone is more expensive. Some cost data were published by Manzione [1991] in relation to a cooling water system with a comparison between chemical additive treatment (unspecified) and ozone. The figures are summarised in Table 14.15.

TABLE 14.15

Cost-benefit analysis of ozone treatment

| Operating expense | Current using chemical treatment $ | Ozone treatment $ |
|---|---|---|
| Chemicals | 6,166 | 0 |
| Blowdown water | 8,332 | 416 |
| Energy costs due to fouling | 18,537 | 10,500 |
| Labour | 4,658 | 0 |
| Water softening | 10,751 | 0 |
| Ozonisation lease and service cost | 0 | 17,952 |
| Total annual cost | 48,444 | 28,868 |

The annual saving based on this analysis would be $19,576. Furthermore because of the lower water make up demand using ozone treatment, there is a saving of $1.5 \times 10^7$ litres per annum.

It would appear that ozone has a significant potential as an additive to combat fouling, in particular as a biocide to reduce biofouling. More detailed work is required however to improve the technology and in particular, to optimise dosing requirements to improve cost effectiveness. Investigations of the use of ozone in relation to other additives, e.g. corrosion inhibitors are also necessary.

*Glutaraldehyde*

The properties of glutaraldehyde have been described [Eagar *et al* 1986, Grab *et al* 1990]. The molecule consists of a three carbon aliphatic chain with aldehyde groups at either end, i.e.

$$\underset{O}{\overset{H}{\diagdown}}C\text{-}CH_2\text{-}CH_2\text{-}CH_2\text{-}C\underset{O}{\overset{H}{\diagup}}$$

(14.37)

The aldehyde groups are chemically reactive and in biocide application reaction is with primary amine groups. It is generally applied in aqueous solution where glutaraldehyde exists in a complex equilibrium mixture of different hydrated forms.

$$
\begin{array}{ccccccc}
\text{O} & & \text{OH} & & \text{OH} & & \text{OH} \\
\| & & | & & | & & / \\
\text{CH} & & \text{CH-OH} & & \text{CH-OH} & & \text{CH}_2 \!-\! \text{CH} \\
| & +\text{H}_2\text{O} & | & +\text{H}_2\text{O} & | & -\text{H}_2\text{O} & \\
\text{CH}_2 & \underset{-\text{H}_2\text{O}}{\rightleftharpoons} & \text{CH}_2 & \underset{-\text{H}_2\text{O}}{\rightleftharpoons} & \text{CH}_2 & \underset{+\text{H}_2\text{O}}{\rightleftharpoons} & \text{CH}_2 \quad \text{O} \\
| & & | & & | & & \text{CH}_2\!-\!\text{CH} \\
\text{CH}_2 & & \text{CH}_2 & & \text{CH}_2 & & \quad\quad \text{OH} \\
| & & | & & | & & \\
\text{CH}_2 & & \text{CH}_2 & & \text{CH}_2 & & \\
| & & | & & | & & \\
\text{CH} & & \text{CH} & & \text{CH-OH} & & \\
\| & & \| & & | & & \\
\text{O} & & \text{O} & & \text{OH} & &
\end{array}
$$

(14.38)

(14.39)

The biocidal activity of glutaraldehyde is based on the ability of aldehydes to undergo alkylation reactions. The reactivity of glutaraldehyde with amino groups is primarily with ammonia and primary amines; reaction with secondary amines can also occur, but in general the reaction is slower. The bifunctional nature (the two aldehyde groups) means that each end of the molecule can react with different amino groups, so that the glutaraldehyde forms a bridge or cross link between the groups [Hughes and Thomas 1970]. The reaction involves two stages, i.e.

$$
\begin{array}{c}
\text{H} \\
\diagdown \\
\;\;\;\; \text{C-R} + \text{NH}_2 - \text{R}^1 \rightarrow \\
// \\
\text{O}
\end{array}
\quad
\begin{array}{c}
\text{H} \\
\diagdown \\
\;\;\;\; \text{C-R} \\
\diagup \;\;\; \diagdown \\
\text{HO} \;\;\;\; \text{NH-R}^1
\end{array}
$$

(14.40)

$$
\begin{array}{c}
\text{H} \\
\diagdown \\
\quad \text{C - R} \\
\diagup \diagdown \\
\text{HO} \quad \text{NH - R}^{1}
\end{array}
\longrightarrow
\begin{array}{c}
\text{R} \\
\diagdown \\
\quad \text{C} = \text{NR}^{1} + \text{H}_2\text{O} \\
\diagup \\
\text{H}
\end{array}
$$

$$(14.41)$$

All cells contain amino acids and thus have sites suitable for reaction with glutaraldehyde. The exact structure of cell walls and membranes will differ from one species to another and this may affect the vulnerability of the cell to the action of the biocide, but the susceptibility remains.

System variables that affect the efficacy of glutaraldehyde as a biocide include temperature, flow velocity, and glutaraldehyde concentration, but the most important factor is *pH*.

The effect of temperature is shown on Fig. 14.21 [Henry 1993]. Increasing the temperature from 20°C to 55°C reduces the time to eliminate the activity of *Bacillus subtilis* spores from 20 hours to something like 10 minutes for a given concentration of glutaraldehyde and cells.

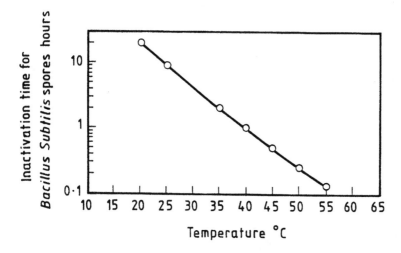

FIGURE 14.21. Time required to inactivate *Bacillus subtilis* spores with glutaraldehyde as a function of temperature

The effect of *pH* on the reduction of viable *Escherichia coli* using 44 *mg/l* of glutaraldehyde at a temperature of 20°C and different *pH* values is shown on Fig. 14.22. The data show that the kill rate is much faster at the higher levels of *pH*. An active cell concentration of 5 x 10⁷ is reduced to zero in 1 hour when the *pH* is 8.5 but it takes 22 hours at a *pH* of 5.0 to achieve the same result [Grab *et al* 1990].

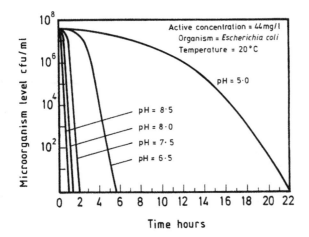

FIGURE 14.22. The effect of *pH* on rate of kill of *Escherichia coli* with glutaraldehyde

The effect of concentration on a mixed culture at different *pH* was demonstrated by Grab *et al*[1990] and is shown on Fig. 14.23. a combination of *pH* 8.1 and a glutaraldehyde concentration of 200 *mg/l* gave a remarkable rate of kill, i.e. reducing viable cell numbers from around 5 x 10⁵ to near zero in about 1 hour.

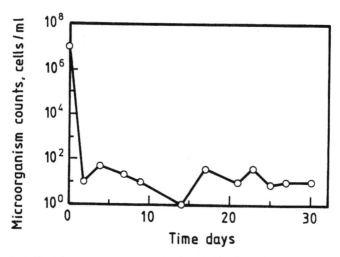

FIGURE 14.23. Activity of glutaraldehyde versus a mixed culture

A comparison of glutaraldehyde with other non-oxidising biocides has been made [Union Carbide 1991]. Each biocide was added to 10 *ml* of sterilised cooling water the *pH* adjusted to 8.5 and the solutions inoculated with a mixture of *P. aeruginosa*, *E. aerogenes*, *B. cereus* and *E. coli*. Aliquots were removed at the indicated times and cell counts made using standard conditions. The data reported in Table 14.16 demonstrate the effectiveness of the glutaraldehyde used biocide in relation to the other additives tested for the conditions of the test, but alternative conditions might be favourable to other biocides in the table.

TABLE 14.16

Biocide efficacy at *pH* 8.5 in cooling tower water

| Biocide | Biocide concentration *mg/l* product | Micro-organism population, *cfu/ml* after | | |
|---|---|---|---|---|
| | | *1 h* | *3 h* | *7 h* |
| AQUCAR* Water Treatment Microbiocide 550 glutaraldehyde based | 80 | $2.3 \times 10^4$ | $8.1 \times 10^3$ | $1.0 \times 10^3$ |
| | 160 | $3.1 \times 10^2$ | 0.0 | 0.0 |
| 2,2-Dibromo-3-nitrilopropionamide | 10 | $1.8 \times 10^7$ | $6.6 \times 10^6$ | $3.5 \times 10^5$ |
| | 25 | $6.5 \times 10^6$ | $7.0 \times 10^5$ | $5.0 \times 10^3$ |
| Methylene bis thiocyanate | 50 | $6.7 \times 10^6$ | $8.0 \times 10^6$ | $1.3 \times 10^6$ |
| Isothiazolinone | 100 | $7.3 \times 10^6$ | $5.8 \times 10^6$ | $1.1 \times 10^6$ |
| | 200 | $6.1 \times 10^6$ | $5.1 \times 10^6$ | $9.2 \times 10^5$ |
| Sodium dimethyl dithiocarbamate and Disodium ethylene bisdithiocarbamate | 500 | $7.8 \times 10^6$ | $8.0 \times 10^6$ | $1.5 \times 10^6$ |
| | 1000 | $6.5 \times 10^6$ | $7.0 \times 10^6$ | $3.0 \times 10^6$ |
| Alkyl dimethyl benzyl ammonium chloride | 50 | $1.3 \times 10^6$ | $2.8 \times 10^5$ | $2.6 \times 10^5$ |
| | 100 | $9.1 \times 10^3$ | $6.1 \times 10^3$ | $6.2 \times 10^3$ |
| Poly[oxyethylene-(dimethyliminio) ethylene-(dimethyliminio)ethylene] dichloride | 9 | $9.3 \times 10^6$ | $8.8 \times 10^6$ | $1.1 \times 10^6$ |
| | 18 | $2.1 \times 10^6$ | $9.1 \times 10^5$ | $3.2 \times 10^5$ |
| Control | 0 | $2.7 \times 10^7$ | $2.7 \times 10^7$ | $2.6 \times 10^7$ |

*AQUCAR is a registered trademark of Union Carbide Chemical and Plastics Technology Corp.

The effect of velocity on biofilms subject to glutaraldehyde treatment, will be similar to its effect with other biocides described in this chapter, i.e. the higher the fluid stress the higher the effectiveness of a given concentration of glutaraldehyde.

14.2.5.4 The future use of biocides

Concern for the environment and the need to introduce a harmonised scheme throughout the European Union has prompted the European Commission to issue a proposal [European Community 1993] concerning biocides and related additives. It is not just concerned with additives in cooling water but a wide range of substances (and organisms) that are used in such areas as preservatives, public hygiene, pest control and antifouling coatings. Biocidal products are defined as "active" substances and preparations containing one or more active substances, put up in the form in which they are supplied to the user, intended to destroy, deter, render harmless, prevent the action of, or otherwise exert a controlling effect on any harmful organism.

An excellent summary of the proposal and a brief review of its implications have been made by Foster and Wilson [1994]. It would seem that the proposal attempts to be all-embracing and general, so that the particular needs of one industry may not be fully considered. The draft proposal covers the following aspects:

1. Biocidal products to be authorised in each Member State before the product can be placed on the market or used in that country.

2. Authorisation of products to depend on inclusion of active substances in a list that will be agreed by the Community and which will form an annex to the directive and composed of representatives of Member States and the European Commission.

3. Applications for authorisation to be supported by data demonstrating that, subject to proper use, the product will not harm human or animal health or the environment.

4. Applicants to comply or justify non-compliance with extensive data requirements set out in the annexes to the directive; the requirements include toxicology, ecotoxicology and efficacy testing.

5. Mutual recognition of Member States' authorisations, subject to certain exemptions and safeguards.

6. Initially authorisations to apply to new products containing active substances not on the market when the directive comes into operation.

7. Existing active substances to be subject to a 10-year review programme; reviews to be shared between Member States and to lead to decisions on inclusion or otherwise in the list annexed to the directive.

8. Transitional arrangements to allow Member States to apply existing national controls to biocidal products containing existing active substances for up to 10 years till those substances are reviewed.

Foster and Wilson [1994] while welcoming the spirit under which the draft proposal has been formulated, raise concerns regarding scope and control of individual biocides, workability of the scheme, transitional arrangements and the order in which individual substances will be added to the register or otherwise. The main concern however, is the cost of implementing the proposal in full. The cost to manufacturers of biocidal products could be prohibitive. It may mean that development of new and perhaps less environmentally damaging products, is prevented because of the associated costs in demonstrating the acceptability of a new product. In cooling water treatment for instance, industry may be confined to existing products and associated costs, with no opportunity for improvement. The next 10 - 20 years may see the demise of the current open once through and recirculating cooling water systems in favour of closed circuits and a greater emphasis on air cooling. There could be implications for the efficiency of cooling and hence on product cost not least the cost of electricity generation, and a consequent impact on global warming from the need for higher fossil fuel consumption for a given output

## 14.3   GAS SYSTEMS

The development of additives for alleviating gas phase fouling problems is far less advanced than for liquid systems and is generally confined to treating flue gases during combustion. The discussion therefore, will deal principally with additives for combustion operations.

The fouling associated with flue gases is extremely complex involving particulate deposition, chemical reactions, vapour condensation and corrosion, as a result the approach to combating fouling is empirical and very much a trial and error procedure. More detail on the origins of deposits in flue gases may be found in Chapter 16.

Fireside additives have been developed primarily to combat fouling in fossil fuel combustion, but more recently, the technology has been transferred to problems associated with other fuels, e.g. industrial domestic and agricultural waste. Additives have also been used to overcome corrosion and acidic emissions, but a discussion of these problems is outside the scope of this book except where they may affect heat transfer.

The use of additives was initially confined to the combustion of fuel oils, presumably because the system lent itself to the injection of suitable chemicals and compounds. In more recent times the use of additives has been applied to coal combustion and at present treatment of flue gases derived from waste combustion is being developed. For instance fluidised bed combustors for waste incineration are very suitable for fouling mitigation using additives. The mixing effect of the fluid bed ensures rapid dispersion of the additive. Chemical addition to natural gas combustion is not generally necessary since deposit formation is not a problem. One of the principal difficulties in discussing the use of additives in connection with the combustion of fuel, is the wide variability of the fuels in use in terms of composition. The spectrum of impurities in a given fuel will have a marked effect on the deposition characteristics displayed by the fuel during combustion, as discussed in Chapter 16. Furthermore the fouling problem will be significantly influenced by the combustion conditions that include temperature and excess air.

Additives may be solid or solutions, or suspensions in a suitable solvent, or liquid. Where solids are employed their particle size is usually of the order of 20 - 30 $\mu m$ with direct injection into the combustion space in the vicinity of the burner where pulverised coal or liquid fuels are employed. For solid fuel combustion on a grate, the additive may be mixed with the fuel prior to combustion or injected in some way to the flame region. Mixing with the fuel may present difficulties of producing a uniform distribution, particularly for solid fuels. The additive is usually separately applied to avoid these problems.

The location of the injection point (or points) will be important as far as the effectiveness of the additive is concerned, incorrect application may even produce negative results. Furthermore effectiveness will also depend on good mixing with the combustion gases possibly in the flame region. The availability of the additive as a powder or liquid solution or suspension, will also have an effect on additive efficiency in reducing fouling and slagging problems. Liquids are likely to be easier to handle than pneumatically transported particles and it may be necessary to have a close particle size range for powders to ensure satisfactory injection.

A wide range of additive concentration is to be found for different industrial applications; much depends on the quality of the fuel, but concentrations in excess of 50 $mg/kg$ of fuel could be considered to be uneconomic.

Some additives and blends of additives are protected by patents. In the use of proprietary formulations the technical assistance of the supplier will almost certainly be necessary to ensure satisfactory results.

In general the principal requirement of the additive is that it will modify the deposit that accumulates on heat transfer surfaces and other parts of the combustor, so that they are more easily removed by the cleaning techniques described in Chapter 15. At the same time some additives react with the acid gases such as $SO_2$ produced in the burning process, thereby reducing the potential for corrosion. Radway [1985] produced a very comprehensive abstract and index of published work related to corrosion and deposits from combustion gases. The volume contains a large number of references to the use of additives. The use of additives to facilitate on-load cleaning of utility boilers has been discussed by Hazard *et al* [1983]. Radway and Hoffman [1987] have produced a definitive assessment of additives used in combustion operations. The following discussion is largely based on their work.

*Magnesium oxide*

Magnesium oxide may be applied as a powder or as a suspension in oil and it has found use in a wide range of coal and oil fired equipment [Locklin *et al* 1980]. The chemical is an alkaline hygroscopic powder and it is usually one of the ingredients of many proprietary blends of combustion additives. It would appear that the origin of the magnesium oxide and the preparation technique influence its performance. Conditions within the combustion space will also affect its activity. The specific surface area of the additive will also represent a factor in its efficiency.

An alternative to the use of magnesium oxide as the additive is to inject magnesium hydroxide into the high temperature zone to provide a highly reactive form of magnesium oxide, i.e.

$$Mg(OH)_2 \rightarrow MgO + H_2O \qquad (14.42)$$

The technique of $Mg(OH)_2$ injection is favoured in the UK and Europe.

Dolomite that may be considered to be $CaMg(CO_3)_2$, contains a relatively high proportion of $MgO$, but its effectiveness appears to be lower than the purer forms of the compound.

Magnesium salts have been used to combat corrosion resulting from the presence of vanadium in oil fired equipment. High melting point magnesium sulphates and vanadates are produced [Matveevicheva 1969].

*Limestone*

Limestone is a widely distributed mineral containing a large percentage of calcium carbonate. The quantity and quality of impurities contained in limestone will depend very much on location. Most powdered limestone additives contain magnesium carbonate.

Under the high temperature conditions prevailing during combustion, the limestone is converted to the oxide ($CaO$) with a very small particle size. In the combustion of fuel oil where sulphur may be present the reaction of $CaO$ with $SO_2$ may produce $CaSO_4$ that may cause deposition problems in the convection sections in a steam raising facility.

$$2CaO + 2SO_2 + O_2 \rightarrow 2CaSO_4 \qquad (14.43)$$

For this reason limestone is not often used in conjunction with oil combustion [Radway and Hoffman 1987]. The presence of high ash in coals however, tends to modify the sulphate deposition so that it is possible to use limestone as a deposit modifier in the combustion of certain coals.

Addition of limestone has been used to reduce slag viscosity in wet bottom coal fired plant and pulverised coal combustion, to facilitate slag and deposit removal.

*Silica*

The mineral silica is available from a number of sources ranging from natural deposits to byproducts from chemical processing. It is generally available as a finely divided powder which facilitates its use as an additive. It has been employed to reduce deposition and corrosion problems associated with vanadium in fuel oil. In some coal fired equipment silica has been added with copper oxychloride to control slagging [Radway and Hoffman 1987].

*Vermiculite*

Vermiculite consists of a hydrated magnesium aluminium silicate that under the application of heat (in the temperature range 450 - 1200°C), can expand up to 15 times its original volume by the loss of the water of hydration as steam. Rapid incorporation of the aditive into a deposit weakens the deposit by the expansion of the vermiculite thereby facilitating deposit removal by soot blowing (see Chapters 15 and 16).

*Borax*

Borax containing minerals have been used to lower slag viscosity in wet bottom and cyclone combustors to aid removal. Compounds of borax have also been used to facilitate ash disposal from boiler equipment.

*Aluminium hydroxide and oxide*

Slag modification can be achieved by the use of aluminium hydroxide and aluminium oxide that may be called "activated alumina". Radway and Hoffman [1987] conclude that aluminium trihydrate $(Al_2(OH)_6)$ is to be preferred to aluminium oxide because it is "softer" and less abrasive causing less wear on burner tips. Aluminium compounds have been used in conjunction with copper oxychloride to modify slag in coal firing operations.

*Manganese additives*

Manganese compounds have found use not only to improve combustion but also to reduce problems associated with slagging and fouling. The presence of manganese reduces the ignition temperature of the fuel by catalytic action. A range of manganese chemicals is available including chloride, sulphate, napthenate and sulphonate. Application may be in the form of powder, dispersions in water or oil, water solutions of inorganic compounds or oil solutions of the organo manganese compounds.

Manganese compounds have been used in coal as well as oil firing equipment to alleviate slagging and fouling problems. Low temperature corrosion has also been reduced by the use of these chemicals. The relatively high cost of manganese compounds may reduce their universal applicability.

## Rare earth compounds

Reductions in corrosion and deposit formation in oil fired equipment have been achieved with these chemicals in addition to improved combustion. The availability includes powders, water solutions and organo-metallic compounds.

## Copper oxychloride

James and Fisher [1967] reported the use of copper oxychloride (3 $CuO.CuCl_2$ $4H_2O$) as a coal additive to reduce slagging in pulverised coal boilers. Its use in a UK power station resulted in a dramatic reduction in slagging by the production of a soft friable deposit which was easily removed from the furnace walls.

An extensive and well controlled full scale trial, was carried out on a power station in Holland by Sanyal [1990] using copper oxychloride. A dose rate of about 2 $ml$ (as an oil suspension) showed a marked influence on slagging propensity. The equipment effectiveness was equivalent to that obtained immediately after cleaning. Sanyal was not able to determine an optimum dosing level in his research.

Sanyal and Coleman [1982] compiled data on successful applications of copper oxychloride and their observations are recorded in Table 14.17.

Investigations into the action of copper oxychloride to reduce deposition problems were made by Livingstone et al [1983]. They suggest that under the high temperature conditions, the additive volatilises to form $CuCl$ and $Cu_2O/Cu$ in the gas phase. These products of decomposition then condense on the surface of the particles (fly ash) moving forward with the flue gas and on the surfaces of the large open pores in the sintered region of the slag deposits. The sintering process associated with deposit formation, is arrested by the crystallisation of an iron rich phase at temperatures in the range 1200 - 1400°C. The result is a lower rate of slag formation, and the production of a solid deposit with an open structure and lower mechanical strength.

The literature reveals that additives can be effective in reducing problems of fouling and slagging on heat transfer surfaces located in flue gas streams. A difficulty in their use arises from the fact that it is not possible to carry out effective laboratory tests to determine effectiveness for particular fuels in different designs of combustor and under differing operating conditions. The equipment operator is left with no other choice than to carry out extensive plant trials. There is no guarantee of success and there have been occasions when the additive chosen for trial has in fact, made the fouling problem even worse. The best option would be a trial on a pilot scale combustor but this kind of equipment is not readily available.

TABLE 14.17

Applications of copper oxychloride to reduce fouling and slagging in power stations

| Power station | Nature of slagging difficulties | Effects of the additive |
|---|---|---|
| Cottam (UK) | Platen superheaters slagging | No details but a reduction in heavy ash build-up |
| Ironbridge (UK) | Heavy slag build-up on hopper slopes | Assisted in coping with Silverdale coal. Reduced ash crusher maintenance |
| Champagne-sur-Oise (France) | Furnace slagging | Increased friability of bottom ash. Reduced need to drop load to shed slag. Reduced at temp. spray rates |
| Vitry (France) | Screen tube deposits | Reduced deposits. Reduction in soot blowing requirements |
| Les Aserevilles (France) | Not reported | Increased friability of bottom ash. Reduced hopper slope bridging |
| Bouchain (France) | Not reported | 'Probable' reduction in furnace deposits |
| Pont-sur-Sambre (France) | Furnace slagging | Slag deposits more friable |
| Hanasaari 'B' (Finland) | Furnace slagging | Marked improvement in ease of controlling slag formation. Slag deposits more readily removed by sootblowing |
| PLEM Buggenum (Holland) | Severe furnace slagging and bottom ash | Reduced deposition. Marked improvement in ease of bottom ash removal |

The use of additives for deposit control in flue gas systems has a potentially excellent future but a great deal of fundamental research is required to provide a better understanding of the chemical and physical mechanisms involved. The problem is complicated by the complex composition of most fuels.

REFERENCES

Atkinson, A., Gerrard, A.F. and Harris, A., 1979, Development of additives to control inorganic fouling in industrial water systems, in: Fouling - Science or Art? Inst. Corr. Sci. and Tech., Inst. Chem. Engrs., University of Surrey.

Bahadur, A., 1993, Chromate substitutes for corrosion inhibitors in cooling water systems. Corrosion Reviews, 11, 105 - 122.

Baldwin, L.V., Feeney, E.S. and Blackwelder, R., 1985, The investigation and application of ozone in cooling water treatment. Proc. 46th Annual Meeting. Int. Water Conf. Pittsburg.

Bell, J., 1988, Ozone product technology. Proc. Conf. Ozone in Water Quality Management. Int. Ozone Assn. London, 187 - 204.

Betz, 1976, Handbook of industrial water conditioning. Betz Laboratories Inc., Trevose.

Bird, T.L., 1988, Corrosion of mild steel in ozonised air conditioning cooling tower water. Waterline, Ind. Water Soc. September, 39.

Blenkinsopp,S.A., Khoury, A.E. and Costerton, J.W., 1992. Electrical enhancement of biocide efficacy against *Pseudomonas aeruginosa*. App. and Environ. Microbio. 58, 11, 3770 - 3773.

Bott, T.R., 1990, A comparison of chlorine and ozone for cooling water treatment. Waterline, Ind. Water Soc. June, 37 - 46.

Bott, T.R., 1992, The use of biocides in industry, in: Melo, L.F. *et al* eds. Biofilms - Science and Technology, Kluwer Academic Publishers, Dordrecht.

Bott, T.R. and Kaur, K., 1994, Water treatment using ozone. Eurotherm Seminar 40, Aristotle University of Thessaloniki.

Crawford, J.D. and Miller, R.M., 1963, Refinery fouling effect of trace metals and remedial techniques. Proc. 28th Mid-year Meeting Am. Pet. Inst. Division of Refining, Philadelphia.

Crittenden, B.D., 1984, Corrosion fouling, in: Fouling and Heat Exchanger Efficiency, Continuing Education Course. Inst. Chem. Engrs., University of Leeds.

Donoghue, J.P., 1993, Ozonation of cooling tower water: a utility's perspective. Commercial and Industrial Marketing Workshop, San Francisco.

Eagar, R.G., Leder, J. and Theis, A.B., 1986, Glutaraldehyde: Factors important for microbial efficacy. Proc. 3rd Conf. on Progress in Chemical Disinfection. Binghamton, NY.

Edwards, H.B. and Sellers, R.L., 1991, Industrial cooling water treatment with ozone. Conf. Proc. Ozone in the Americas. Additional symposium on ozone treatment of cooling waters. Int. Ozone Assn. Toronto, 433 - 468.

Edwards, H.B., 1983, The economical advantage of ozonating the cooling tower water system. Proc. 6th Ozone World Congress. Int. Ozone Assn. Washington.

Edwards, H.B., 1987, Ozone - an alternative method of treating cooling tower water. Proc. Cooling Tower Institute Annual Meeting, New Orleans.

EPRI, 1984, Dechlorination technology manual, EPRI Research Report CS3748, Palo Alto

EPRI, 1985, Condenser-targeted chlorination design, EPRI Research Report CS5279, Palo Alto.

Epstein, N., 1983, Fouling in heat exchangers, in: Taborek, J. and Hewitt, G.F., eds. Heat Exchanger Theory and Practice.

ESDU, 1988, fouling in cooling water systems. Item No. 88024, ESDU Intl.

Grab, L.A., Emerich, D.E., Baron, S.J. and Smolik, N.A., 1990, Glutaraldehyde: a new slimicide for papermarking. Proc. Papermakers Conf. TAPPI, Atlanta.

Haluska, J.L. and Shaw, D.L., 1975, Antifoulant test shows fuel savings. Oil and Gas Journal, September, 93 - 100.

Harris, A. and Marshall, A., 1981, The evaluation of scale control additives, in: Progress in the prevention of fouling in industrial plant, Syposium University of Nottingham.

Hawthorn, D., 1992, The scaling index is dead, long live the scaling equation, in: Bohnet, M., Bott, T.R., Karabelas, A.J., Pilavachi, P.A., Séméria, R. and Vidil, R. eds. Fouling Mechanisms Theoretical and Practical Aspects. Editions Europeénes Thermique et Industrie, Paris.

Hazard, H.R., Krause, H.H. and Sekercioglu, I., 1983, Use of additives to facilitate on load cleaning of utility boilers. CS3270 Research Project 1839 - 2. EPRI, Palo Alto.

Henry, B.D., 1993, Union Carbide Europe SA, private presentation.

Hollerbach, G.H. and Krauss, J.C., 1987, Using carbon dioxide to treat water and process fluids. BOC Group Technology Magazine, 6, 26 - 32.

Hughes, R.C. and Thomas, P.F., 1970, Cross linking bacterial cell walls with glutaraldehyde. Biochem. J. 119, 925.

Hugo, W.B., 1987, in: Hugo, W.B. and Russel, A.D. eds. Pharmaceutical Microbiology, 4th edition. Blackwell Scientific Publications, 282.

Humphrey, M.F., French, K.R. and Howe, R.D., 1979, Ozonation of cooling tower waters. US Patent 4, 172, 786.

Jamialahmadi, M. and Müller-Steinhagen, 1991, Reduction of calcium sulphate scale formation during nucleate boiling by addition of EDTA. Heat Trans. Eng. 2, 2, 19 - 26.

Jamilahmadi, M. and Müller-Steinhagen, H., 1993, Scale formation during nucleate boiling - a review. Cor. Review, 11, 1 - 2, 25 - 54.

Kaur, K., Leadbeater, B.S.C. and Bott, T.R., 1988, Ozone as a biocide in industrial cooling water treatment. Proc. Int. Conf. Fouling in Process Plant, University of Oxford.

Kaur, K., Leadbeater, B.S.C. and Bott, T.R., 1991, Effects of ozone as a biocide in an experimental cooling water system. Ozone Sci. and Eng. 14, 517 - 530.

Kemmer, F.N., 1988, Editor, The Nalco Water Handbook. McGraw Hill Book Co. New York.

Khalanski, M. and Bordet, F., 1980, Effects of chlorination on marine mussels. Water Chlorination, Environmental Impact and Health Effects, 3, Ann Arbor Science, 557 - 568.

Knudsen, J.G., Tijitrasmoro, E. and Dun-gi X, 1984, The effect of zinc chromate inhibitors on the fouling characteristics of cooling tower water, in: Suitor, J.W. and Pritchard, A.M. eds. Fouling in Heat Exchange Equipment. HTD vol. 35, Am. Soc. Mech. Engrs. 11 - 18.

Langelier, W.F., 1936, The analytical control of anticorrosion water treatment. J. Am. Water Works Assn. 28, 1500 - 1521.

Langelier, W.F., 1946, Chemical equilibrium in water treatment. J. Am. Water Works Assn. 38, 169 - 178.

Lewis, B.G., 1985, Mussel control and chlorination. CEGB Report No. TRPD/L/2810 R85.

Livingston, W.R., Sanyal, A. and Williamson, J., 1983, The role of copper oxychloride in reducing slagging in pulverised fuel fired boilers. Proc. Conf. The Effectiveness of Fuel Additives. Inst. Energy, Leatherhead.

Locklin, D.W. *et al*, 1980, Electric utility use of fireside additives, CS1318 Research Project 1035 - 1, EPRI, Palo Alto.

Manzione, M., 1991, Water conservation by ozonation of cooling tower waters - practical experiences in Northern California. Conf. Proc. Ozone in the Americas. Additional symposium on ozone treatment of cooling waters. Int. Ozone Assn. Toronto, 374 - 390.

Matveevicheva, V., 1969, X-ray diffraction and infra red spectroscopic study of the phase composition of the oxide system $V_sO_5$ - $MgO$. Rus. J. Phys. Chem. 43, 143.

Mayo, F.R., Stavinoha, L.L. and Lee, G.H., 1988, Source of jet fuel thermal oxidation tester deposits from oxidised JP8 fuel. Ind. Eng. Chem. Res. 27, 362 - 363.

McCoy, J.W., 1980, Microbiology of cooling water. Chemical Publishing Company, New York.

Mitchell, P.K., 1985, Bromination: two methods available for sanitizing. Pool and Spa News, 15th April 128 - 129.

Müller-Steinhagen, H.M. and Branch, C.A., 1988, Comparison of indices for scaling and corrosion in heat exchangers. Can. J. Chem. Eng. 66, 1005 - 1007.

Nesaratnam, R.N. and Bott, T.R., 1984, Effects of velocity and sodium hypochlorite derived chlorine concentration on biofilm removal from aluminium tubes. Proc. Biochem. 19, 14 - 18.

Neufeld, A. *et al*, 1987, Industrial Corr. 5, 6.

Nichols, W.W., 1994, Biofilm permeability to antibacterial agents, in: Wimpenny, J., Nichols, W., Stickler, D. and Lappin-Scott, H. eds. Bacterial Biofilms and their Control in Medicine and Industry. Bioline, School of Pure and Applied Biology, University of Wales College of Cardiff, 141 - 149.

Parkiss, B.E., 1977, Biotechnology of Industrial Water Conservation. M & B Monographs, Mills and Boon Ltd., London.

Radway, J.E., 1985, Corrosion and deposits from combustion gases - Abstracts and Index. Hemisphere Pub. Corp. Washington.

Radway, J.E. and Hoffman, M.S., 1987, CS5527 Research Project 1839 - 3, EPRI, Palo Alto.

Rice, R.G. and Wilkes, J.F., 1991, Fundamental aspects of ozone chemistry in recirculating cooling water systems. Corrosion '91 Paper 205, Nat. Assn. Cor. Engnrs. Houston.

Rice, R.G. and Gomez-Taylor, M., 1986, Occurrence of byproducts of strong oxidants reacting with drinking water contaminants - scope and problem. Environ. Health Perspectives, 69, 31 - 44.

Ryznar, J.W., 1944, A new index for determining the amount of calcium carbonate formed by water. J. Am. Water Works Assn. 36, 472 - 486.

Sanyal, A., 1990, Copper oxychloride as a fireside additive in coal fired utility boilers, CS6751 Research Project 1839 - 4, EPRI, Palo Alto.

Sanyal, A. and Coleman, C.R., 1982, The control of slagging in coal fired utility boiler furnaces: a case history of the use of copper oxychloride as a fireside additive. ASME Paper 82-JPGC-Ful.

Stephenson, W.K. and Rowe, C.T., 1993, Fouling control in ethylene recovery systems. AIChemE. Symp. Series 89, 335 - 340.

Sugam, R., Guerra, C.R., Del Monaco, J.L., Singletary, J.H. and Sandvik, W.A., 1980, Biofouling Control with Ozone at the Bergen Generating Station. Report CS1629, EPRI, Palo Alto.

Union Carbide Chemical and Plastics Co. Inc., 1991, AQUCAR Water Treatment Microbiocide 550: Comparative biocide efficacy in *pH* 8.5 cooling tower water. Union Carbide Chemical and Plastics Co. Inc. Speciality Chemicals Division, Danbury CT.

Veale, M.A., 1984, Control of deposits in cooling water systems, in: Fouling and Heat Exchanger Efficiency. Continuing Education Course. Instn. Chem. Engrs., University of Leeds.

Watkinson, A.P., 1988, Critical review of organic fluid fouling. Final Report AWL/CNSV-TM-208 Argonne National Laboratory, Illinois.

Weinland, B.W., Miller, R.M. and Freedman, A.J., 1967, Reduce refinery fouling. Materials Protection, 6, 2, 41 - 43.

Whitehouse, J.W., Khalanski, M., Saroglia, M.G. and Jenner, H.A., 1985, The Control of Biofouling in Marine and Esturine Power Stations. CEGB Northwestern Region.

# CHAPTER 15

## Heat Exchanger Cleaning

### 15.1   INTRODUCTION

In order to maintain or restore efficiency it is often necessary to clean heat exchangers. Methods of cleaning may be classified into two groups; on-line and off-line cleaning.

The use of additives discussed in Chapter 14, is a technique for reducing or eliminating the deposition of fouling material on heat exchanger surfaces during operation and may be regarded in broad terms, as on-line cleaning using chemicals. Other on-line techniques are used that depend on physical and mechanical mechanisms and do not involve the application of chemicals. On-line cleaning may be on a continuous basis or intermittent. In general these techniques involve additional equipment included at the design stage provided the potential fouling problem is recognised at the time the design is made. If the fouling problem only becomes apparent after a period of operation it is possible to retrofit on-line cleaning equipment. It is usually more expensive however to make these modifications than to have included the equipment in the initial design.

An alternative to on-line cleaning is to stop operations and to clean the heat exchanger either chemically or by mechanical means. It may be possible to restore performance without the need to dismantle the heat exchanger, but usually it will be necessary to have access to the inside surfaces. Where frequent off-line cleaning is anticipated it would be prudent to consider the installation of a "standby" heat exchanger to be brought on-stream when the on-line heat exchanger performance becomes unacceptably low, thereby providing the opportunity to clean the fouled heat exchanger while at the same time, maintaining production. Cleaning was reviewed by Bott [1991].

### 15.2  ON-LINE CLEANING

A number of techniques are available for reducing the effects of potential fouling without the need to take the heat exchanger off-stream. Some of the methods lend themselves to certain operations or heat exchanger designs, while others are more universal in application.

#### 15.2.1  Circulation of sponge rubber balls

The use of recirculating sponge rubber balls (for instance the Taprogge system) has found wide application in the power industry for the maintenance of condenser efficiency as discussed further in Chapter 16. It has also been used with some success to reduce scale formation in multi-stage flash desalination plant. The technique is capable of preventing the accumulation of particulate matter, biofilm formation, scale and corrosion product deposition. It is only applicable to flow through the inside of tubes.

According to Aschoff *et al* [1987] the first sponge rubber ball cleaning systems were introduced to the European market in 1957. In the mid-1970s the Taprogge patent expired opening the way for other suppliers to enter the market. The dominant supplier however, remains Taprogge, due to wide experience in the application of the technology.

The technique involves slightly oversized sponge rubber balls continuously passed through the tubes of a heat exchanger by the water flow. A schematic diagram of the system is shown on Fig. 15.1. The oversize ensures that the ball is "squeezed" as it passes through the tube so that the ball is pressed against the tube wall thereby wiping the surface. The balls are extracted from the outlet flow from the exchanger by a suitable strainer, drawn off and pumped back to the inlet section of the heat exchanger to be recycled through the equipment. In general the balls are soft and resilient so as to avoid excessive wear on the heat exchanger surfaces. Balls with an abrasive surface maybe used for tenacious deposits or for preliminary cleaning following retrofitting of the technology, but their long term use should be avoided to prevent excessive tube wear. Limited use of these balls however, does have the advantage that the surface becomes "polished" with attendant reduction for the deposit adhesion (see Chapter 6). The opportunity for abrasion is exemplified by the relative hardness of the carborundum coating used on the balls as demonstrated in Table 15.1 (based on Eimer 1985). Wear may be reduced by the use of plastic granulate coated balls.

TABLE 15.1

Hardness of tube material and abrasive sponge rubber balls

| Material | Vickers hardness |
|---|---|
| Carborundum coated ball | 3400 |
| Plastic granulate coated ball | 95 |
| Titanium | 145 |
| Typical stainless steel | 142 |
| Brass | 84 |
| Cupro nickel (90/10) | 75 |
| Cupro nickel (70/30) | 95 |

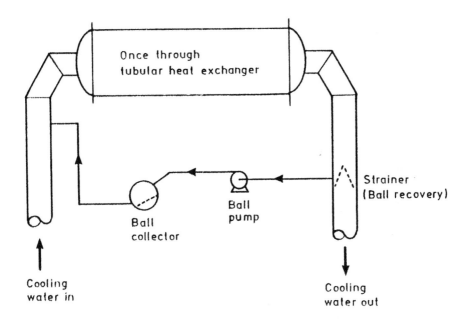

FIGURE 15.1. The Taprogge system for the circulation of sponge rubber balls

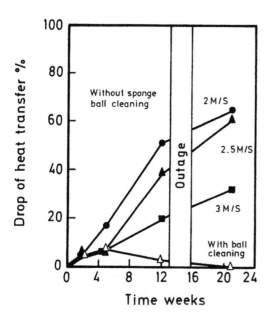

FIGURE 15.2. The effectiveness of the Taprogge sponge rubber ball circulation system

It is apparent that the technique is extremely successful [Aschoff 1987 and Bott 1990] for cooling water applications. The effectiveness is demonstrated by data prepared by Eimer [1985]. Fig. 15.2 shows how the presence of the circulating balls maintained heat transfer coefficients in a condenser associated with a nuclear power station, in comparison to the situation without the ball circulation system in place.

The passage of the balls through the heat exchanger is a random process, but provided there are sufficient balls in the system then the frequency of balls passing through each tube will be sufficient to maintain clean surfaces. Eimer [1985] demonstrates how the ball frequency $f_k$, (balls/$h$) may be determined. If the ball charge is 10% of the number of parallel tubes in the exchanger then

$$f_k = \frac{0.1 \times 3600}{t_u} \text{ balls} / h \tag{15.1}$$

where $t_u$ is the average ball circulation time in seconds.

If $t_u = 30s$ then

$$f_k = \frac{0.1 \times 3600}{30} = 12 \text{ balls} / h \tag{15.2}$$

i.e. each tube "sees" a ball 12 times per hour. According to Eimer [1985] this is a typical frequency.

These calculations assume that the tube plate is clear of obstruction. If there are any deposits or accumulation of debris within the header box, the effectiveness of the cleaning system may be impaired. Under these circumstances some tubes may not receive balls so that they become very susceptible to fouling. Good pre-screening is required to ensure that maldistribution is not a problem. Problems may sometimes occur due to the retention of a ball within a tube.

The cleaning system need not be operated continuously but based on intermittent application. For these conditions the cleaning period, i.e. the time the balls are in circulation and the interval from the beginning of a cleaning cycle till the beginning of the next cycle, must be taken into account. An example is provided by Eimer [1985]. If the system is operated for 1 hour daily with a ball frequency of 12 balls/$h$ and the cleaning cycle is 24 hours, the same cleaning effect could be obtained by taking half the ball charge, i.e. reducing the ball frequency by 50% to 6 balls per hour and increasing the cleaning period to 2 hours.

Eimer [1985] concludes that in the extreme case, i.e. with continuous operation but with $\frac{1}{24}$ of the normal ball charge would result in a ball frequency of 0.5 balls/$h$ and would be considerable deviation from the typical value of 12 balls/$h$ given by Equation 15.2 and is not recommended.

As the example shows it is necessary to optimise the mode of operation to maximise the beneficial effects. The following need to be taken into consideration:

Water quality
Material of construction
Protective film formation, corrosion (see below)
Fouling mechanisms and extent of likely fouling
The need for pre-screening to avoid clogging of the tube sheets

Computer programmes are used by the system suppliers (e.g. Taprogge) to optimise ball type, hardness, diameter, frequency and cleaning period for given conditions.

Corrosion resistance is often achieved by the presence of a film or layer on the surface to be protected (see Chapters 10 and 14). It is imperative therefore that any on-line cleaning technique does not destroy the ability of the film to protect the surface.

Continuous ferrous sulphate addition has been used for many years to reduce water side corrosive attack of steam condenser tubes. It is usually applied to once through cooling water systems because of its low cost, to provide an iron-rich protective film on the tube surface. For recirculation systems other more expensive, corrosion inhibitors are generally employed. Two phases of the ferrous sulphate treatment programme may be recognised. The first phase involves the initial laying down of the protective film. The second phase involves the maintenance of the film, which would be otherwise destroyed by the shear effects of flow.

In general the thicker the film the greater the resistance to corrosion, but the protective film also represents a heat transfer fouling resistance. The thickness of this layer therefore must be such that it is adequate for protection without the accompaniment of excess reductions in heat transfer. Optimum plant operation therefore depends on obtaining a compromise between these conflicting factors. The circulation of sponge rubber balls may assist in this optimisation process. The sponge rubber ball removes the bulk of the voluminous soft deposit produced by the treatment, but at the same time, leaving on them a lacquer-like iron-rich film on the surface. The result can be good corrosion resistance with minimal reduction in heat transfer efficiency, but the choice of operating conditions needs careful consideration.

A comparison [Aschoff 1985] shows that in general the sponge rubber ball cleaning system appears more expensive to operate than a chlorination addition for the equivalent plant to combat biofouling. These authors also show however, that where dechlorination is required (see Chapter 14) to meet discharge requirements the overall cost will be higher than the mechanical cleaning system. The use of the sponge rubber ball system generally gives stable operating conditions, whereas the use of chlorine may give rise to erratic and sometimes an unreliable response, to the chlorine dosing regime depending on the changeable potential for biofouling of the recirculating cooling water [Bott 1990]. A more extensive comparison is made in Chapter 16.

The cost of installing a circulating ball system may be quite high but short payback times of less than one year have been reported. It would appear that a typical payback time in 1985 would have been 3 - 5 years based on fossil fuel plants with moderate fouling problems [Aschoff *et al* 1985]. It is unlikely that there has been a substantial change since that time.

It has to be remembered that the condenser represents only part, albeit a large part, of the equipment associated with a cooling water system and, although the use of sponge rubber balls may eliminate the need for treatment for the condenser, additives may be needed for the remaining equipment. For instance it may be possible to dispense with chlorination in respect of the condenser, but the prevention of biofouling in pipe lines, cooling tower and basin will still be necessary. Furthermore scale inhibitors may be required even in the condenser fitted with a circulating sponge rubber ball system. Much depends on the water quality available for cooling and the current fuel cost.

The sponge rubber ball system is only suitable for cleaning the inside of tubes and therefore its use is restricted to tubeside single pass or possibly U tube designs for two passes in shell and tube heat exchangers. The system cannot be used for the shell side or for plate heat exchangers and other non tubular designs.

15.2.2 Brush and cage systems

The brush and cage system of on-line cleaning of heat exchangers is similar in concept to the sponge rubber ball technology for the prevention of deposit accumulation on the inside of tubes. It carries the same restrictions on application as the sponge rubber ball system, i.e. it cannot be used for anything other than straight tubes.

The principle of the brush and cage system is that a brush fabricated from suitable metal wires or polymer filaments, is passed through the tubes by the liquid flow. At either end of each exchanger tube there is a cage located so as to arrest the brush projectile and prevent its passing into the header box. After a given period of time, determined by the mechanism and rate of the anticipated fouling, the flow is reversed and the brush returns along the tube to the other end to be caught in the cage located at that end. The shuttling of the brushes through the tubes maintains the cleanliness of the tubes. Flow reversal is achieved during operation usually automatically, on a predetermined time cycle. Some power plant cooling water systems have flow reversal facilities for back washing and this could be used to execute the oscillating flow.

One of the advantages claimed for the technology in comparison with the sponge rubber ball system, is that it requires less maintenance. A major disadvantage perhaps in some situations more than others, is the momentary interruption in flow while the flow reversal is accomplished. The disruption of steady state conditions in the plant operation may be difficult to accommodate and could have implications for product quality. The cost of the complex piping arrangements and the control system could be expensive, particularly in a retrofit application. For reasons of cost and integration into the plant as a whole installation is best for new plant. The technique is not generally used for power station condensers and is more useful for use with individual heat exchangers in process plant. Careful design and operation is required to avoid problems of wear to heat exchanger tubes.

### 15.2.3 Air or gas injection

Air injection, sometimes called "rumbling" may be used to reduce deposit formation on heat transfer surfaces. The technique involves the use of slugs of air in liquid flows (i.e. a two phase mixture) to create regions of high turbulence near the wall as the slugs of air pass through the equipment. The method is usually applied to areas where accessibility is difficult, e.g. the shell side of shell and tube heat exchangers. The method is generally not considered to be elegant and may be quite ineffective for tenacious deposits that require much larger forces for removal, than can be provided by the passage of slugs. Care has to be exercised where flammable volatile liquids are being processed to avoid the possibility of forming explosive mixtures. Under these circumstances it might be necessary to consider the use of an inert gas such as nitrogen, but the cost may be high and prohibitive.

### 15.2.4 Magnetic devices

The use of magnetic fields to reduce or eliminate scale formation in pipes has been attempted for many years. It could be supposed that slightly soluble compounds such as $CaCO_3$ existing in solution as charged ions, would be affected by the application of an electric field, and that this could form the basis of a technique to alleviate fouling. As long ago as 1958 one manufacturer of magnetic devices claimed to have sold several hundred thousand units [Eliassen *et al* 1958], and similar devices are constantly advertised for domestic use in locations where hard water persists. Duffy [1977] reported that magnetic devices had not been used extensively for industrial applications in the United States, due to the strong criticism which they have received from engineers. He described work that demonstrated no influence of magnetic fields on the precipitation of $CaCO_3$. The scepticism is still apparent in industry today, despite the fact that successful applications have been reported. Donaldson [1988] for instance, reports a successful use of a magnetic device. In the steel industry pretreatment phosphating systems involve solutions of zinc, manganese, iron and chromium phosphate. In spray and dip operations a sludge and scale are formed, which result in jet blockage, restrictions in pipe flow, scaling of heat exchangers and the inner walls of tanks. The use of hard water to rinse the treated steel can also give rise to precipitation of calcium phosphate and consequent scale formation. In a particular application dealing with the production of motor van bodies, a magnetic unit was installed in the hard water rinse system following an alkali rinse bath. The use of the device resulted in the removal of the calcium phosphate scale that had formed on the dip tank walls and associated surfaces.  The original scale was of the order of 6 - 8 *mm* thick on the walls of a tank with a volume of 1.63 *m³*.  The estimated saving in acid cleaning was £22,000 per annum.

The use of a device in the hard water circuit prior to a plate heat exchanger resulted in the deposition of a soft scale easily removed by brushing, in contrast to the hard scale formed in the absence of the magnetic unit.

Donaldson [1988] also reports two examples where magnetic devices were installed in zinc phosphate spray systems where evidence was found for:
1. A reduction in the amount of sludge formed with a consequent reduction in phosphate consumption.

2. Reductions in jet blockage

The financial saving was estimated to be £9,500 per annum.

Other examples of scale reduction and associated benefits are reported by Nordell [1961], Donaldson and Grimes [1987], Donaldson [1988] and Donaldson and Grimes [1988].

The evidence of these examples cannot be ignored in the face of the criticism of the technology that has been referred to above.  An argument used to discredit the results of industrial trials, is that under the test conditions there is good control of the particular process with attendant benefits that would have occurred without the presence of the magnetic device, if good control was permanently exercised.  On the other hand Donaldson [1988] reports carefully controlled laboratory tests where a pipe badly scaled with calcite gradually became clean over prolonged exposure to a magnetic field under flowing water conditions.

Donaldson [1988] attributes the beneficial effects of magnetic devices to a number of closely related changes brought about by the application of magnetic fields.
1. Crystal size is increased with less opportunity for incorporation in a scale structure.

2. Crystallinity of growing crystals is affected with a distinction between single crystals and aggregate particles in precipitates.

3. Morphology of growing crystals is changed through the changes in the rate of growth in different planes of the crystal structure.

4. Changes in the fluid-precipitate equilibrium that alter the relative stabilities of two phases with relatively similar lattice energies, e.g. X-ray diffraction data show differences in zinc phosphate phases with and without magnetic treatment.

5. Evidence from research with zinc phosphate solutions suggests an increase in solubility in the presence of a magnetic field.

It would appear that the technique shows promise but more work needs to be done on the use of magnetic devices for the reduction of scaling problems before the technology can be used with confidence. Furthermore the possibility of inducing electric currents in metallic structures, that could result in enhanced corrosion (see Chapter 10), must be carefully considered.

### 15.2.5 Soot blowers

The technique most often employed in combustion systems to maintain efficient heat transfer across surfaces exposed to the flue gas, is the technique of soot blowing. The method was originally devised, as the name implies, to remove accumulations of soot from the internal surfaces of combustion plant, but it now has a more universal application for the removal of any deposits that may interfere with the heat transfer process. The use of additives (see Chapter 14) to modify the structure of slags and foulants, may improve the effectiveness of soot blowing equipment.

A soot blower consists of a jet of steam or air that impinges on a deposit or causes the flue gases in the region of the deposit to impinge on it. The force created by the change in momentum of the gases as they strike the deposit weakens and "knocks" the deposit off the surface on which it has accumulated. Because of the relative closeness of heat transfer surfaces (e.g. banks of tubes) and their extent, it will generally be necessary to use a number of soot blowers. Manual or more usually automatic, rotating and traversing will be required to ensure that all heat transfer surfaces are subject to the soot blower action.

The major designs of soot blowers are of two types; the rotary and long retractable arrangements. The rotary soot blower consists of a multi nozzled element capable of rotation but permanently located within the heat transfer region to be cleaned. An arc of jets therefore accomplishes the removal of encrustations of deposit. Because the soot blower is usually permanently and completely exposed to the flue gases, rotary designs are generally limited in their application to locations in the combustor where the temperature is below 500°C. Retractable devices may be employed where the temperatures are high so that exposure is limited. Long retractable soot blowers may be needed in order to reach deep into the flue gas space, particularly in the region of radiant superheaters. Adequate cooling of these soot blowers by circulating air or steam will be necessary since they are subject to radiation from the flame.

The principal designs of soot blower fall into three types [ESDU 1992].

1. Single nozzle soot blowers can be retractable or non-retractable. The latter are used in the intermediate or low temperature zones, where retraction is not necessary and the nozzle assembly is not likely to suffer undue damage. Retractable versions are necessary in high temperature regions so as to avoid subjecting the equipment to aggressive conditions for longer than is necessary. A modified retractable design may be required to clean refractory walls.

2. Lance blowers are essentially improved versions of the nozzle designs. Three versions are in use and include retractable, partly retractable and oscillating.

   The standard retractable lance soot blower usually travels at right angles to the tubes to be cleaned, the jets describing a helical pattern passing across and between, the individual tubes. It is usual to have opposed steam ports and the lance travels the width of the surfaces to be cleaned. Where large units are concerned lances are fitted on either side to reduce the length of lance. This version is suitable for use in the highest temperature zones.

   If the space availability outside the unit is limited it may be necessary to fit partly retractable soot blowers. As the design requires that part of the lance is left permanently in contact with the flue gases, it cannot be used in high temperature regions.

   An oscillating soot blower is used where complete rotation is unnecessary or undesirable for any reason, such as accelerated erosion in certain regions.

3. Multijet blowers have non-retractable elements made from high quality materials than can withstand high temperature flue gases. They may be located either transverse or parallel to the tubes to be cleaned and are rotated during the soot blowing cycle. The major application is for the removal of loosely held deposits. Because of the severe conditions to which these blowers are subject, replacement rates may be higher than for the corresponding retractable units.

   Multijet soot blowers also find application in the cooler parts of the unit, such as in the convection sections, extended surface economisers and regenerative air heaters.

   The simplest method of activating soot blowers is to use steam from the superheater. If compressed air is used as the medium a compressor and air receiver will be required with additional maintenance and the associated costs.

   High pressure water jets may be used in preference to compressed air or steam, and because of the greater density of water the impact force of the water jet is much higher than air or steam. In addition to the removal force produced by the jet itself the thermal shock on the deposit due to the cooling effect, will also facilitate deposit removal. There is no general acceptance of the water jet activated soot blower however, since there is a risk that the large forces involved particularly from thermal shock, will damage the heat exchangers and the associated structures. Furthermore there is also a risk that large pieces of deposit on becoming dislodged may fall through the structure and heat exchangers, to cause damage. An alternative is to use water jets in conjunction with steam or air so that the risks of structural damage are reduced. On the other hand a cloud of evaporating droplets that impinge on surfaces in steam raising plant, e.g. the superheaters, may cause severe erosion.

   It is not usual or necessary in most boiler plant to operate soot blowers on a continuous basis, but rather on a regular cycle. A period between operations might be eight hours, but the frequency and extent of operation must be determined in relation to the plant design and the fuel being burnt (see Chapter 16), since this would determine the character of the heat exchanger deposits.

Because of the damage principally erosion, that may arise from the use of soot blowers it is good practice to check the surfaces swept by the jets. If erosion does appear to be a problem it may be necessary to modify the cycle and duration of the soot blowing operation, or to reduce the steam or air pressure till acceptable conditions have been achieved.

Soot blower equipment for new designs are generally based on developing the technology from experience, particularly in terms of the fuel to be burnt in the equipment. Different impingement pressures to accommodate differing rates of deposit accumulation may be obtained by the use of different sized nozzles.

As the discussion demonstrates the design of soot blowing equipment is entirely empirical based on an evolutionary process. Nevertheless it is a technique that can be made effective in maintaining heat transfer efficiency in large scale steam raising plant.

### 15.2.6 The use of sonic technology

The principle underlying the application of sound is that the vibration created by the energy associated with the transmission of sound will disturb and dislodge deposits on surfaces, i.e. "to shake" the deposit free. Cavitation produced by the propagation of sonic waves in the continuous phase near the deposit surface, can also assist the removal process.

In general the technique has been used for problems in gas systems, particularly flue gases from combustors, i.e. very similar to the application of soot blowers. According to Garret-Price [1985] considerable attention has been applied to the application of low frequency sound emitters (horns) to the relief of fouling problems on heat exchangers, but both high and low frequency sound has been used. Good results are possible in dealing with loose, friable deposits, e.g. dust accumulations in the convection section of boiler plant, but the use of sound is much less effective in respect of sticky and tenacious deposits that are generally associated with slagging. In contrast to the intermittent use of soot blowers some devices are used on a near continuous basis.

The sonic device or horn, generally operates at an audible frequency of around 220 *Hz* producing a sound not unlike that of a ship's foghorn with an intensity of around 130 *dB*. Low frequency resonant sound in the region of 0 - 20 *Hz* may be used to create a sound field for the removal of soot. An example of equipment for this duty consisted of a 5 *m* mild steel horn with a weight of 250 *Kg*. An activating air supply of 300 - 1200 *m³h* at about 5 bar was necessary for effective operation. The cost of such equipment is usually considered to be similar to the more traditional soot blowing facilities.

It would appear that most installations of sonic technology to control heat exchanger fouling, fall into the category of retrofit applications. The use of the technique may be to supplement the effectiveness of conventional soot blowers, or as a substitute for ineffective equipment.

The use of sonic technology is not widespread. There is some reluctance to its use on account of the possible structural damage that might accrue from unwanted vibration. There may also be a problem of noise pollution in the vicinity of the equipment. It has to be said however, that the technique has considerable merit but further research and development is required to provide a better understanding of the mechanism by which deposits are removed by sound waves.

### 15.2.7 Water washing

The manual application of water jets for on-line cleaning has been a technique used for many years to restore heat transfer efficiency in flue gas systems. In some plant permanent water jets or sprinklers, have been installed, but the method is less sophisticated than the use of soot blowers described in Section 15.2.5. Apart from the force applied by the water jets, the thermal shock on the deposit will often cause cracking and subsequent spalling of the foulant. Intermittent use of jets and sprays may be sufficient to maintain a reasonable heat transfer efficiency. Developments and improvements in the effectiveness of soot blowing equipment has resulted in the declining use of water washing. In many situations the use of on-line water washing is a last resort where other techniques have been unsuccessful. Good quality water is to be preferred because water with a high solids content is likely to be a fouling fluid in its own right, due to deposition from the evaporating water jet. If sea water is employed for instance, it will be necessary to give a final wash with clean water to remove salt deposits from the heat transfer surfaces.

The effect of cold water sprayed onto heat exchange surfaces in boiler plant can have the effect of reducing the steam temperature that is in effect a reduction in overall thermal efficiency. Furthermore excursions in steam temperature (and heat transfer surface temperature) can result in damage to the equipment, particularly the boiler tubes. It is necessary to ensure that the amount of water is kept to a minimum in terms of quantity and time of application.

Single jet water lances may be employed where the surfaces to be cleaned are readily accessible but multinozzled jets are usually required where the surfaces are not visible from inspection doors, e.g. the superheater surfaces. The number and size of the jets used for water washing will very much depend upon the nature of the deposit to be removed, the location within the combustion space and the likely effect of obstructions to the manipulation of the lances. Obstructions encountered may be internal or external to the flue gas space.

In the regions of lower temperature in the flue gas space, water washing rather than jets may be employed. The technique is applicable to economisers with horizontal tubes and air preheaters, and involves the installation of a water distribution system above the top bank of tubes to provide a curtain of water. Soluble deposits are more likely to be removed by this method than more tenacious and fused deposits that generally require the greater force of water jets. Consideration has to be given to the nature of any liquid effluent that may be produced by water washing. For instance it is possible to produce acidic liquids from the solution of acidic components contained in the flue gas, e.g. sulphur dioxide, that are corrosive. There could also be problems associated with the disposal of liquid effluents produced by water washing.

A modification of on-line water washing is to use chemical solutions with the aim of softening the deposit or changing its structure in some way, so as to facilitate removal. Again it is necessary to consider the implications of the use of particular chemicals, e.g. mineral acids, in respect of potential corrosion problems.

### 15.2.8 Shot cleaning

A technique to remove deposits from heat transfer surfaces located in flue gas streams, is to use the kinetic energy of shot falling under the action of gravity. The method involves the release of metallic shot above heat exchangers positioned in a vertical flue gas duct, so that it cascades down through the equipment and is collected at the bottom of the duct before transfer back to the release point.

The philosophy of the use of shot cleaning is not so much to remove deposits but to prevent their formation. Indeed, if attempts are made to remove existing deposits by the use of shot it is possible for the deposit, particularly if it is molten or soft, to capture the shot and this could add to the fouling problem rather than alleviating it.

It is necessary to ensure by regular inspection, that heat transfer surfaces subject to shot cleaning are not being damaged. If localised damage is found to occur it may be necessary to install some form of protection shield or redistributor to deflect the falling shot.

Shot cleaning has been used principally for economisers and recuperative air heaters but it is suitable for use in regions where temperatures are below 800°C.

### 15.2.9 Galvanic protection

In Chapter 10 the electrochemical nature of corrosion is discussed. Since corrosion of metals involves the passage of electric currents it is logical to adopt an electrical method to prevent corrosion. In Chapter 14 the use of both anodic and cathodic inhibitors (barriers) to prevent the flow of electrons is considered. It is possible to achieve similar protection by the use of electric currents.

*Cathodic protection*

In Chapter 10 the corrosion of iron is explained by the removal of a metal ion from an anodic site with the production of electrons. The corrosion if iron is described by Equation 10.2.

$$Fe \rightarrow Fe^{2+} + 2e^{-} \tag{10.2}$$

The corresponding reaction at the cathodic site is explained by Equation 10.5 where electrons are consumed, i.e.

$$2H^{+} + 2e^{-} \rightarrow 2H \rightarrow H_{2} \tag{10.5}$$

As the corrosion proceeds all the electrons generated by the anodic reactions are consumed in the cathodic reactions.

Applying Le Chatelier's Principle to Equations 10.2 and 10.5 it is to be expected that by electrical intervention in the corrosion reactions, that the rates of the anodic and cathodic reactions could be altered. If electrons are removed from the system the rate of the anodic reaction increases to yield more electrons. At the same time the rate of the cathodic reactions decreases so as to consume fewer electrons. Under these circumstances the iron will corrode more rapidly, the hydrogen production rate will decrease and the electrode potential of the metal will rise. On the other hand if electrons are imposed on the metallic iron from an external source, the cathodic reaction rate will increase and that of the anode will decrease. Under the imposition of this external source of current the iron will dissolve more slowly, the rate of hydrogen production will increase and the electrode potential will fall. The use of an external source of electrical current (electrons) in this way, to reduce or prevent metallic corrosion is generally called cathodic protection.

A metal surface subject to general overall corrosion, is usually considered to have similar numbers of anodic and cathodic sites. With time these sites will move over the surface so that more or less uniform corrosion is established across the surface. The application of cathodic protection to the surface eliminates or at least reduces, the electrochemical potential between the anodic and cathodic areas so that the entire surface has a uniform potential and is cathodic in character [Fontana 1986]. The original concept of the protecting mechanism was that according to Kirchoff's laws, a cathodic current applied to a metal subject to corrosion opposes the current leaving the surface. It is possible therefore, provided the applied cathodic current is sufficiently large, that the anodic currents can be reduced to zero. When the surface is just underprotected there is a slight penetration occurring uniformly over the whole surface. An alternative explanation suggests that the applied cathodic current causes a progressive reduction in the size of the active anodes, rather than a reduction in the anodic current density [LaQue and May 1965]. When the surface is just underprotected it would be anticipated that the surface becomes roughened due to the very localised attack, since the movement of the anodic areas would be restricted.

There are two implications associated with these two concepts of the mechanism of cathodic protection in respect of underprotection. If the older theory applies then it is to be expected that there will be more or less uniform corrosion over the surface. The later theory of LaQue implies that pitting corrosion will occur. There is evidence however, that suggests that the earlier theory of uniform protection is applicable [Dexter *et al* 1985].

There are two ways in which electrons may be supplied to the metal to be protected.

*Sacrificial anode technique*

A sacrificial anode can be any metal substantially anodic to the metal to be protected. When an electrical circuit is established, a galvanic cell is formed between the metal to be protected and the sacrificial anode providing the opportunity for electrons to flow from the sacrificial anode to the protected metal. The sacrificial anode is so named since in the process of providing electrons it is corroded according to an equation of the form:

$$M \rightarrow M^{Z+} + Ze \qquad (15.2)$$

where $M$ is any metal

and $Z$ is an integer

A schematic diagram of the arrangement is shown on Fig. 15.3. The reaction at the cathode is:

$$2ZH^+ + 2Ze \rightarrow ZH_2 \qquad (15.3)$$

Other reactions are possible at the cathode including oxygen reduction depending on the chemistry of the environment.

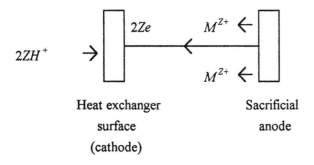

Heat exchanger
surface
(cathode)

Sacrificial
anode

FIGURE 15.3. Simple explanation of the action of a sacrificial anode

Common sacrificial anode metals include alloys of magnesium, zinc and aluminium. The performance of these anode materials is closely related to their alloy chemistry and the environment and must be chosen with these factors in mind.

### Impressed current technique

The principle of an impressed current system involves the supply of a protective current from some direct current power source, e.g. storage batteries, rectifiers or d.c. generators, through an auxiliary anode. Fig. 15.4 illustrates the arrangement.

The reactions at the cathode will be similar to those when a sacrificial anode is employed. If the electrode is a noble metal or is electrochemically inert, oxidation of the environment will occur preferentially. For an aqueous environment the reaction could be:

$$2H_2O \rightarrow O_2 + 4H^+ + 4e \tag{15.4}$$

or in sea water cooled heat exchangers the reaction could be:

$$2Cl^- \rightarrow Cl_2 + 2e \tag{15.5}$$

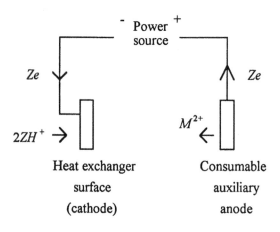

FIGURE 15.4. Simple explanation of the action of an impressed current

The choice of anodic material depends on a number of factors but principally cost and the required life. Consumable electrodes in general, are less costly but must be replaced as they are consumed, and they have the limitation that they cannot sustain high current loadings. A major difficulty may be the contaminating effects of the dissolution products. Non consumable alloy electrodes such as silicon-iron, lead-antimony-silver, platinum-titanium and platinum-tantalum avoid some of the limitations of consumable anodes, but they are expensive.

It is possible with impressed currents to "overprotect". Overprotection may give rise to problems comparable to the corrosion which the system seeks to prevent. Moderate overprotection of steel may not represent too much of a problem except the use of unnecessary current and the consumption of the auxiliary anodes. Higher levels of overprotection however, may give rise to excessive hydrogen at the cathodic surface with consequent deleterious effects. With metals such as aluminium, zinc, lead and tin excess alkali generated at the surface by overprotection may cause damage to the structure.

In Chapter 10 it is stated that corrosion of a surface may be related to the velocity of the corroding fluid across the surface; the higher the velocity the higher the rate of corrosion. For this reason the magnitude of the current density required to maintain a particular cathodic potential will increase with electrolyte velocity. Due to changing cross-sectional area for a given volumetric flow, the necessary current is likely to vary from location to location within a heat exchanger.

In Chapter 12 micro-organisms in suspension were referred to as "living colloids". It is to be expected therefore, that microbial fouling will be influenced by cathodic protection. Maines [1993] however, suggests that the picture is far from clear with respect to electrode material, types of organism and the technique adopted for cathodic protection. It would appear that in some situations biofouling is increased and in others it is reduced. Where a biofilm is present it would be expected to modify the electrochemical conditions at the metal surface and thereby influence the metal dissolution reactions [Dexter *et al* 1989].

A comparison of the sacrificial and impressed current techniques is given in Table 15.2 and is based on Ashworth [1986].

Cathodic protection has been demonstrated as a viable means of reducing corrosion in power station cooling water systems. Good design is essential, but regular monitoring of the system is essential to ensure that conditions have not changed with time that could result in accelerated corrosion [Burbridge 1986].

In Chapter 8 crystallisation and scale formation are discussed and the effect of *pH* was demonstrated as being a factor in deposit formation. Furthermore the equilibrium of ions in solution is likely to be affected by the presence of electric currents and the current density. As a consequence, a characteristic feature of cathodically protected metallic surfaces in contact with solutions containing mineral ions, notably sea water, is the formation of deposits or cathodic protection scale. The use of controlled scale formation is mentioned as a method of corrosion control in Chapter 14. Cox [1940] proposed the deliberate formation of calcium deposits on steel by the imposition of large cathodic currents to act as anti-corrosive self healing layer.

A metal or alloy in contact with an electrolyte (i.e. a solution containing ions) completely immune from corrosion by cathodic protection, will have current entering over the entire surface. The surface of the metal performs as a region of reduction. For oxygen containing water the following reactions are possible:

$$O_2 + 2H_2O + 4e \rightarrow 4OH^-$$                  (15.6)

and $2H_2O + 2e \rightarrow H_2 + 2OH^-$                  (15.7)

TABLE 15.2

A comparison of sacrificial and impressed current techniques for cathodic protection

| Technique | Advantages | Disadvantages |
|-----------|-----------|---------------|
| Sacrificial anodes | Independent of external supply of electric power. Simple to install. simple to augment if initial protection proves inadequate. Cannot be incorrectly attached to the structure. does not require control. Difficult to overprotect. | Small driving force may restrict application. |
| Impressed current | Possible to employ a large driving voltage so that large areas can be protected. Few anodes required. Voltage may be adjusted to suit the particular system. | A reliable source of d.c. current is imperative. Danger of overprotection. May be difficult to provide a satisfactory potential profile over a heat exchanger. Corrosion may be accelerated by careless connection, i.e. making an anode of the structure to be protected. |

It can be seen that the consequence of these reactions is to increase the concentration of hydroxyl ions at the metal surface, and may be taken as being equivalent to a corresponding reduction in $H^+$ concentration. The result is an increase in the *pH* at the surface being protected. The *pH* at the surface has been found to be directly proportional to the applied cathodic current [Guillen and Feliu 1966] and calculated to be in the range 10.75 - 11.25 [Wolfson and Hartt 1981].

The *pH* of an aqueous solution is controlled by the reactions described in Chapter 8, i.e.

$$H_2O + CO_2 \rightarrow H_2CO_3$$                  (8.2)

$$H_2CO_3 \rightarrow H^+ + HCO_3^- \tag{8.3}$$

and $HCO_3^- \rightarrow H^+ + CO_3^{2-}$ (8.4)

The presence of cathodically generated hydroxyl ions according to Equations 15.5 and 15.6 will lead to further reactions.

$$OH^- + HCO_3^- \rightarrow H_2O + CO_3^{2-} \tag{15.8}$$

$$CO_3^{2-} + Ca^{2+} \rightarrow CaCO_3 \downarrow \tag{15.9}$$

Because the $OH^-$ ions are close to the metal wall the precipitation of the $CaCO_3$ will occur on or close to, the surface. Colloidal precipitates may be formed in the viscous sub-layer near to the metal surface. In general, natural waters, e.g. sea water, will include other mineral salts in solution so that other compounds are likely to be included in the deposit. For instance if $Mg^{2+}$ ions are present they are likely to react with the excess of $OH^-$ ions present at the metal surface, i.e.

$$Mg^{2+} + 2OH^- \rightarrow Mg(OH)_2 \downarrow \tag{15.10}$$

In the presence of the protecting electric current, deposition and growth of mixed scale on the metal surface is a complex phenomenon. For all *pH* values up to 12, colloidal $Mg(OH)_2$ is positively charged and is therefore attracted to the negatively charged surface produced by the imposed electric current. On the other hand $CaCO_3$ in colloidal suspension is negatively charged and will be repulsed from the metal surface under protection [Pathmanaban and Phull 1982]. Complex mixtures of salts forming the adherent scale are possible. Edyvean [1984] observed layers close to the surface as $Mg(OH)_2$ and $MgCO_3$ mixtures, further out in the deposit the predominance of $MgCO_3$ was observed within $CaCO_3$ at the outer layer of the deposit. Edyvean also noted that the scale thickness was not uniform. The resistance offered by the scale will also affect the current flow for a given electrical potential. The influence of electrolyte velocity on current demand, $OH^-$ ion production and the changing *pH* in relation to distance from the protected surface, has implications for the nature and extent of the electrically induced deposits.

The calcareous deposit formed on cathodically protected metal surfaces will also be affected by the presence of micro-organisms and macro-fouling. Physical disruption and alteration of structure, composition and crystal form will influence the response to cathodic protection. The consequent economic implications of changes in current density requirement for protection, and the engineering design implications and associated economics may be significant [Maines 1993]. Work is therefore required to improve the understanding of the relationship between cathodic protection, biofouling and scale formation in cathodically protected systems.

*Anodic protection*

Anodic protection relies on the formation of protective films on metal surfaces by means of externally applied anodic currents. It is a relatively new development in comparison to cathodic protection, with at present, limited practical application.

The technique may be understood in terms of metallic passivity, i.e. the loss of chemical activity experienced by certain metals and alloys under particular environmental conditions as a result of surface film formation. Equations 15.2 and 15.3 suggest that the application of an anodic current to a metal should tend to increase metal dissolution and decrease hydrogen production. Metals that display passivity, such as iron, nickel chromium, titanium and their alloys respond to an anodic current by shifting their polarisation potential into the passive region. Current densities required to initiate passivity are relatively high [Uhlig and Revie 1985] but the current density to maintain passivity are low, with a consequent reduction in power costs [Scully 1990].

15.2.10 The use of inserts

Enhancement has been used for many years to improve heat transfer, particularly for viscous liquids. The principle involved is to disturb the viscous sub-layer near to the heat transfer surface thereby reducing the resistance to heat flow. Since the problem of fouling is very concerned with the transport of foulants across the viscous sub-layer it is to be expected that any attempt to reduce the resistance to heat flow will also affect the propensity towards fouling. Hewitt *et al* [1994] have given some background to the use of heat transfer enhancement techniques.

Two basic methods for heat transfer enhancement are possible:
1. Modification of the surface, e.g. roughening the use of integral fins, or machined grooves.

2. The use of inserts generally applicable to the inside of tubes in shell and tube heat exchangers, e.g. twisted tapes or wire matrices. An example of a wire insert is shown on Fig. 15.5.

Rough surfaces tend to enhance fouling as described in Chapters 7 - 12 where the mechanisms of fouling are discussed. The effect of finning on particulate deposition from air flows was discussed in Chapter 7. The use of machined surfaces is expensive and may also encourage deposit formation, although it does not appear that much work has been done in this area to determine the influence of regular "roughness" on fouling. Although the prime consideration has been to improve heat transfer efficiency, it has been found that inserts can reduce fouling problems where chemical reactions are involved [Gough and Rogers 1987], and where wax and other organic precipitates arise [Gough 1994, Gardner 1994].

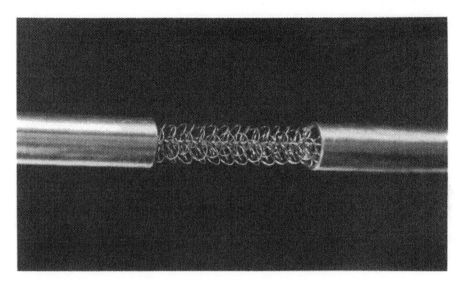

FIGURE 15.5. An example of a wire matrix insert (courtesy of Cal Gavin Ltd.)

Some inserts are primarily designed to reduce fouling, e.g. wire spring-like configurations that are free to oscillate and move under the influence of the flow conditions. The movement of the insert against the wall knocks or rubs off, any deposit as it is formed. A problem of this approach is that repeated local stressing brought about by the movement may give rise to fatigue fracture and disintegration of the insert. Pieces of the tubulator that are freed in this way may constitute an additional fouling problem. A more robust design is the one shown on Fig. 15.5. Here the insert is held tightly within the tube by a "push fit" as well as anchorage at the end with many points of contact between the wire matrix and the tube inner surface.

Gough and Rogers [1987] report (and discussed by Bott [1990]) the use of wire matrix tubulators inside the tubes of a shell and tube heat exchanger used for heating tar oil. The tar oil passed through the tubes with steam at the appropriate pressure condensing in the shell. The data obtained from the operation of the heat exchanger showed a rapid reduction in overall heat transfer coefficient over a period of 4 months (see Table 15.3).

The data show a rapid deterioration in performance to an intolerably low level over the four month period. The inefficiency was attributed to deposit accumulation on the inside of the tubes since fouling on the steam side is unlikely to change so rapidly. The fall in overall heat transfer coefficient was almost 50% over the four month period.

TABLE 15.3

Plant data obtained from a tar oil heater

| Period from beginning of test months | Tar oil flow rate kg/s | Overall heat transfer coefficient $W/m^2K$ |
|---|---|---|
| 0 | 1.69 | 149 |
| 2 | 1.69 | 84 |
| 4 | 1.47 | 77 |

The exchanger was stripped and cleaned by high pressure water washing (see later in this chapter) and wire matrix turbulence promotors were inserted into the tubes. A further four month test was made. The essential data are included in Table 15.4. It can be seen from Table 15.4 that the overall heat transfer coefficient fell by a relatively small amount (just over 4%) and may be compared with the 50% fall with no inserts in place.

TABLE 15.4

Performance data with inserts in place

| Period from beginning of test months | Tar oil flow rate kg/s | Overall heat transfer coefficient $W/m^2K$ |
|---|---|---|
| 0 | 1.68 | 229 |
| 4 | 1.71 | 219 |

Based on these trials new exchangers specifically designed to utilise wire wound turbulators, were installed. As a result the exchangers were able to operate for 4.5 years without cleaning, i.e. a considerable improvement in terms of fouling propensity. The calculated fouling resistance at the end of the operating period (4½ years) was only about 30% of that allowed for in the design for plain tubes (with no inserts), and yet using plain tubes the design fouling resistance was reached after only 2 months.

Laboratory studies using wire matrix inserts have been made by Crittenden *et al* [1993] using a light Arabian crude oil containing 10% waxy residue. The experimental data show that with inserts in place, asymptotes on the fouling resistance - time curves are reached within 10 hours of start up. Furthermore the values of the fouling resistance at the asymptote are only between about 2 and 7% of the recommended TEMA values recently revised [Chenoweth 1990].

Crittenden *et al* [1993] conclude that the disturbance of the hydrodynamics by the turbulators is largely responsible for the beneficial effects on fouling. The additional turbulence close to the heat transfer surface provides four potentially advantageous effects:

1.  A reduction in the residence time of fouling precursors close to the hot surface.

2.  A reduction in the volume of liquid which is heated to a temperature in excess of that of the bulk liquid.

3.  Elimination of nucleation at the surface.

4.  An increase in the rate of removal or release of deposits at or close to the wall.

The use of inserts is not without penalty since the increased turbulence requires additional energy. The presence of the turbulator increases the resistance to flow so that pumping costs are increased for a given volumetric flow rate.

As demonstrated by the work of Gough and Rogers [1987], the use of inserts for improving the performance of existing heat exchangers subject to reaction fouling, can be beneficial. Where chemical reaction fouling is anticipated in a new heat exchanger installation, the use of inserts should be considered in the initial design concept.

Little fundamental work appears to have been carried out into the use of inserts to combat fouling from other fouling mechanisms. It has to be remembered that in one sense, the insert is a "filter" and may therefore enhance fouling where "sticky" particles are concerned, e.g. micro-organisms. The effect may not be so much in respect of heat transfer but in regard to excessive pressure drop.

A great deal more work is required to provide a better understanding of fouling reduction by inserts, with particular reference to the fouling mechanisms discussed in Chapters 7 - 12.

### 15.2.11 Special features and designs

There are a number of techniques that may be used in conjunction with heat exchangers, their design and basic concept, to assist in the maintenance of clean conditions during operation. The technologies are not widely employed but they will be briefly discussed for the sake of completeness.

*Surface treatment*

In Chapter 6 and in subsequent chapters in dealing with different fouling mechanisms, the importance of surfaces in the fouling of heat exchangers is discussed. Low energy surfaces are likely to reduce the tendency of particles to attach. Changes in the surface characteristics therefore, have the potential to reduce the incidence of fouling. In addition to reducing fouling the surface may also prevent corrosion. It may be difficult to separate these effects. In microbial fouling operations it may be possible to include a biocide in the coating, that is slowly released over the lifetime of the coating. The technique has been used for many years to prevent biofouling of ships' hulls, by the use of biocidal paint (see later in this section). The coating itself will represent a resistance to heat transfer in exactly the same way as any deposit residing on the surface.

The advantages of successful lining of heat exchangers subject to fouling and corrosion may be listed as follows:

1.  Carbon steel and cheaper alloy steels can be used in contact with aggressive waters.

2.  Fouling is reduced on alloy tubes.

3.  The possibility of galvanic corrosion is reduced due to the electrical barrier provided by the coating.

4.  In general surfaces are smoother thereby reducing frictional losses.

5.  Off-line cleaning is facilitated.

Careful application of linings is required to provide a reasonably uniform thickness and to give complete coverage. Substantial variations in thickness will affect the local heat transfer and may affect the overall performance of the heat exchanger. Incomplete coverage may not substantially affect fouling reduction, but where corrosion resistance is required, breaks in the lining membrane can lead to severe localised corrosion.

*Vitreous enamels*

Vitreous enamels have been used for many years as liners for reactors and other chemical processing equipment but it is only relatively in recent times that the technology has been used in conjunction with heat transfer equipment. Conventional acid resisting vitreous enamels soften at around 450°C but developments have enabled enamels to be used at much higher temperatures. Page [1980] has reviewed the opportunities for the application of enamels to heat exchangers, particularly for use in acidic conditions at high temperatures.

Combinations of ceramic and metal systems - sometimes referred to as "cermets" have excellent resistance to mechanical damage (one of the limitations associated with the use of vitreous enamel for process plant). The combination of ceramic and metal in surface coatings also provides oxidation and corrosion resistance to the underlying metal substrate. The technique is particularly applicable to regions of low temperature in air preheaters in combustion plant, i.e. near or below the acid dew point. Page [1980] reports some extensive tests. Over a five year period some 40 corrosion resisting steels, alloy steels and organic and inorganic coatings were tested in combustion systems. The corrosion losses over a 12 month period were of the order of 10 - 27% for the stainless and low alloy steels, but various vitreous enamels showed only 1 - 3.8% corrosion loss. An improved enamel showed only a 0.5% loss over a similar period. The treated surfaces were also found to be less susceptible to soot deposition, and any deposits were easily removed.

The corrosion resistance of enamel is generally attributed to a surface layer of silica. Acid first dissolves the surface alkali oxide leaving a hydrated porous layer which is mainly silica, that acts as a barrier through which corrosive agents must diffuse for further attack to occur. As the affected layer deepens, the diffusion process slows down so that the enamel becomes protected by a skin of acid resisting silica.

Two major problems of using enamels for heat exchanger protection are:

1. The difficulty of ensuring a uniform continuous coating. For instance if "pin holes" remain in the enamel layer they will provide opportunities for condensed acids (or other corrosive agents) to attack the substrate metal. Gases generated at the enamel/metal interface, e.g. hydrogen, in attempting to escape, can cause flaking of the enamel sometimes accompanied by a rapid deterioration of the enamel structure, and consequent detrimental effects on the integrity of the whole structure.

2. The enamelling process requires "firing" in a suitable high temperature environment and this can impose a severe restriction on the size of components capable of being coated. It is possible to design the equipment in small units to be combined in the final assembly, but this imposes its own problems particularly in respect of the integrity of the joints. Furthermore the cost of manufacture will be increased above that for non-modular assembly.

Other problems may arise from differential expansion between the rigid enamel and the substrate metal.

*Plastic and polymer coatings*

The internal lining of shell and tube air cooled heat exchangers to reduce corrosion and fouling and prevent abrasion, has been used for many years. The scope of the technology has been generally widened to include many different applications. Because of the effects of temperature on many of the polymers their use is restricted to relatively low temperatures. Coatings are available that will be satisfactory over a range of operating conditions including temperature and *pH*.

The coatings used in heat exchanger technology are usually based a phenol formaldehyde and phenol epoxide resins. Application to a heat transfer surface is by a multi-coat flooding process. Partial polymerisation of each coating is obtained by placing the heat exchanger (or components) in stoving (heating) ovens. The thickness may be monitored electrically, so that when the desired thickness has been achieved the polymerisation process may be completed by final stoving at a temperature of around 250°C. The resultant homogeneous coating of 200 - 250 $\mu m$ thickness is hard, glossy and smooth with hydrophobic properties. At the same time, the coating retains sufficient flexibility so that it can withstand expansion and contraction, to be expected in fixed tube shell and tube heat exchangers working up to temperatures approaching 200°C. The extension to crack is generally in the range 1 - 4.5%. Some typical applications of resins used for heat exchanger lining are given in Table 15.5.

The principal application for polymer coated heat exchangers as Table 15.5 demonstrates, is to reduce corrosion but the smooth glossy surface will provide reduced friction loss and any fouling, e.g. micro-organisms or sediment, is more easily removed by increasing velocity or by hosing during off-line cleaning operations, than from the substrate metal.

### Thin coatings and paints

Callow [1990] has described the use of antifouling paints in conjunction with ships' hulls to reduce microbial and macro-fouling, but the principle could be applied to heat exchangers in suitable environments. She defines three major types of antifouling paints.

1.  Soluble matrix - also known as conventional paints.

2.  Contact leaching.

3.  Self polishing or ablative paints.

In the soluble matrix type cuprous oxide and sometimes other biocides, is incorporated into a resin binder. The acidic resin slowly reacts with the sea water (*pH* approximately 8.0) and dissolves continuously exposing the cuprous oxide particles that are gradually released at the water/solid interface. The newly created surface provides antifouling properties. Depending on thickness the life of these paints on ships' hulls, in contact with sea water, is 1 - 2 years.

Contact leaching involves a higher copper content and an active life of 15 months. The matrix itself is insoluble and as copper dissolves from the surface, underlying particles of copper are exposed. As time passes the rate of release of copper gradually falls, due to the mass transfer resistance of the outer layers of matrix from which the copper has already been leached.

Self polishing co-polymers are tributyltin and methyl methacrylate co-polymers into which copper salts and other biocides, together with appropriate pigments are incorporated. In contact with sea water hydrolysis occur, releasing tributyltin and producing a smoother surface in contact with the water. The rate of hydrolysis depends on the temperature and *pH* and the pigments incorporated in the resin.

TABLE 15.5

Applications of resins to reduce fouling and corrosion

| Resin | | Application with fluids |
|---|---|---|
| Phenol formaldehyde | | |
| | Pigmented | Strong acid to weak alkali |
| | | Salt solutions |
| | | Cooling water |
| | | Gases |
| | | Organic liquids |
| | Filled | Weak acids and alkalis |
| | | Aqueous solutions |
| | | Vapours |
| Phenol epoxide | | |
| | Pigmented | Strong acid to strong alkali |
| | | Cooling water |
| | | Brackish and sea water |
| | Filled | Weak acids to alkalis |
| | | Aqueous solutions |
| | | Vapours |
| Substituted phenol epoxide | | |
| | Pigmented | Strong acid to alkali |
| | | Salt solutions |
| | | Cooling waters |
| | | Gases |
| | | Organic liquids |
| | Filled | Acids and alkalis |
| | | Aqueous liquids |
| | | Vapours |

The use of tributyltin and copper biocides may not be a matter of concern for ocean going vessels where the dilution is infinite, but the use of these biocides in conjunction with say cooling water systems, is not recommended because of the toxic effect and environmental risks. In order to utilise the technique for heat exchangers therefore, it would be necessary to develop suitable biocides that may be incorporated in the resin matrix and still be effective in controlling biofilm formation and growth, without detrimental effects.

Santos *et al* [1992] studied the effectsof thin layers of polymers on biofouling under different velocity conditions, in comparison with glass surfaces. They found that biofouling due to *Pseudomonas fluorescens* was not only a function of the polymer applied, but also the velocity of water across the surface. The surface treatments used were:

1. Glass clad HP (GCHP) a heparin modified siloxane resin giving hydrophilic properties to the surface.

2. Glass clad IM (GCIM) is a polyethylene imine modified resin used to provide glass surfaces with a greater affinity for microbial cell retention.

Some data are recorded on Fig. 15.6. Absorbence of infra red (vertical axis) is a measure of the biofilm thickness. The importance of flow velocity is demonstrated. The results show that the Glass clad HP provides a surface that is similar to glass (i.e. less hospitable to biofilm retention than common construction metals).

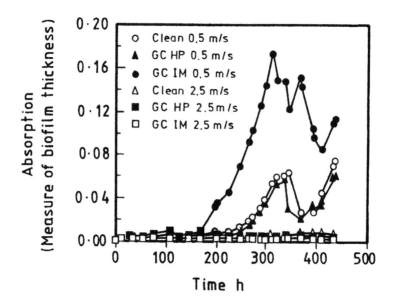

FIGURE 15.6. The effects of polymers compared with a glass control tube at 0.5 and 2.5 *m/s* water velocities

A problem with the use of thin coatings and paints is the maintenance of the surface coverage. Over a period of time the effectiveness is likely to fall and coating replacement will be necessary. In all but the simplest heat exchangers, this may be difficult and expensive.

## Non-metallic heat exchangers

Glass heat exchangers, shell and tube and coil in tank designs have been used for many years for corrosive applications. The smooth surface is less hospitable to deposit retention than the traditional metal construction. More recently heat exchangers manufactured from polymers and plastics have found application in the chemical and process industries. Redman [1989] has reviewed the use of non-metallic heat exchangers. In general, polymers chosen for heat exchanger use have excellent corrosion resistance. The relatively low cost compared to high quality alloys coupled with lightness make them attractive. Other advantages are the ease of cleaning and resistance to fouling. Among the disadvantages are the low melting points of thermoplastics and low thermal conductivities. Table 15.6 compares the thermal conductivities (and densities) of some common construction materials [Redman 1989].

TABLE 15.6

Thermal conductivities and densities of some comon materials of construction for heat exchangers

| Material | Thermal conductivity W/mK | Density g/cm³ |
|---|---|---|
| Copper | 390 | 8.96 |
| Mild steel | 50 | 7.8 |
| Stainless steels | 15 | 7.8 |
| Ni/Cr/Mo alloys | 8 | 9.0 |
| Titanium | 16 | 4.5 |
| Tantalum | 53 | 16.6 |
| Borosilicate glass | 1.16 | 2.23 |
| Polytetrafluoethane | 0.19 - 0.23 | 2.17 |
| Fluorinated ethylene propylene | 0.20 | 2.15 |
| Polypropylene | 0.17 | 0.89 |
| Polyethylene | 0.34 | 0.89 |

Mott and Bott [1991] compared the accumulation of biofilm on simulated heat exchanger tubes. The equilibrium or plateau value of biofilm deposit, has the highest for 316 stainless steel tubes with lower amounts on electropolished stainless steel, glass and fluorinated ethylene propylene. Table 15.7 summarises their data. The tests were carried out using *Pseudomonas fluorescens* under identical conditions. The velocity of flow through the tubes was 1 *m/s*.

TABLE 15.7

The accumulation of biofilm on various surfaces

| Material | Biofilm accumulation $g/m^2$ |
|---|---|
| Glass | 169 |
| Fluorinated ethylene propylene | 150 |
| 316 stainless steel as received | 206 |
| electropolished 316 stainless steel | 159 |

It would appear that the fluorinated ethylene propylene had approaching 30% less biofilm compared to that on the as received 316 stainless steel tube under similar conditions. The glass had 18% less biofilm.

The usefulness of plastic tube heat exchangers was demonstrated by Hughes *et al* [1988]. Evaporating sea water from thin falling films in low energy polyamide tubes, the extent of the fouling was quite small. The principal constituent of the deposit was $Mg(OH)_2$.

The use of bundles of thin walled plastic tubes have been used in shell and tube heat exchangers. Apart from the resistance to deposit accumulation afforded by the polymeric material, the oscillation of the flexible tubes under the influence of the flows in the shell as well as within the tubes assists in keeping the surface clean. A recent innovation is the use of ceramic heat exchangers for handling corrosive fluids, e.g. heating corrosive gases to 580°C prior to incineration at 950°C and also as an acid condenser for sulphuric acid at 250°C [Redman 1989].

*Scraped surface heat exchangers*

Scraped or wiped surface heat exchangers, were developed to improve heat transfer to or from, viscous liquids either in sensible heat transfer (e.g. margarine coolers) or for evaporation (e.g. polymer solutions). Hewitt *et al* [1994] have reviewed the technology. The unit usually consists of a relatively large diameter tube that carries an axially mounted shaft onto which blades are attached. The blades may be spring-loaded or "free floating". As the blades rotate they either scrape away the viscous sub-layer or disturb it under wiping conditions. The blade action particularly scraping, maintains clean heat transfer surfaces. Fig. 15.7 is a simplified sketch of the exchanger which may be mounted vertically where the liquid usually falls as a thin "film" down the exchanger walls (as in Fig. 15.7), or in the horizontal position, where the exchanger tube runs full of liquid. The technique is particularly useful for heat transfer to thermally sensitive liquids, e.g. foods, since the time of contact of liquid with the hot surface is relatively short. Apart from reducing the incidence of fouling from chemical reaction, the action of the blades removes any deposit as it is formed, e.g. scale formation in evaporation or particle sedimentation or freezing fouling.

Limitations of the technology include low capacity, and generally higher maintenance costs associated with rotating machinery.

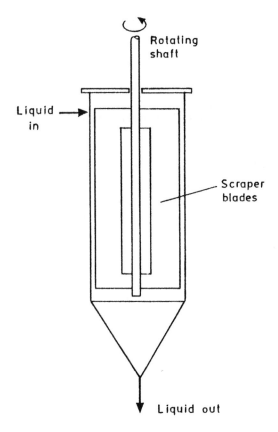

FIGURE 15.7. Schematic diagram of the action of scraped surface heat exchangers

*Chemical cleaning*

On-line cleaning may be achieved by the injection of chemical solutions into the process stream. A major difficulty arises from product contamination which may persist even after the dosing has been stopped. It is essentially an adaption of off-line cleaning discussed in the next section.

*The use of radiation*

Radiation sterilisation of microbial-laden water is a possible method of controlling biofilm formation and as such may be regarded as a cleaning technique. the use of ultra violet light has been used for a number of years for this purpose. The application is limited however, for three reasons:

1.  The principal difficulty is that the ultra violet light is only capable of interacting with the micro-organisms it "sees" in the water, i.e. generally the planktonic cells. It is unlikely that micro-organisms attached to a surface will be much affected. Once the surface has become contaminated cleaning will have to be directed to the water/biofilm interface.

2.  Optical access is needed requiring suitably positioned transparent sections.

3.  Large systems will require large ultra violet light installations with generally high cost.

The process may be acceptable for high quality product manufacture, e.g. food processing where absolute sterilisation is imperative.

Gamma rays may be used in the same way as ultra violet light, but the necessary safety requirements are greater and therefore the costs will be even larger than the equivalent ultra violet techniqe.

## 15.3 Off-line cleaning

It is inevitable that sooner or later many heat exchangers will require off-line cleaning to restore their heat transfer efficiency even though steps have been taken to eliminate or minimise potential fouling problems, at the design stage and during subsequent operation. Ideally this could be carried out at the annual shutdown or at some other planned time. Emergency shutdown of heat exchange equipment for cleaning is disruptive and expensive. Cleaning and associated maintenance costs are discussed in Chapter 3. There are a number of options available for off-line heat exchanger cleaning, but care has to be exercised not only in the choice of method, but also in its implementation to prevent damage to the equipment, to protect all employees involved against accident, injury or health risks, and to protect the environment from detrimental effects of the cleaning process. Vigilance is required since the cleaning operation is generally on an infrequent basis and is considered to be a "temporary" exercise. For these reasons there is a temptation to use a "lash up" rather than something more robust that takes time and effort to install. Excessive pressures by operators to execute a "quick turnround" of equipment must not compromise safety.

15.3.1 Manual cleaning

Manual cleaning can be carried out dry or in the wet condition. It is the simplest technique involving wiping, brushing or scraping heat exchanger surfaces to remove deposits. Because it is labour intensive the cost of manual cleaning may be high. It is generally applicable to plate heat exchangers where access to individual plates is possible by dismantling the plate array. If solvents are employed the potential risks of explosion and safety risks from toxicity to personnel will require special attention.

In large combustion equipment manual cleaning may be employed to remove friable deposits from surfaces. It will be necessary for personnel to enter the equipment and this will require special precautions to ensure that the risk to operatives is minimised, e.g. the use of portable air blowers and hoses for cooling and to supply fresh air. Protective clothing, goggles and efficient masks or respirators are likely to be essential.

In some equipment, e.g. jacketted vessels, it may be neessary to chip off hard deposits such as might be encountered in a phenol formaldehyde resin batch reactor fitted with heating and cooling coils and jackets.

15.3.2 Mechanical cleaning

Mechanical cleaning techniques are common throughout industry. Specialist firms are often employed to clean heat exchangers since they have wide experience and knowledge.

*Water, steam and air lances*

Water washing can be effective in the removal of deposits, but for tenacious deposits it may be necessary to use high pressure water or steam lances. Detergents may be added to the water to improve the cleaning process. It may be necessary to use blast cleaning for tenacious deposits involving the use of air laden with suitable abrasive material propelled at high velocity to impinge on the deposit to be removed. The nature of the abrasive material depends on the deposit strength and the risk to the equipment. Abrasive material may include sand, nut shells, or shot. In large boiler plant for instance, it will be necessary to ensure that there is sufficient space available to collect projectile material and pieces of deposit. The effectiveness of the technique depends very much on the availability of access to the fouled surfaces. Care is required in the use of high pressures to avoid damage to plant and accidental injury to personnel. Lester and Walton [1982] have listed foulant characteristics, the applicable mechanical technique for removal and typical heat exchangers. Table 15.8 summarises these authors observations.

TABLE 15.8

Foulants and common methods of removal [Lester and Walton 1982]

| Foulant | Typical heat exchanger | Mechanical technique | Operating pressure bar |
|---|---|---|---|
| Airborne contaminants, e.g. dust, grit | aluminium air coolers | Jet washing | 2 - 4 |
| Soft deposits, mud, loose rust, biological growths | Shell and tube exchangers, film coolers | Jet washing | 40 - 150 |
| Waxes, grease | Condensers, etc. | Hydro steaming | 30 |
| Heavy organic deposits, polymers, tars | Condensers | Jet washing with or without pre-treatment with chlorinated or aromatic solvents | 300 - 400 |
| Boiler scale, water side and fire side | Boilers, economisers, preheaters | Jet washing, pneumatic percussive or pneumatic abrasive techniques including crushed olive stones | 300 - 700 |
| External deposits on heat exchangers, e.g. paint, rust | All types where applicable | Wet sand blasting | Depends on equipment design |

Bott [1990] has reported the beneficial effect of water washing. Due to the accumulation of deposits, in heat exchangers placed in the exhaust stream of large diesel/generators, the temperature drop in the exhaust gases fell from the design figure of 214°C to around 190°C in about 1 - 3 days. The heat exchangers used for steam raising (evaporator and superheater), consist of banks of finned tubes with a total heating area of 648 $m^2$. A series of water jets fitted above the heat exchangers, is used to restore heat transfer efficiency on a periodic basis after shutting down the diesel engines. The wash water is collected in a tank below the heat exchangers. In general the accumulations from diesel exhausts are not tenacious, unlike deposits often found in the high temperature regions of coal combustors, so that the cleaning process is not difficult, provided there is adequate coverage of the heat transfer surface. Fig. 15.8 shows how effective the water wash was in removing the deposits.

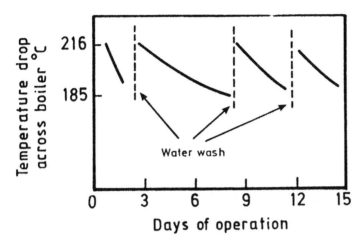

FIGURE 15.8. Improvements in thermal performance as a result of water washing

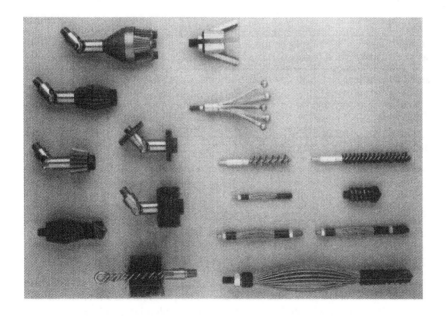

FIGURE 15.9. Some devices for clearing fouled tubes (courtesy Goodway Tools)

*Tube drilling and rodding*

For difficult to remove deposits, e.g. carbonaceous deposits on the inside of crude preheater tubes, it may be necessary to drill along the axis of the tubes to restore the full flow area. Other techniques, e.g. chemical cleaning or water jetting, are not possible because of the difficulty of penetration into the mass of deposit from the ends of the tubes as the deposit often occupies almost the full cross sectional flow area, or because the deposit is extremely dense, hard and intractable. Various devices may be applied to the rotating shaft including drills, cutting and buffing tools and brushes that may be made from different materials, e.g. steel, brass or nylon, depending on the tube material and the nature of the deposit. Fig. 15.9 shows some of these devices and Table 15.9 some of their applications.

TABLE 15.9

Tube cleaning devices

| Item | Application |
|------|-------------|
| Hollow drills | Severely or completely blocked tubes |
| Stainless steel wire cutting/buffing tools | Light to medium scale deposits finishing previously drilled tubes |
| Stainless steel brushes | Soft deposits and light scales in ferrous tubes (e.g. carbon steel, stainless steels) |
| Brass brushes | Soft deposits and light scales in non-ferrous tubes |
| Nylon brushes | Soft deposits of slime, mud, etc. in non-ferrous tubes (e.g. copper, brass, plastic) |

The drilling may be accomplished by a compressed air motor and a water supply will be necessary to help lubricate the drilling operation and to flush away the deposit as it is removed. Fig. 15.10 illustrates the principle.

For the effective use of these techniques good access is required at either end of the tube bundle. The work has to be carried out by good quality labour to reduce the risk of damage to the heat exchanger tubes. The problems are intensified with hard deposits. Where completely blocked tubes are involved it would be necessary to drill a pilot hole through the deposit to facilitate the use of the larger drill.

Drilling and rodding is generally only applicable to the inside surfaces of tubes. It may be possible to use a rodding technique for the outside of tubes on square pitch array but this would be difficult or impossible for tubes on triangular pitch.

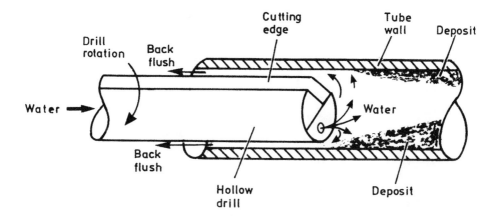

FIGURE 15.10. The technique of drilling out a heat exchanger tube using a hollow drill

## *The use of bullets, scrapers or scrubbers*

An alternative to drilling is to use a bullet or scraper that is "fired" along the length of the tube to scrape away the deposit. The projectile is propelled through the tube by a stream of high pressure air or water or both. The technique is only applicable to the inside of tubes. Examples of scrapers or bullets are shown on Fig. 15.11. It will be seen that in general the cleaners have a core with a surrounding open scraper structure that facilitates deposit removal and carries debris forward.

The cleaners shown on Fig. 15.11 have a series of "scrapers" graded so that any deposit is progressively removed as illustrated on Fig. 15.12. The water gun for firing the projectiles is shown on Fig. 15.13. It operates at pressures between 1.4 x $10^3$ and 2.75 x $10^3$ $kN/m^2$.

The mechanical tube cleaners may be fabricated from metals rubber or plastics. One of the concerns in the use of these projectiles is the likely wear on the inside of heat exchanger tubes. Hovland *et al* [1988] have demonstrated that the erosion effect of using these cleaners is negligible. The tube wear by cleaners fabricated from steel after 100 shots through 90/10 copper nickel alloy was in the range 12.7 - 22.9 $\mu m$. Three different procedures were used in the tests and reflected the number of shots (or passes) the cleaners performed before being replaced. The authors conclude that tube wear with harder materials, e.g. stainless steels or titanium would be similar or less.

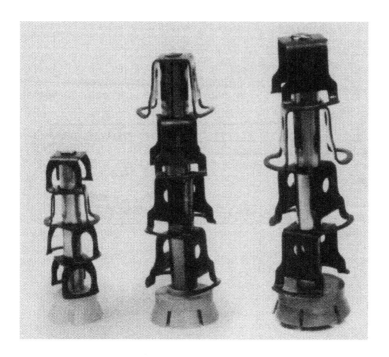

FIGURE 15.11.  Mechanical tube cleaners (courtesy of Conco Systems Inc.)

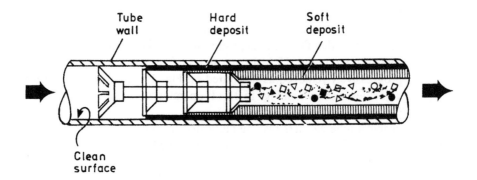

FIGURE 15.12.  The mechanical tube cleaning action (courtesy of Conco Systems Inc.)

FIGURE 15.13. The water gun (courtesy of Conco Systems Inc.)

## *Cleaning with explosives*

Elsewhere in this chapter some of the shortcomings of the use of water jets and abrasive materials for the cleaning of stubborn deposits have been mentioned. The techniques are primarily used to clean heat transfer surfaces in combustion plant exposed to flue gases. An alternative is to use controlled explosions, where the energy to remove the deposit, is transmitted by a shock wave in the air adjacent to the surface to be cleaned, or by the general vibration of tubes brought about by the explosion. It is a relatively new innovation in boiler plant cleaning introduced in the last 5 or 6 years.

The areas which can be successfully cleaned inside boiler plant include superheaters, reheaters, economisers, furnace nose, furnace walls and furnace hopper. It is possible to begin the cleaning process while the structure is still hot (i.e. as opposed to cool conditions required for the entry of cleaning personnel for manual cleaning or for the use of air or water jets.

The advantages claimed for the technology include [JRC International 1994]:
1. Rapid cleaning

2. Excellent improvements in thermal efficiency

3. Improved safety: no high pressure hoses

4. No corrosion

5. No problems with liquid sludge

6. Avoids cementing problems associated with the use of water jets

7. Suitable for regions that high pressure jets cannot reach

8. Lower cost than the use of water jets

Fig. 15.14 shows boiler tubes before and after treatment with controlled explosions.

### 15.3.3 Steam soaking

Steam soaking involves the admission of steam into the heat exchanger so that there is intimate contact between the steam and the surface to be cleaned. The cleaning process is accomplished by a combination of temperature, temperature difference with associated differential thermal expansion (or contraction) and the susceptibility of the deposit to vapour and hot water penetration. It is commonly used in steam raising equipment. Care has to be exercised in the application of steam soaking so as not to damage equipment. For instance it may be necessary to fill heat exchanger tubes with water if steam soaking is used on the outside of tubes. Frequent inspection is required to ensure that no damage occurs. Particular problems include corrosion and possible effects on refractory linings in high temperature (combustion) plant. Where soot blowers are installed it may be possible to use this equipment to allow steam admission, otherwise specially designed and adequate, steam entry ports will be required.

The technique is not recommended for general use where there are risks of damage to the equipment, particularly in combustion equipment where mixtures of construction materials are present, e.g. refractories and different alloys and metals exist together. Even here though, when the deposit is extremely difficult to remove, the technique could be used as a last resort.

Steam soaking could be applied to heat exchangers where biofouling is a problem. If the temperature is raised sufficiently high and for a sufficiently long period, micro-organisms and indeed macro-biological deposits will be killed and subsequent water flushing is likely to remove the biofilm from the surface. In general the technique is not recommended for closed systems for biological fouling problems since, although the biological activity may be arrested, large volumes of dead material and debris may give rise to blockage of heat exchangers and transfer lines downstream.

Figure 15.14. Boiler tubes before and after explosive cleaning (courtesy JRC International)

15.3.4 Thermal shock

Because of the differences in thermal expansion that in general, exist between fouling deposits and metal surfaces changes in temperature particularly rapid changes, are likely to cause cracking of foulant layers residing on metal surfaces, with the possibility of flaking. The technique is closely related to steam soaking. It would be necessary to follow the process with water flushing to carry away the dislodged material and it is likely that the process would have to be repeated several times before reasonably clean surfaces had been obtained. Thermal shock can be applied in two ways, either rapidly cooling a hot surface or rapidly heating a cold surface. The former is more likely to be the preferred route in most applications.

Bridgwater and Loo [1984] experimentally studied the spalling of scale from metal surfaces in an investigation primarily aimed at determining how such deposits are removed from surfaces under flowing conditions. The work has relevance to the use of thermal shock for the removal of foulants from heat exchanger surfaces. According to these authors the key feature is the stored elastic energy due to thermal stresses and the release of this energy due to the propagation of defects in the crystalline structure. Calcium sulphate spalled continuously while the deposition took place, but calcium carbonate scale was insensitive to changes in fluid velocity although it was susceptible to large changes in heat flux. Fig. 15.15 shows how the wall temperature at a point on the experimental tube changed with time due to the effects of scale removal. The tube was heated and a calcium sulphate solution flowed on the outside. It will be seen that on a fairly regular cycle the scale was removed and rebuilt. The sudden drops in temperature are attributed to the scale removal process.

Heat shock may be used for biofouling. If the water side is drained, and the temperature of the heat transfer surface is allowed to rise, provided a sufficiently high temperature has been reached, the micro- and macro-organisms will be killed. The problem arises however as with steam soaking, that when the water flow is restored large quantities of dead biological matter released from the surface may give rise to blockage problems. Furthermore if the temperature is allowed to become too high, the biofilm may be "cooked on" to the surface with increased difficulties of removal.

*Osmotic shock*

In Chapter 12 the function of micro-organisms is demonstrated to depend on the transfer of nutrients (and waste products) through the cell membrane. It is to be expected therefore, that any process that utilises the transport properties of membranes, could be used to control the viability of the cell. Osmotic processes depend on semi-permeable membranes and the phenomenon may be applied to the cleaning of heat exchangers on which a biofilm resides. In simple terms water

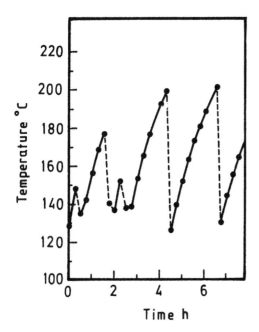

FIGURE 15.15. Variation in wall temperature with time for a point on a heated tube in contact with calcium sulphate solution

migrates across permeable membrane from regions of low concentration (say of a salt) to regions where the concentration is higher. It is possible therefore to apply osmotic shock to either fresh or sea water systems, by subjecting them to salt water or fresh water respectively. As a result the cell may come under the effects of internal pressure followed by cell lysis or alternatively to become water deficient that will result in death. The technique is difficult to apply when using sea water to combat biofouling in "fresh" water systems, due to the likely problems of corrosion. The long timescale for the technique to be effective is also a deterrent in the use of the method.

## 15.3.5 Chemical cleaning

Chemical cleaning of process plant has been practised for many years and the present sophisticated techniques are generally based on the original "fill and soak" methods.

Before chemical cleaning can be applied a number of factors have to be taken into account. Not least amongst these considerations is the risk to the environment and the safe disposal of any effluent. Information on the following is required before any attempt at chemical cleaning can be undertaken.

1. Nature of the deposit so that the appropriate treatment may be specified.

2. Material of construction of the heat exchanger and associated pipework and equipment so that corrosion can be avoided.

3. Ability to heat the equipment so that the effectiveness of the cleaning process can be raised.

4. Volume of the system so that the quantity of the required chemical may be estimated.

5. Degree of cleanliness required as a result of the cleaning operation.

6. Effluent problems associated with the safe disposal of the used cleaning chemicals.

7. Hazards associated with the cleaning process and safety requirements.

8. Cost.

9. Time that may be allowed for the cleaning operation.

Some of the chemicals that have been used for cleaning purposes together with the foulant they are capable of removing are given in Table 15.10 which is based on the observations of Lester and Walton [1982]. These authors have also discussed the compatibility of chemical cleaning agents with some common materials of construction (see Table 15.11).

As with problems of deposition, so effective removal of deposits will depend on velocity and temperature. It may be anticipated that high velocity and elevated temperatures of operation would assist the cleaning process by increasing the shear, and improving the rate of chemical reaction, on which the effectiveness of the chemical cleaning agents depends. Jackson [1984] has given some relevant data on the use of a 2% caustic soda solution to clean a plate heat exchanger fouled with tomato paste. The quantitative data are given on Figs. 15.16 and 15.17. The conclusion is that for a given circulation time and a flow rate below 0.42 $m/s$ an increase in the temperature of the caustic soda leads to a reduction of the deposit remaining on the plates and an increase of its removal rate. Once the velocity is increased above the critical value of 0.42 $m/s$ the deposit remaining decreases for a given circulation time.

TABLE 15.10

Some chemical cleaning agents

| Foulant | Chemical cleaning agent | Comments |
|---------|------------------------|----------|
| Iron oxides | Inhibited hydrofluoric, hydrochloric, citric or sulphamic acid, EDTA | Inhibited hydrofluoric acid is by far the most effective agent but cannot be used if deposits contain more that 1% *w/v* calcium |
| Calcium and magnesium scale | Inhibited hydrochloric acid, EDTA | As for iron oxides |
| Oils or light greases | (a) Sodium hydroxide, carbonate or tri sodium phosphate with or without detergents or | Alkalis are also used for the removal of biological deposits |
| | (b) Solvent - water emulsions, e.g. kerosene - water | Useful for non-ferrous systems which may be attacked by alkalis |
| Heavy organic deposits, e.g. tars, polymers | Chlorinated or aromatic solvents followed by jet washing | Trichloroethylene and perchoroethylene are non-flammable |
| Coke/carbon | Alkaline solutions of potassium permanganate or steam-air decoking | Steam/air decoking is used on furnace tubes where coke is removed by controlled burning |

TABLE 15.11

Compatibility of some materials of construction with common cleaning agents

| Material | Compatible cleaning agent |
|---|---|
| Carbon steels | Inhibited mineral or organic acids, organic acids, organic solvents, alkalis and chelants |
| Austenitic steels | Inhibited hydrofluoric, nitric, sulphuric, phosphoric and organic acids, chelants and non-chlorinated organic solvents |
| Copper, nickel and their alloys | Inhibited sulphuric or organic acids, organic solvents |
| Aluminium | Only weak acids (e.g. citric, sulphamic) should be used, organic solvents |
| Cast iron | Inhibited mineral or organic acids, organic solvents |
| Concrete | Attacked by hydrochloric acid, but to a lesser extent by sulphuric or organic acids |

Generally speaking plant pumps should not be used for circulating chemical cleaning solutions (e.g. acids and alkalis) but they can usually be employed for flushing operations. The circulation of flammable solvents needs particular care as the heat exchanger may not be located in a flame proof area. Particular attention must be given to "dead legs" where cleaning chemicals may accumulate giving rise to problems later during the cleaning procedure or in subsequent plant operation. Adequate flushing must be provided.

No chemical cleaning especially involving acids, alkalis and other aggressive agents, should be undertaken without good analytical control to reduce the potential risk of corrosion.

The relatively complex procedure that may be required and the associated safety considerations for chemical cleaning of heat exchangers is illustrated by a case study published by French [1981]. The deposit on the surfaces of a stainless steel heat exchanger had the aproximate composition given in Table 15.12.

The following steps were taken to clean the heat exchanger:

1. A 10% solution of a high boiling aromatic solvent emulsifier with surfactants was used to remove the bulk of free oil by circulation through the heat exchanger. After circulation the unit was drained and flushed. Care would have had to be taken due to the flammability of the solvent and to avoid unacceptable discharge.

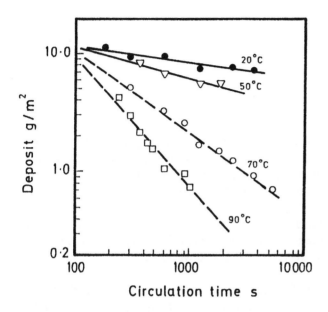

FIGURE 15.16. Soil remaining on plates after cleaning against cleaning time. Solution velocity 0.304 *m/s*

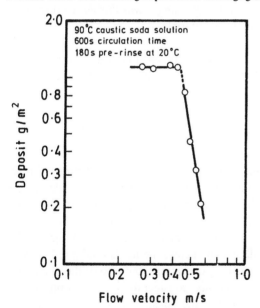

FIGURE 15.17. Soil remaining on plates after cleaning against solution velocity

TABLE 15.12

Deposit analysis

| Component | Weight % |
|---|---|
| Free oil | 12 |
| Carbonised organic matter | 34 |
| Acid insoluble substances | 8 |
| Iron oxides | 41 |
| Iron sulphide | 5 |

2. Treatment with a hot alkaline solution of potassium permanganate to break down the organic and carbonaceous content of the deposit and to lift it off the metal surface. The procedure was followed by adequate flushing. Again care would be required in disposing of the liquid effluents.

3. Washing with an inhibited acid to dissolve the oxides and sulphides of iron. Hydrochloric acid was not used because of the risk of chloride stress corrosion cracking of the stainless steel. Sulphuric acid was used.

   It was necessary because of the sulphide in the deposits, to vent safely the toxic hydrogen sulphide gas. If large quantities had been expected it might have been necessary to vent via a flare to reduce the risk to personnel and the environment. After the acid treatment the exchanger was flushed with water to remove traces of the acid and suspended material.

4. The final stage involved rinsing with a dilute alkaline solution of sodium carbonate and sodium nitrate to leave the heat exchanger not only clean, but in a "safe" condition.

## 15.4   CONCLUDING REMARKS

As this chapter has demonstrated there is a wide range of options for the cleaning of heat exchangers both on-line and off-line. The final choice adopted for the relief of a particular fouling problem depends on many factors but the prime considerations will be that of cost and safety. Cost, not only for the cleaning operation itself and the associated equipment requirement, but also for off-line cleaning, the down time involved, will usually dictate the method adopted. Safety not only of personnel involved in the cleaning operations, but also of the plant itself and the environment must be given serious consideration in relation to the techniques available. The cheapest may not be the safest.

There are many specialist companies with experience in off-line cleaning of heat exchangers and it could be extremely beneficial to call upon their expertise to tackle difficult problems. At the same time adequate supervision of the cleaning operation must be maintained at all times in the interests of efficiency as well as safety.

REFERENCES

Aschoff, A.F., Rose, M.D. and Sopocy, D.M., 1987, Performance of mechanical systems for condenser cleaning. Report CS 5032, Research Project 2300 - 11, EPRI, Palo Alto.

Ashworth, V., 1986, The theory of cathodic protection and its relation to electrochemical theory of carrosion, in: Ashworth, V. and Barker, C.J.L. eds. Cathodic Protection: Theory and Practice. Ellis Horwood Ltd., Chichester.

Bott, T.R., 1990, Fouling Notebook, Institution of Chemical Engineers, Rugby.

Bott, T.R., 1991, Mitigation of fouling in design (and operation), in: Foumeny, E.A. and Heggs, P.J., eds. Heat Exchange Engineering, Vol. II, Ellis Horwood, New York, Chapter 7.

Bridgwater, J. and Loo, C.E.,1984, Removal of crystalline scale: mechanisms and the role of thermal stress. Proc. 1st UK Nat. Conf. on Heat Transfer, Vol. 1, 455 - 463, I.Mech.E./I.Chem.E.

Burbridge, E., 1986, Process plant and cooling water systems, in: Ashworth, V. and Barker, C.J.L. eds. Cathodic Protection: Theory and Practice. Ellis Horwood Ltd., Chichester.

Callow, M., 1990, Ship fouling: problems and solutions. Chem. and Ind. 123 - 127, 5th March.

Chenoweth, J.M., Final report of HTRI/TEMA Joint Committee to review the fouling section of TEMA standards. Heat Trans. Eng., 11, 73 - 107.

Cox, G.C., 1940, US Patent, 2, 200, 469.

Crittenden, B.D., Kolaczkowski, S.T. and Takemoto, T., 1993, Use of in tube inserts to reduce fouling from crude oils. AIChE Symp. Series, 295, 89, 300 - 307, Atlanta.

Dexter, S.C., Sibert, A.W., Duquette, D.J. and Videla, H.A., 1989, Use and limitation of electrochemical techniques for investigating microbial corrosion. Proc. NACE Conf. Corrosion '89. Paper 616.

Dexter, S.C., Moettus, L.N. and Lucas, K.E., 1985, On the mechanism of cathodic protection. Corrosion 41, 598 - 607.

Donaldson, J.D., 1988, Scale prevention and descaling. Tube International 39 - 42, January.

Donaldson, J.D. and Grimes, S.M., 1987, Scale prevention in steel pretreatment by magnetic treatment. Steel Times Int. 44 - 45, December.

Donaldson, J.D. and Grimes, S.M., 1988, Lifting the scales from our pipes. New Scientist, 43 - 46, 18th February.

Duffy, E.A., 1977, Investigation of magnetic water treatment devices. PhD Thesis, Clemson University.

Edyvean, R.G.J., 1984, Interactions between micro fouling and the deposit formed on cathodically protected steel in sea water. Proc. 6th Int. Congress on Marine Corrosion and Fouling. Athens, 469 - 483.

Eimer, K., 1985, Recommendations for the optimum cleaning frequency of the Taprogge tube cleaning system. Technical Report 85 - 26, Taprogge Gesellschaft mbH, Wetter.

Eliassen, R., Skrinde, R.T. and Davis, W.B., 1958, Experimental Performance of "miracle" water conditioners. J. Am. Water Assn. 50, 1371 - 1385.

ESDU, 1992, Fouling and slagging in combustion plant. Data Item 92012, ESDU Int., London.

Fontana, M.G., 1986, Corrosion Engineering, 3rd Ed. McGraw Hill Book Co., New York.

Gardner, G., 1994, No fouling. Chem. Engr. No. 541, 11th August, Supplement S9.

Garrett-Price, S.A., 1985, Fouling of Heat Exchangers, Noyes Publications, New Jersey.

Gough, M.J. and Rogers, J.V., 1987, Reduced fouling by enhanced heat transfer using wire matrix radial mixing elements. AIChE Symp. Series 257, Vol. 83, 16 - 21, Pittsburgh.

Gough., M.J., 1994, Tubeside inserts solve refinery wax crystallisation fouling problem. Chem. Tech. Europe, 1, 1, 21.

Guillan, N.A., Feliu, S., 1966, Contribucion al estudio de algunas variables que afectan a la proteccion catodica del acero. Revista de Metallurgia 2, 519 - 532.

Hewitt, G.F., Shires, G. and Bott, T.R., 1994, Process Heat Transfer. CRC Press Inc. Chapter 28.

Hovland, A.W., Rankin, D.A. and Saxon, E.G., 1988, Heat exchanger tube wear by mechanical cleaners. Proc. Power Generation Conf., Philadelphia, 25 - 29 Sept.

Hughes, D.T., Bott, T.R. and Pratt, D.C.F., 1988, Plastic tube heat transfer surfaces in falling film evaporators. Proc. Second UK Nat. Conf. on Heat Transfer, Vol. 2, 1101 - 1113.

Jackson, A.T., 1984, Cleaning characteristics of a plate heat exchanger fouled with tomato paste using 2% caustic soda. 1st UK Nat. Conf. Heat Transfer, I.Mech.E./I.Chem.E., 1, 465 - 472.

JRC International, 1994, Engineered Explosive Cleaning Services.

LaQue, F.L. and May, T.P., 1965, Experiments relating to the mechanism of cathodic protection of steel in sea water. Proc. 2nd Int. Congress on Metallic Corrosion, NACE.

Lester, G.D. and Walton, R., 1982, Cleaning of heat exchangers. Proc. Conf. Practical Applications of Heat Transfer. I. Mech. E. C55/82, 1 - 9.

Maines, A.D., 1993, Interactions between marine fouling and cathodic protection scale. PhD Thesis, University of Leeds.

Mott, I.E.C. and Bott, T.R., 1991, The adhesion of biofilms to selected materials of construction for heat exchangers. Proc. 9th Int. Heat Trans. Conf., Jerusalem, Vol. 5, 21 - 26.

Nordell, E., 1961, Water Treatment for Industrial and Other Uses. 2nd Ed. Reinhold Pub. Corp., New York.

Page, M., 1980, Vitreous enamels provide cost effective corrosion protection. Corr. Prevention and Control 17, February.

Pathmanaban, S. and Phull, B.S., 1982, Calcaneous deposits on cathodically protected structures in sea water, UK Nat. Corr. Conf. 165 - 168.

Redman, J., 1989, Non-metallic heat exchangers. Chem. Engr. No. 459, 17 - 26 April.

Santos, R., Bott, T.R. and Callow, M.E., 1992, Coated surfaces in relation to biofilm formation, in: Melo, L.F., Bott, T.R., Fletcher, M. and Capdeville, B. eds. Biofilms - Science and Technology, Kluwer Academic Publishers, Dordrecht.

Scully, J.C., 1990, The Fundamentals of Corrosion, 3rd Ed. Pergamon Press, Oxford.

Uhlig, H.H. and Revie, R.W., 1985, An Introduction to Corrosion Science and Engineering, 3rd Ed. John Wiley and Sons, New York.

Wolfson, S.L. and Hartt, W.H., 1981, An initial investigation of calcareous deposits upon cathodic steel surfaces in water. Corrosion, 37, 70 - 76.

# CHAPTER 16

# Fouling Assessment and Mitigation in Some Common Industrial Processes

## 16.1  INTRODUCTION

A discussion of fouling problems associated with a wide range of industrial processes is outside the scope of this book but it is useful to discuss fouling and its mitigation in some common industrial processes. Cooling water is used in many industries, e.g. power generation, refineries and chemical manufacture and although problems are often site specific, it is useful to give a general background to cooling water fouling. Steam is a common heating and working utility in industry and it is often produced by the combustion of fossil and other fuels in suitably designed steam raising plant. Combustion therefore is associated with power generation, and most process industries. Of special interest is the food industry where, in addition to maintaining heat transfer efficiency, the cleanliness of heat exchangers is also a factor in hygiene and product safety and quality.

## 16.2  COOLING WATER SYSTEMS

In general any water available in sufficient quantity, may be used for cooling purposes, but the wide spectrum of water quality will reflect the variety of fouling problems likely to be encountered. The emphasis in cooling water management is on quantity rather than quality. Indeed the use of good quality water is not justified in most industrial cooling systems. Water with a low propensity to foul however, is required for totally enclosed systems where access for treatment to combat fouling is either severely restricted or non-existent.

The two principal ways in which cooling water systems operate are:

1. "Once through" where the water is passed through the cooling system and rejected, usually back to the source (e.g. sea, lake or river). The inlet and discharge locations need to be carefully sited so as to avoid mixing and hence a reduction in cooling efficiency.

2. Recirculating systems where the water is recycled through the system several times before rejection in "blowdown". In order to maintain cooling efficiency it is necessary to reduce the temperature of the water after it has passed through the heat exchangers, by evaporation of some of the circulating water. The evaporative cooling may be carried out in a cooling tower or in a spray pond. In both methods the concept is to present a large wetted surface area to the atmosphere to encourage evaporation thereby removing heat as latent heat.

Cooling the water may be enhanced by forced air passing through a cooling tower of suitable design.

A simple sketch of a recirculating system employing a cooling tower to reduce water temperature is given in Fig. 16.1.

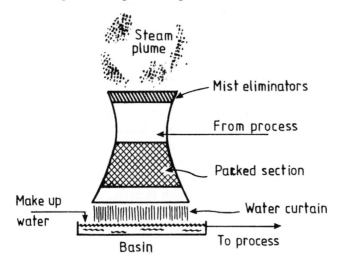

FIGURE 16.1.   A simplified flow diagram of a recirculating cooling water system.

Sources of water and application are presented in Table 16.1.

TABLE 16.1

Sources of water and cooling systems

| Source | Usual application |
| --- | --- |
| Town's and borehole water<br>Steam condensate | Usually recirculating due to the<br>relatively good quality and cost |
| Lake, river, stream or canal water | Often once through because of the<br>variable quality, but may be<br>recirculating |
| Treated sewage and waste water<br>Process water | May be recirculating or once through |
| Sea, esturine or brackish water | Usually once through |

In once through systems the fouling potential is associated with the source of the water which may contain particulate matter and debris of various sorts, micro-organisms, macro-organisms and dissolved solids.   In recirculating systems the contamination may also occur due to contact with atmospheric air which is likely to contain dust particles and micro-organisms and may also contain soluble gases. The result could be a combination of particle deposition, biofilm formation, scaling or corrosion fouling.   Process leakage (via heat exchangers) into the cooling system may also produce contamination.

Water quality in respect of particulate, crystallisation and corrosion fouling will involve:

alkalinity or acidity

hardness

dissolved gases         expressed  as *mg / l*

suspended solids

*pH*

Biological data will cover:

microbial population  *cells / ml*

biological oxygen demand  (BOD)

concentration of organic material *g / l*

Unless the majority of information is available it is not possible to make a proper assessment of the potential fouling problem.

An analysis of a typical cooling water in respect of non biological fouling might be as given in Table 16.2.

TABLE 16.2

Analysis of a typical cooling water

| Constituent | Expressed as | Concentration *mg/l* |
|---|---|---|
| *Ca* hardness | $CaCO_3$ | 125 - 175 |
| Chloride | $Cl^-$ | 50 - 175 |
| Sulphate | $SO_4^{--}$ | 125 - 180 |
| Nitrate | $NO_3^-$ | 25 -   40 |

Other typical properties might be:

| *pH* | 7.5 - 8.2, i.e. slightly alkaline |
|---|---|
| Conductivity | 80 - 120 *mho/mm* |
| Langelier Index | -0.5 to +1.0 |

The Langelier index (see Chapter 14) suggests that the water is near equilibrium but it could have a slight tendency to either scale formation or to corrosion depending on local conditions.

The biological character of the water is likely to be variable particularly in respect of the season, i.e. large differences between summer and winter.

Plant and operating features that influence the potential for fouling are:

Plant design

Residence time of the cooling water in the plant

Water consumption (make up water addition)

Changes in heat load and consequent changes in the interface temperature between the cooling water and solid surfaces, e.g. heat transfer surfaces

Cooling water velocity

## 16.2.1 Cooling system design to reduce fouling potential

There are a number of features in the design of cooling water plant that need attention in order to reduce or even eliminate the incidence of fouling. The extent of the individual mechanisms of particulate deposition, corrosion, scale formation and biofouling, will depend on the quality of the cooling water, the temperature and the material of construction.

### Basic concepts

Simple designs are to be preferred since this reduces the opportunity for fouling to occur and facilitates cleaning. It may be desirable to have back flushing arrangements for heat exchangers, particularly if the cooling water flows are through the shell side of shell and tube heat exchangers. Other aspects that need consideration before design work is undertaken include:

1. The choice of material of construction will influence the extent of possible corrosion and may affect the retention of other deposits, for instance corrosion resistant titanium tubes have been known to foul more rapidly than some stainless steel alloys under similar conditions. It is unlikely in large cooling water systems, e.g. power station condensers, that treatment of the heat exchange surfaces will be considered as a feature for the reduction of fouling, due to the relatively high costs involved including initial cost and maintenance.

2. Where the design of a shell and tube heat exchanger is anticipated a choice has to be made in respect of whether the water flows through the tubes or in the shell. In general the cooling water is passed through the tubes in order to facilitate cleaning. In steam condensers for instance, the condensate side of the exchanger is generally quite clean. On the other hand, if the process fluid is under a relatively high pressure, in order to reduce capital cost it may be necessary to pass the process stream through the tubes and the cooling water through the shell; such an arrangement avoids the necessity and high cost, for a

thick walled shell to contain the high pressure of the process fluid. Under these design requirements it will be necessary to take careful note of the increased cleaning difficulties that will undoubtedly occur and to allow for them in the design.

In order to reduce the potential fouling on the inside of tubes through which cooling water is flowing a velocity in the region of 1 or 2 *m/s* will be required. Lower velocities are likely to encourage fouling and higher velocities may result in excessive pumping costs, and dependent on equipment design and water quality, erosion may be encouraged. If it is necessary for the cooling water to flow through the shell side a velocity of 1 *m/s* based on the maximum flow area, is recommended. Low velocities in the shell may give rise to heavy tenacious deposits that are difficult to remove, even by the use of chemicals, because of the restricted access.

At the initial design stage it is necessary to consider the range of velocities necessary over a twelve month period to maintain the required cooling capacity. During the winter cooling water temperatures will be generally much lower than in the summer months and, as a result, velocities through heat exchangers in the winter are likely to be much lower than in the summer for the same duty. To some extent this is self compensating, i.e. biological activity is less in winter compared to summer operation, but other mechanisms such as particulate deposition, may be higher in winter due to the generally higher suspended solids content of natural waters (e.g. rivers) resulting from higher "run off" and soil erosion.

The basic concept of the plate heat exchanger and the associated high turbulence of water flowing between the plates, reduces the potential for fouling. The unique design also facilitates cleaning. Where large quantities of heat have to be removed such as in power station condensers, the use of plate heat exchangers is not generally possible due to the limitations of unit size.

3. The operating temperature of the system, touched on under 2 above, requires careful attention to reduce the potential for fouling. The fouling mechanisms discussed in Chapters 7 - 12 are all to a greater or lesser extent, influenced by temperature. In the temperature range generally encountered in cooling water systems fouling is likely to be encountered at the higher temperatures, but the extent of the fouling will be dependent on the water quality as described earlier in this chapter. It is difficult to give precise recommendations in respect of temperature of operation due to the wide variation of water quality and operating conditions, but an outlet temperature of the cooling water from the heat exchanger no more than 40°C will help to restrict the fouling. A further restriction on temperature to reduce deposition problems would be to design for a water/metal interface temperature of less than 70°C. Limitations on temperature may be imposed by suitable choice of velocities not only in respect of the cooling water, but also the process fluid being cooled.

*General design of the cooling water circuit*

The design of the cooling water system external to the heat exchanger can influence the potential for fouling in the associated heat exchanger.

The distribution of the cooling water around the site requires specific attention. Two basic types of system are possible, namely radial distribution and a ring main. The former is susceptible to settlement and possible biological activity and corrosion in "legs" that are closed because the process plant served is not operating. Settlement may result in blocked lines and difficulties in restarting, with consequent effects on heat exchanger performance. In a ring main, provided a number of process units served are operating, there is a positive flow that reduces the possibility of settlement and deposition. It is essential to maintain flows wherever possible and regions of no flow (dead legs) avoided.

It is valuable to provide connections on the cooling water system, that may be used for flushing to dislodge any accumulations of deposits that might occur. The opportunity to isolate heat exchangers by the inclusion of bypass systems, must be included so that off-line cleaning can be carried out without the need to shut down whole sections of the total cooling water circuit.

Consideration at the design stage should be given to the possible need for filtration. Much depends on the quality of the water available, and any trends in water quality that are likely to occur in the future. Filtration may be required at the point of abstraction (e.g. from a river or lake) for once through systems and to reduce the load on the main filters in recirculating systems. In the latter, filtration of the total flow is likely to prove uneconomic, so that side stream filtration should be considered to maintain acceptable particulate concentrations. Sidestream filtration of up to 10% of the total flow is a compromise. Most filter systems will require some kind of back flushing facility with clean water (mains or filtered) for cleaning and throughput restoration.

*Cooling towers*

Although the design and operation of cooling towers are not in general, directly associated with the fouling of heat exchangers, they may influence conditions within the system as a whole, that could affect fouling mechanisms. It is useful therefore to discuss cooling towers to complete the background to likely fouling in cooling water systems.

In order to encourage evaporative cooling there is a need to bring atmospheric air into intimate contact with the water to be cooled. The cooling tower generally contains suitable packing material that provides extended surfaces for the mass transfer of water vapour from the liquid phase into the air stream (see Fig. 16.1). It is usual for the water to fall under gravity through the tower countercurrent to the upward flow of air. In some designs there is crossflow of the air through the downflow of water. Modern cooling tower designs and retrofit situations, employ plastic vacuum formed packings as an alternative to the earlier use of wooden slats.

Wood, particularly when wet, is heavy and susceptible to biological attack; plastic offers an inert and lighter structure at a generally lower cost.  A disadvantage of plastic tower fills however, is that if large accumulations of deposits on the packing do occur, distortion and even collapse of the packing may result, due to the combined weight of the deposit and associated water content.  Good water quality control can do much to prevent these problems.  In a large installation such as on power stations, the air flow through the cooling tower may be achieved by natural draught.  On smaller units in the process industries for instance air flow may be achieved by a suitable fan.  If the air is "blown" through the tower the technique is usually referred to as "forced draught", but if the fan is located on the outlet air side it is generally referred to as "induced draught" operation.  Mist eliminators will almost certainly be required to prevent discharge of water droplets from the top of the tower.  Where the use of a fan is envisaged it is usual to employ the forced draught technology for two reasons:

1.  It is generally difficult to eliminate all the droplets leaving with the air flow so that in induced draught designs, these droplets are likely to enter the fan system with the possibility of erosion and corrosion and damage to the fan.

2.  The volumetric flow through the cooling tower increases due to the vaporisation of water so that a greater load is placed on the fan.

Design of towers with natural and fan assisted draught should incorporate:

1.  Adequate basin volume under the tower to cope not only with the normal water flow but also surges of flow that may occur from time to time.

2.  Traps or sumps in the basin into which silt or sludge can accumulate so that these deposits may be removed from the system by sludge handling facilities.

3.  Good water distribution over the top of the packing to provide the maximum opportunity for good mixing and mass transfer.

4.  Adequate access around the tower to facilitate cleaning and maintenance.

5.  Tower location relative to the prevailing wind direction to minimise contamination problems from dust, gases and other airborne components that may occur.

It may be necessary to take account of problems due to freezing in winter, either when the plant is not operating or under the reduced flow conditions associated with winter operation.

The detailed design of cooling towers has been described by Hewitt *et al* [1994].

*Environmental and safety considerations*

It is incumbent on the designers and operators of industrial plant that they are environmentally acceptable and safe. There are certain aspects of large scale cooling plant that require attention in this regard. The very nature of the operation of cooling towers generally results in a plume of visible steam leaving the top of the tower which may lead to rain-like deposition of water droplets downwind. The result may be one or a combination of the following:

1. Drizzle and fog in the vicinity of the tower (from the top and base of the tower).

2. Wetting of surrounding areas that may freeze in winter.

3. Contamination from dissolved salts (and possibly additives) when the water evaporates.

4. Corrosion of surroundings.

A major consideration has to be possible incidence of *Legionella* since the moist conditions, the formation of droplets and the temperature (20 - 45°C) are conducive to the growth of this bacterium. It is not usually a hazard unless it grows in large numbers [Butcher 1991], but it thrives on pipe scale, in biofilms, slimes and sludges (i.e. the same contaminants that give rise to heat exchanger fouling). There are several *Legionella* species of which *L. Legionella* is the most pathogenic. Legionnaire's Disease is the result of infection by *Legionella* and something like 12% of cases are fatal. The bacterium is also responsible for other less serious illnesses. Infection is from airborne water droplets which are generally inevitable in association with cooling towers (and spray ponds). Although the production of water droplets may be reduced by attention to water distribution in the tower, the fitting of mist eliminators, reducing air velocity especially in forced draught systems, reducing the water loading and by the fitting of side screens to reduce drift, the possibility of droplet and mist formation cannot be neglected.

Jacob [1992] reports on the implementation of new guidelines for the control of *Legionella* that came into effect on 15th January 1992 [HSE booklet HS(9)70]. The prevention or control of Legionellosis (including Legionnaire's Disease) applies to any water system (not exclusively to cooling water systems) which can release a spray or aerosol during operation or maintenance. The code lists four main requirements that must be undertaken by companies or operators:

1. Identify and assess sources of risk from *Legionella*.

2. Prepare a scheme for preventing or minimising the risk.

3. Implement and manage the scheme of precautions.

4. Keep records of the precautions implemented.

The principle method of eliminating the problem of *Legionella* is the use of a biocide. Chlorine is adequate [Butcher 1991], but it is essential that the dose levels are maintained within the water in the cooling tower.

*Dosing equipment*

In Chapter 14 the use of additives to combat potential fouling was discussed. In the use of chemical treatment for cooling water there has to be an emphasis on effective and rapid dispersion since the concentration of the additives employed must be low, i.e. a few *mg/l* where possible, to minimise cost and to reduce potential pollution problems. In general the additive formulation will be based on the need to limit corrosion (i.e. the use of corrosion inhibitors), scale formation (i.e. the use of crystal modifiers, dispersants or threshold chemicals or a combination) and biofouling (i.e. the use of biocides and dispersants). In many installations additives are injected on the suction side of the main pump so that turbulence within the pump will provide rapid mixing. In very large cooling systems multiple injection nozzles will be required to enhance distribution.

In addition there are of criteria that should be observed to provide the best opportunity for fouling control. It is important that when using oxidising biocides particularly, it is usual for there to be a system demand that requires the following to be maintained:

1. The rate of injection must be adjustable so that depending on the state of the system adequate biocide residuals are always present.

2. The rate of injection of material must be monitored, to ensure the dose is sufficient for the estimated duty and to reduce the possibility of excessive costs.

3. It is imperative that dosing takes place downstream from the cooling tower (or spray ponds) to avoid losses due to stripping by the air stream.

4. Suitable tankage for the storing of additives must be provided. The volume required will of course depend on the expected demand, but it may also require agitators or mixers to ensure a homogeneous blend. In this regard solutions rather than slurries are to be preferred in order to reduce the possibility of settlement.

5. Adequate safety features related to the particular additive must be in place.

It is sometimes necessary to retrofit dosing equipment. It goes without saying that any such addition must be properly designed, engineered and installed so as to minimise the risk of malfunction and accident.

*Modern designs*

In an attempt to reduce capital costs the volume of cooling water plant has been systematically reduced, generally by eliminating the tower basin. The total volume of the system may be relatively small, but with a high rate of circulation, i.e. a circulation time of 1 - 2 mins. For a given heat load the evaporation requirements will be identical to those of a larger traditional system where the half-life of the water in the system may be days, whereas for the small volume system it may be of the order of minutes. With such rapid flow rates and small volume, control is difficult. Serious fouling or corrosion can occur without the installation of sophisticated control and dosing equipment [ESDU 1988]. The operation may be compared with the older designs where the large volume of cooling water takes a relatively long time to react to changes in *pH*, biocide addition, corrosion inhibitor or scale reducing chemical concentration, allowing control to be practised on an incremental basis.

### 16.2.2 Blowdown in recirculating systems

In open recirculation systems water is lost from the system in a number of ways including windage loss (i.e. droplets carried out of the top of the cooling tower (or from the sprays in spray ponds) or from the basin area (drift loss), and leakage from the system, but the principal ways water is removed from the system is due to evaporation in the tower (or pond) to effect the cooling. The general effect is a concentration of the dissolved salts so that unless action is taken, the concentration will build up to such an extent that scale formation would become a major problem. The dilution provided by make up water will to some extent offset the effect, but it will not be a complete solution. It will be necessary therefore to discharge water containing a relatively high concentration of dissolved salts to be replaced with water from the source (make up) with a lower concentration of dissolved solids. The deliberate discharge of water is generally referred to as "blowdown" and may be continuous or intermittent.

A concentration factor *n*, may be defined by:

$$n = \frac{\text{dissolved salts concentration in blowdown water}}{\text{dissolved salts concentration in make up water}} \qquad (16.1)$$

Fig. 16.2 illustrates the various losses encountered in a cooling water system.

The make up water $M = E + W + B + L$ \qquad (16.2)

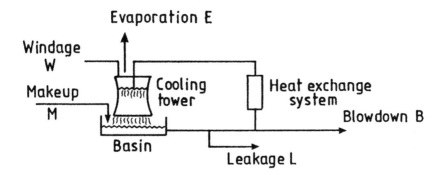

FIGURE 16.2. Water loss in cooling water system with a natural draft cooling tower

If $W + L$ are considered to be small compared to $E$ and $B$

$$M = E + B \qquad (16.3)$$

so that making a dissolved salt balance assuming no salts in the evaporative loss

$$c_m = c_b \qquad (16.4)$$

where $c_m$ and $c_b$ are the dissolved solid concentration in the make up and blowdown respectively

Therefore $\dfrac{c_b}{c_m} = \dfrac{M}{B}$ $\qquad (16.5)$

Substituting for $M$

$$\frac{c_b}{c_m} = \frac{B+E}{B} \qquad (16.6)$$

But $\dfrac{c_b}{c_m} = n$ from Equation 16.1

so $n = \dfrac{B+E}{B}$ $\qquad (16.7)$

or $B = \dfrac{E}{n-1}$ $\qquad (16.8)$

In general $n$ will be in the range 3 to 5 so that the amount of blowdown may be estimated. The precise value of $n$ will depend on a number of factors including:

1. The water temperature likely to be encountered.

2. Climatic conditions (i.e. winter and summer) and associated air temperatures and humidity.

3. The chemical nature and concentration of the dissolved salts in the make up water.

4. The changes in temperature between the inlet and outlet water of the cooling tower.

5. Residence time in the system.

6. The nature and concentration of added chemicals.

An approximation of $n$ may be obtained by assuming that the evaporation loss may be taken as 1.5% of the circulation rate $C$ (Fig. 16.2) per 10°C temperature change in the cooling tower [ESDU 1988].

16.2.3 Problems of pollution

In Chapter 14 the use of additives was discussed, in particular the application of biocides to the control of biofouling. The presence of these chemicals in the blowdown from recirculating systems or discharge from once through systems, may represent a pollution problem with a potential risk to the environment. There may be a legal requirement to treat the water before discharge which will add to the operating costs (see Chapter 14).

Quite apart from the added chemicals, the presence of an increased concentration of dissolved salts (due to evaporative cooling) particulate matter, and agglomerates may also give rise to environmental problems. The consequences may involve:

1. Interaction with ground water dissolved solids to give precipitation.

2. Disturbance of the nutrient balance in natural water including ground water to the detriment of primitive fauna and flora.

3. Oxygen depletion leading to ecological damage.

Apart from the chemical effects discharge of warm water, particularly from once through systems, may give rise to thermal pollution. If the discharge temperature is too high the environment may be affected:

1. An effect on fauna and flora in the river, lake or stream into which the discharge is made.

2. There may be surface steaming which could be regarded as undesirable.

3. Reduced oxygen solubility with effects on living matter.

The operation of the cooling system including the heat exchangers, must be such that the temperature of the discharge is kept within reasonable limits. It is possible that for particular locations there may be legal requirements to this effect. Consideration must also be given to noise pollution that may affect personnel or people working or living in the vicinity of the site. Noise may arise from:
1. Fans operating on forced draught cooling towers.

2. Noise of water falling into the basin of a cooling tower.

It is necessary therefore to minimise the effects of sound pollution to within at least, the legal limits that may prevail at a particular site.

Visual pollution may arise from excessive steam plumes from the top of cooling towers.

### 16.2.4 Management concepts and water utilisation in recirculating systems

In the light of the earlier discussions in this section it is possible to provide suggestions on the process design and management of recirculation cooling water systems. EPRI [1982] issued a comprehensive manual providing design and operating guidelines for the treatment of recirculating cooling water. This section is broadly based on the recommendations of the EPRI manual since they have been compiled from first hand experience of a wide range of different cooling water systems. The design, management and water utilisation will be concerned with a number of requirements in mind, not least the problem of fouling by the cooling water on heat transfer surfaces. The discussion will be centred on power generation plant, but the same principles may be applied to other cooling water systems.

The sources of water available for cooling purposes were listed in Table 16.1, and before a cooling water system can be designed the origin of the water must be known for the particular site since its quality will affect the fouling propensity and hence the treatment requirements. The following should be considered:
1. The water availability in the vicinity of the proposed plant. There may be limitations on the use of water from rivers, lakes and aquifiers governed by rainfall, evaporation, water rights and so on.

2. The quality of the raw water as defined in Table 16.2.

3.  Discharge restrictions that may be in force to limit discharge of substances considered to be environmentally unacceptable (see earlier).

4.  The philosophy adopted by the organisation responsible for the installation. The approach may be driven by considerations of public relations.

   Three alternative modes of operation may be considered in the light of the preliminary considerations, i.e.

1.  Zero discharge involves the reuse or safe disposal of all waste water on site. Factors that impinge on choosing not to discharge include the water availability and the economics. It may be more cost effective to treat and reuse cooling water blowdown rather than treat it to meet stringent discharge regulations.

2.  Treatment of waste water (blowdown) before discharge. The criteria are generally based on economics. As implied in (1) above it may be more cost effective to treat the discharge water to meet required standards rather than to prepare the water for reuse. If more than one stream of waste water is produced it may be economical to keep the streams separate before discharge, i.e. the stream with low contamination may be fit for discharge without treatment, whereas the stream that is highly contaminated may require treatment. It is possible of course that mixing lightly and heavily contaminated streams may product a discharge stream that is acceptable.

3.  It may be allowed to discharge some streams without treatment. In general this applies to cooling water systems where there is a plentiful water supply and which operate at low cycles of concentration without a significant addition of treatment chemicals. Discharge without treatment may also be possible where the receiving water has a large volume in relation to the discharge volume, and rapid mixing is assured to reduce the overall concentration to acceptable levels. In some locations "indirect discharge" may be possible where the waste water is used for say irrigation purposes. Care has to be exercised however, to prevent soil contamination from the build-up of pollutants from the discharge. Ground water may also be affected.

   Once the management options for water utilisations have been determined, there is a need to consider the alternative treatment opportunities that are available in the light of these preliminary assessments. EPRI [1982] provide sixteen treatment options. The treatment options are principally aimed at reducing or eliminating the potential for fouling of the water cooled condensers. Efficient heat transfer across the condenser is an essential contributor towards power production at minimum cost. Table 16.3 lists some of the advantages and disadvantages of these treatment options and Figs. 16.3 - 16.5 provide flow sheets of three common systems.

TABLE 16.3

Some advantages and disadvantages of different treatment programmes

| Treatment | Advantages | Disadvantages |
|---|---|---|
| Chemical treatment. | Low capital investment | Potentially high annual operating costs |
| Single stage make up softening. | Relatively low investment costs. Established technique. Well defined chemical dosing requirements. | High annual operating cost due to high chemical requirements. Rapid changes in make up water quality may be difficult to accommodate. |
| Single stage side stream softening. | Improved silica removal. Reduced plant size and space requirements. May be coupled to side stream filtering. | High annual operating cost due to high chemical requirements. Rapid changes in make up water quality may be diffult to accommodate. Difficulties of maintaining consistent operation. |
| Make up plus sidestream softening. | Greater removal of $Ca$, $Mg$, $Si$ and suspended solids than can be achieved by either system alone. Established techniques. Well defined chemical dosage requirements. | High annual operating cost due to high chemical requirements. Rapid changes in make up water quality may be difficult to accommodate. Difficulties of maintaining consistent operating conditions. |
| Split stream make up softening[+]. | Low capital investment. | High annual operating costs due to the high chemical requirments. Rapid changes in make up water quality may be difficult to accommodate. |

[+] This treatment option uses a conventional make up clarifier-softener system operated in parallel with a second treatment system that may itself have alternatives. The simplest arrangement involves the treatment of a portion of the make up water in a clarifier-softener and blended with the bypassed water to produce the desired hardness and alkalinity. Other alternatives are more complex with attendant operating difficulties.                    (cont'd)

TABLE 16.3 (continued)

| Treatment | Advantages | Disadvantages |
|---|---|---|
| Two stage make up softening. | Greater removal of silica and magneisum than can be achieved in a single stage softener. Established technique. Well defined additive requirements. | Higher capital investment than for a single stage process. High annual operating costs due to high chemical requirements. Rapid changes in make up water quality and flow rate may be difficult to accommodate. |
| Ion exchange softening of make up. | Hardness salts and silica can be controlled. Blowdown can be reused prior to final disposal. | High capital investment, maintenance and operating costs. Unknowns associated with sidestream treatment using aluminium sulphate. |
| Reverse osmosis or electrodialysis treatment of sidestreams. | Concentrations of all ions and suspended solids can be controlled. | High capital investment, maintenance and operating costs. Problems of membrane fouling. |
| Single stage make up softening plus sidestream treatment with reverse osmosis. | Concentrations of all ions and suspended solids can be controlled. | High capital investment, maintenance and operating costs. Rapid changes in make up water quality may be difficult to accommodate. Problems of membrane fouling. |

(cont'd)

TABLE 16.3 (continued)

| Treatment | Advantages | Disadvantages |
|---|---|---|
| Activated alumina treatment of sidestream. | Simple design. Ease of operation. | Problems of fouling of the activated alumina. Little experience of the technique for large cooling water systems. |
| Combined make up and sidestream softening. | Improved silica removal due to increased temperature and ion concentrations in the reactor vessel. Reduced chemical requirements. | Wide variations of temperature in reaction vessels may be difficult to accommodate. High annual operating costs due to high chemical requirements. Rapid changes in make up water quality may be difficult to accommodate. |
| Sulphate recovery. | Optimisation of chemical use. Lower combined capital investment and annual cost than other systems capable of the same result. | Little experience of the technique for large cooling water systems. High annual operating costs due to high chemical requirements. Rapid changes in make up water quality may be difficult to accommodate. Difficulties of maintaining consistent operating conditions. |

(cont'd)

TABLE 16.3 (continued)

| Treatment | Advantages | Disadvantages |
|---|---|---|
| Waste water concentration. | Ease of operation and maintenance. Reduction of the discharge volume. Produces high quality water for reuse. | High capital investment and annual costs. |
| Waste water cooling tower. | Reduction of the volume of blowdown. Reduction of the heat load on the main cooling tower system. Low operating costs. | Little experience of the technology. Waste water is not recovered for reuse. |
| Continuous mechanical cleaning. | Low capital investment. Ease of operation. Low space requirements. | High annual cost of replacing balls or brushes (see Chapter 15). Potential for abrasion of tube surfaces. "Difficult" scales may not respond to the treatment. |

Fig. 16.3 gives the flow sheet of a cooling water system commonly used in power generation that is also used in the process industries. Chemicals such as acids, biocides, scale and corrosion inhibitors and dispersants are added to control problems of fouling in the recirculating system (described elsewhere in this book). The technique is generally applicable where the make-up water is generally of low hardness and silica and low concentration factors are employed.

The location of the biocide injection is carefully chosen so that the biocide will be active in the tower as well as in the condenser tubes. Active biocides in the tower will reduce the accumulation of biofilms on the tower internals and eliminate problems from *Legionella*. The *pH* of the cooling water may also be adjusted by acid (or alkali) addition in the water entering the tower. The point of addition of scale and corrosion inhibitors is shown to be immediately before the water enters the condenser, i.e. to give maximum protection to the heat exchanger surfaces in the condenser.

The principal advantage of basing the mitigation of fouling entirely on chemical treatment is the relatively low investment cost of the dosing equipment. On the other hand there is potentially, a high annual operating cost.

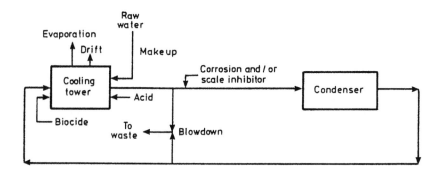

FIGURE 16.3. Simplified flow sheet of a power station cooling water circuit using chemical additives

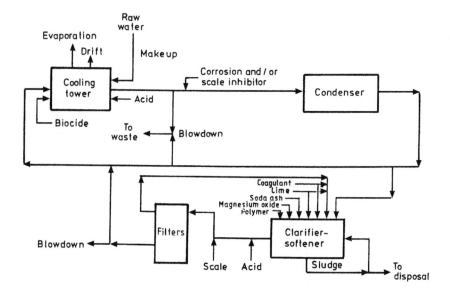

FIGURE 16.4. Simplified flow sheet of a power station cooling water circuit with side-stream softening

Fig. 16.4 gives the flow sheet of a cooling water system employing a single stage side stream softening. The technology uses a clarifier-softener system to treat a sidestream from the closed cycle cooling water system. The system involves a clarifier softener with the ability for sludge recycling followed by suitable filters. Adequate storage facilities for the additive chemicals is also a requirement of the system. A waste sludge disposal system is also required. The principle of the system is that concentrations of calcium, magnesium, bicarbonate alkalinity, silica and total suspended solids built up in the circulating system, are reduced. A smaller volume of water is used than would be the case if the total volume of recirculating water had to be treated. Silica removal is improved since the system operates at elevated temperatures with increased concentrations of silica and magnesium, so that the technology is very suitable to systems where silica removal is a prime requirement.

Fig. 16.5 is a simplified flow sheet of a power station equipment with a sponge rubber ball cleaning system (e.g. Taprogge system). The technique is described in some detail in Chapter 15. The technique reduces the problem of biofouling, scale formation and particulate deposition. The system requires suitable pumps, filter screens and storage for cleaning devices. In addition to the mechanical treatment (only applicable to the condenser tubes) chemical treatment is also required to combat fouling in other parts of the system.

The advantages of using a mechanical system include a relatively low investment cost with low space requirements. The system is relatively simple to operate in comparison to a chemical dosing system that requires constant surveillance. The disadvantages are represented by the high cost of ball (or brush) replacement, the possibility of tube erosion by the abrasive surfaces of balls (or brushes). The effectiveness of the system may be impaired by the presence of hard scales and inadequate filtering of the feed water.

The relative installation costs of a sponge rubber ball system and simple chlorination and chlorination with dechlorination for different power stations are illustrated in Fig. 16.6. Dechlorination may be increasingly required where there are environmental problems associated with discharge. The cost data are historial but the general relationship is likely to hold with today's prices. Sponge rubber ball systems are more expensive than chlorination installations for all plant sizes, but there is a definite cost advantage when dechlorination is required.

16.2.5 The cost penalties of cooling water fouling

Because the performance of a power station is very dependent upon the steam pressure within the condenser, cost analysis is often based on the achievable condenser pressure. In general terms the lower the condenser pressure the lower the unit production cost of electricity.

Bott [1990] has demonstrated how the overall fuel consumption for power generation increases as the condenser pressure increases. Data for a 500 *MW* power station are presented on Fig. 16.7.

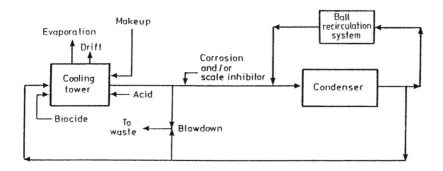

FIGURE 16.5.  Simplified flow sheet of a power station cooling water circuit with sponge rubber ball cleaning

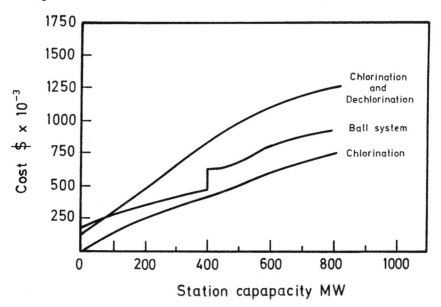

FIGURE 16.6.  A cost comparison of mechanical cleaning and chlorination/dechlorination installations

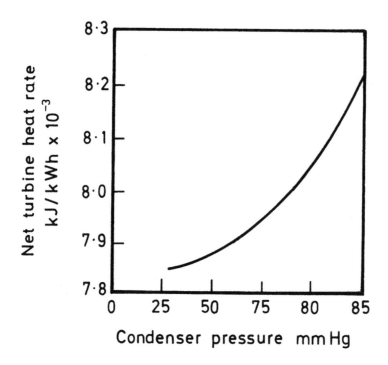

FIGURE 16.7. The effect of condenser pressure on heat requirements for power generation

The heat consumption required to generate 1 *kWh* of electricity rapidly rises as the absolute pressure in the condenser increases. The data show that if the condenser pressure is allowed to rise to 75 *mm Hg* from a base of 25 *mm Hg* the increase is fuel cost is approximately 1.3%.

EPRI [1984] produced data that show how the output from a power station falls as the condenser pressure increases. Fig. 16.8 and Fig. 16.9 illustrate the effects in relation to a 1150 *MW* nuclear fuelled and a 600 *MW* coal fired power station respectively.

Data from the figures are contained in Table 16.4 and show roughly the same reduction in capacity for a similar increase in condenser pressure. Although the percentage reductions are relatively low, nevertheless they are significant in terms of the net cost of power generation. According to EPRI [1984] a 1% loss of efficiency represents in excess of $10⁶ per annum (at 1982 prices) for a typical 600 *MW* coal fired station. The annual additional energy requirement for an increase in condenser pressure above about 40 *mm Hg* absolute for the nuclear and coal fired units is shown on Fig. 16.10.

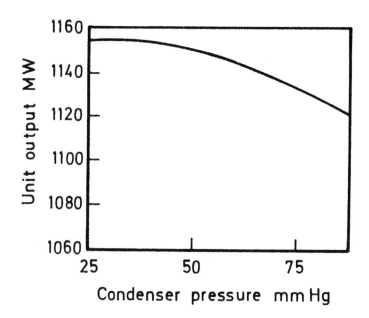

FIGURE 16.8.  Capacity in relation to condenser pressure for a 1150 *MW* nuclear fuelled power station

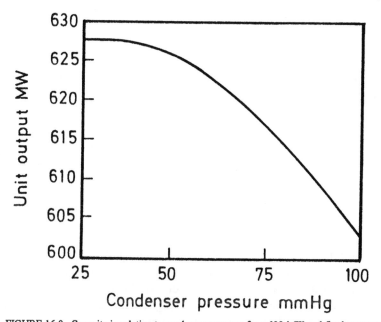

FIGURE 16.9.  Capacity in relation to condenser pressure for a 600 *MW* coal fired power station

TABLE 16.4

The reduction in output due to raised condenser pressure

| Station | Condenser pressure rise *mm Hg* | Reduction in output % |
|---|---|---|
| 1150 *MW* nuclear fuelled | 40 → 90 | 2.9 (based on nominal plant capacity) |
| 600 *MW* coal fired | 40 → 90 | 2.7 (based on the 628 *MW* output achievable in the unit) |

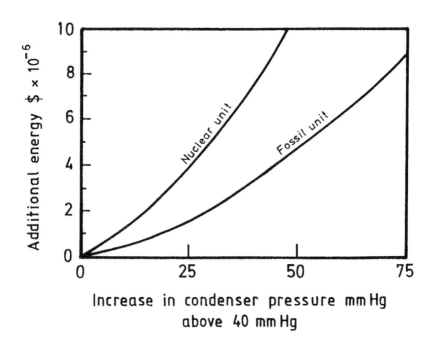

FIGURE 16.10. Additional energy costs as a function of condenser pressure above 40 *mm Hg* absolute

The figures unfortunately are historical and recognising that with increasing costs these figures are likely to be higher at current prices, they do demonstrate the enormous cost that can be attributed to condenser fouling. It has to be remembered however, that the cost of improving condenser efficiency with mitigation techniques will offset some of the savings in fuel cost that the improvement provides.

Fig. 16.11 compares the payback time for the installation of sponge rubber ball mechanical cleaning of heat exchangers [EPRI 1987]. Although these data were again published in 1987 and will therefore include historical costs, they do indicate economic advantage in respect of the system.

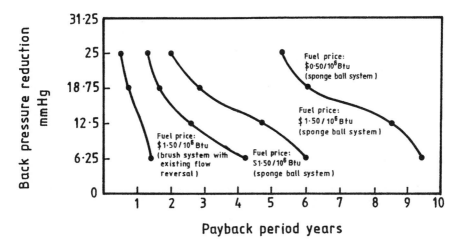

FIGURE 16.11. Payback times for mechanical condenser cleaning systems

The shorter payback times will generally be associated with higher value fuels where the economic penalties for inefficient condenser operation are greatest. The use of natural gas as opposed to residual fuel oils for instance, would provide the opportunity for short payback times.

Bott [1990] reports a long term trial aimed at comparing traditional chlorine treatment of cooling water on a 500 *MW* power station and the use of the Taprogge system of mechanical cleaning (sponge rubber balls). The comparison was based on a twelve month period of the chemical treatment followed by a further twelve month period using the mechanical system. Continuous twelve month tests were employed so that as far as possible, the effects of full seasonal changes on biofilm formation would be taken into account, i.e. the increased biological activity during the summer months and the low levels of micro-organisms encountered during the winter period. The results of the trial were reported in terms of deviations of a target condenser pressure. The target condenser pressure is, as the name suggests, the condenser pressure that is desirable to achieve acceptable efficiency. Condenser pressures above the target mean higher unit production costs, whereas condenser pressures below target represent a bonus over and above what is considered acceptable efficiency. Figs. 16.12 and 16.13 provide the comparison.

It will be seen from Fig. 16.12 that virtually the whole of the twelve month period with chlorine treatment the target condenser pressure was not attained, although it has to be said that there would appear to be a long term trend towards achieving the target. It has to be remembered in assessing the effectiveness of chlorine treatment, that this technique had been employed on the power station for many years and therefore the expertise of the operating staff would have been fully developed, and the results the best that could be obtained with the system.

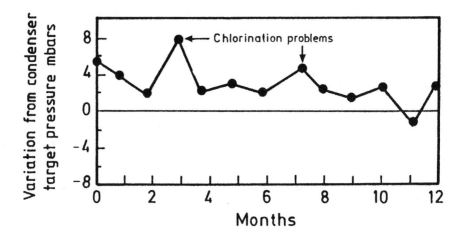

FIGURE 16.12. Condenser pressure variation on a 500 *MW* station over a 12 month period using chlorine

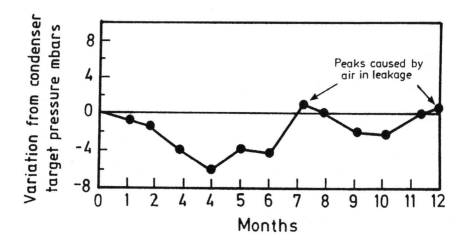

FIGURE 16.13. Condenser pressure variation on a 500 *MW* station over a 12 month period using a sponge rubber ball cleaning system

For almost all the time after the Taprogge system had been installed, the condenser pressure was below the set target, i.e. there was a "better than expected" efficiency (see Fig. 16.13). The short period when the condenser pressure was slightly above the target pressure were attributed to the ingress of air into the steam space. It is reported that after the trial, inspection of the tubes revealed that they were clean and of bright appearance. The overall heat transfer coefficient for the mechanically cleaned tubes was 5 - 10% better than the corresponding coefficients for chlorine "cleaned" tubes. Even allowing for the cost of installing the Taprogge system the improvements in the production costs were appreciable.

From this brief discussion it is apparent that the effective use of mechanical cleaning of condensers to reduce the effects of biofouling, is well established. The use of higher grade fuels and the need to reduce detrimental effects on the environment from the use of chemicals and additional flue gas emissions, will encourage the installation of the technology.

## 16.3    COMBUSTION SYSTEMS

In general, although not exclusively, industrial combustion of fuel is used to raise steam and this will be the basis for the discussion in this section. Other processes involving combustion are furnaces for high temperature processes in refinery operations and chemical manufacture.

In this book the term fouling has been applied to any deposit that accumulates on a heat transfer surface and impedes the heat transport process. In combustion operations the term fouling has a more specialised meaning and may be contrasted with the term slagging. Slagging and fouling in combustion plant have been reviewed [ESDU 1992]. The following are general definitions of these two terms as they refer to combustion systems:

1. *Fouling*: may be defined as the formation of loose, discrete particles or loosely sintered deposits usually in the convection zone of the boiler, for example, in the secondary superheater pendants. In general it is particulate matter carried forward by the flue gas to the cooler parts of the combustor.

2. *Slagging*: may be defined as the formation of fused or sintered deposits usually in the radiant zone of a boiler (the location of the highest temperatures), that is, the furnace walls, ash hopper and radiant superheater. In general the term slag refers to material that is or has been, molten.

There are regions in the combustion space where it is difficult to establish whether or not the problem is one of fouling or slagging and indeed it may be necessary to change the definition at a particular location as the deposition process proceeds. For instance on water tubes, because of the relatively low temperature of the surface at the beginning of the deposition process, the deposit may reside as individual particles. As the accumulation of deposit continues the outer surfaces

*Fouling of Heat Exchangers*

may reach temperatures approaching that of the high temperature flue gases; under these conditions the material may begin to melt and fuse possibly aided by the presence of fluxing agents derived from the mineral matter originally in the fuel but carried, and possibly transferred chemically, in the flue gases.

Fig. 16.14 shows the location of fouling and slagging in a typical pulverised coal steam boiler.

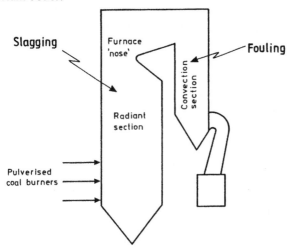

FIGURE 16.14. Location of zones of fouling slagging in a conventional pulverised coal steam boiler

Almost any combustible material may be used as a fuel in steam raising applications but the common fuels include coal, oil and natural gas. There is an increasing trend to use the heat released from the incineration of waste materials, e.g. domestic refuse, for raising steam. The mechanisms by which heat transfer surfaces accumulate deposits in combustion equipment include particle deposition, corrosion, and chemical reaction. The consequences of fouling and slagging in combustion plant are generally similar to that for other heat exchanger fouling problems, i.e. loss of thermal efficiency, reduced throughput and additional maintenance requirements, but there are other detrimental effects including:

1. The risk that large pieces of dislodged deposit may fall into the plant causing possible damage to heat transfer tubes and structure and problems of ash hopper bridging.

2. The presence of deposits may promote corrosion of metal surfaces. In a pure oxidising atmosphere the establishment of an oxide on the surface of metals is often encouraged as it provides a protective layer that limits the rate of metal attack. Most fuels contain chemical compounds or impurities that, during combustion, can give rise to corrosive agents. In the presence of deposits and in combination with particular constituents of the flue gases, for example chlorine and sulphur compounds, the corrosion effects can be much greater than in the presence of a pure oxidising atmosphere. Under such conditions the

protective oxide layer becomes involved with the chemical reactions and is destroyed or severely damaged.

3. The exit flue gas temperatures are raised with the risk of overheating of air preheaters with the potential for equipment damage. (Air preheaters are usually situated in the flue gas stream immediately before the gases are discharged to atmosphere). In contrast if the temperature of the flue gases at the exit from the equipment is too low, problems of corrosion fouling of the air preheaters can occur. In addition to the aggressive agents mentioned in 2 (above) combustion gases will contain water vapour either originating from moisture in the fuel or the combustion of hydrogen containing fuels. As the flue gas temperature is progressively lowered by the transfer of heat, condensation will occur at some specific temperature. The resulting condensate is generally acidic in character and will be highly corrosive. The temperature at which condensate appears is usually termed the acid dew point and is generally around 120°C, although its particular value will very much depend upon the fuel burnt.

The mitigation of fouling in combustion equipment therefore, is of primary concern for designers and operators with special reference to the fuel. The problem of fouling and slagging is an important consideration in the choice of combustion process and for convenience the various fuels will be discussed separately as they generally generate different problems.

## 16.3.1 Coal combustion

Three major types of coal combustor are in common use and include chain grate stokers, pulverised fuel burners and fluidised beds. The residence time of the fuel in the combustion zone of each plant is very different and leads to very different deposition potentials even using coal of similar properties.

The chain grate stoker is essentially a bed of burning coal carried on a moving grate. The principle is illustrated on Fig. 16.15. In concept the coal is fed onto one end of the moving grate and is carried by the grate through the combustion zone. By the time the coal substance has reached the other side of the plant it is reduced to ash; all the combustible material having being burnt to release heat. Air for combustion is usually admitted below the grate (primary air) and secondary air may be introduced above the grate to ensure that all volatile matter burns before it passes into the flue gas system. Tertiary air may also be employed to ensure complete combustion in the gas phase.

On a chain grate stoker the fuel and residues may remain in the high temperature region for up to 30 or 40 minutes. The majority of the ash (up to 80%) remains on the grate with only the least dense ash particles blown off the grate and carried forward in the flue gas. This is the origin of the term "fly ash", to

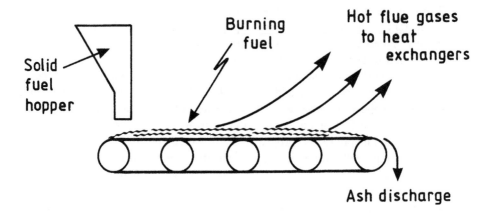

FIGURE 16.15.  The principle of a chain grate stoker fired boiler

describe the mineral matter transported by the flue gases.  In addition carbonisation and distillation of organic compounds takes place on the grate.  Under poor combustion conditions the residue can retain a relatively high carbon content (up to 20%), i.e. a serious loss of thermal efficiency.

In pulverised coal firing the coal is reduced in size to provide about 90% below 75 $\mu m$ [Highley and Kaye 1983] mixed with air in suitable burners firing into the combustion space (as generally illustrated in Fig. 16.14).  Secondary air may be admitted to ensure complete combustion is achieved.

In a pulverised fuel operation the fuel and residues remain in the high temperature region for only 1 to 2 seconds.  The majority of the ash (at least 80%) passes through the flue gas system for removal in precipitators.  The remainder of the ash goes out of the bottom of the pulverised fuel boiler.  Evolution of organic volatile compounds into the gas stream is minimised due to the short residence time of the fuel and incombustibles.

A modification of the pulverised coal plant has a combustion chamber in the form of a cyclone.  Molten slag is collected on the walls of the chamber and flows downwards and out of the bottom through a specially designed port.  The design is generally known as a "slagging combustor" or "cyclone combustor".

In fluidised bed combustion particles of coal generally in the size range that may be up to 25 *mm*, are burnt within a bed of inert particles suspended in an air stream.  The gas flow is adjusted so that all but the finer particles are retained in the bed which behaves as a fluid.  Good mixing and even temperature distribution are the hallmarks of fluid bed combustors.  Fig. 16.16 illustrates the principle of fluidised bed combustion.

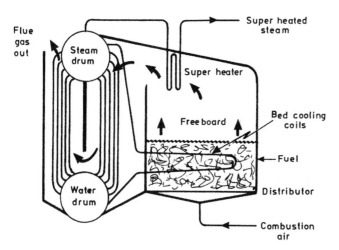

FIGURE 16.16. Schematic diagram of a fluidised bed combustor

*Mineral content of coal*

The spectrum of the non-combustible mineral deposits in coal depends on the geology of the site where the coal is located, its depth in the ground and the age of the deposits. The elements contained in the mineral matter are diverse and can include metals such as *Fe, Al, Ca, Na, K, Ti* and *Mg*. Other elements found in coal include *O, Cl, S, P* and *Si*. Mineral compounds found in coal generally include alumino-silicate (or clay) material, along with quartz ($SiO_2$), and together these impurities represent 60 to 90% of the total mineral matter in coal substance [Raask 1985]. Potassium, sodium and calcium alumino-silicates are common, but clays containing iron and magnesium are also found in coal. Anions usually associated with metals such as iron, magnesium, potassium, calcium and possibly copper and zinc include *Cl, CO₃* and *SO₄*.

Differences between the mineral contents of coals from different mines produce different slagging and fouling characteristics when burnt in an identical combustor. The physical and chemical changes that these materials experience when they are rapidly heated in a flame, have a direct bearing on the deposition process.

The method of firing (that is stoker, pulverised fuel, or slagging combustor) will also affect the extent of particular chemical reactions because each of these techniques will have different temperature/time characteristics.

The mineral matter associated with the coal burnt in a combustor may be broadly classified:

1. Inherent mineral components that are in intimate contact with the coal substance or even chemically bonded with it and finely distributed throughout the dense phase. The sodium and chlorine ions tend to be associated with the coal substance and usually evenly distributed throughout the coal mass, but not

necessarily combined as the salt *NaCl*. In general the ratio of *Na:Cl* is not 1:1 as it would be for *NaCl*.

Since the inherent mineral matter is intimately associated with the coal particles it usually attains a higher temperature during combustion than the surrounding flue gases. Under such conditions, fusion and chemical reaction are likely to occur. Conditions near the flame will probably cause the particles to experience a reducing environment for relatively long periods, that will also influence the character of the chemical reactions, particularly those involving the variable oxidation states of iron.

2. Adventitious mineral matter loosely attached to the coal substance or mixed with the coal during the mining process. A large proportion of the *Al* and *Si* in the coal appear in the adventitious material generally in the form of alumino silicate (associated with clays). Adventitious mineral material will tend to behave as isolated particles in the combustor and may experience oxidising or reducing conditions, depending on the location either within the flame or in the cooler regions.

The high temperatures experienced in the flame above a chain grate stoker or in a pulverised combustor are likely to soften or melt some mineral constituents. Changes in the shape of the mineral particles that depend upon the surface tension and the viscosity of the ash residues at the particular temperature are likely to occur as a result of the rise in temperature due to combustion. The originally irregularly sharp-edged particles are likely to assume a more rounded or even spherical form. The water loss and transformation reactions may not have gone to completion by the time these deformations have occurred. It is likely that the smallest quartz particles will fuse and become spherical in the flame and, at temperatures of 1500°C and above, will react with the other ash-forming particles.

The temperatures encountered in normal combustion conditions are generally not high enough to volatilise any silica, but it is possible that where higher temperatures do exist a small degree of silica vaporisation could occur, via the reduction of silica with carbon to $SiO$ which later reoxidises to $SiO_2$, creating a fume.

Sulphur is present in coal either organically bound or as part of some inorganic species, usually as pyrite $(FeS_2)$. In general, inorganic sulphur content is usually higher than that associated with the organic molecules, but the organic sulphur content of British coals is likely to represent of the order of 50% of the total sulphur. Under the conditions of combustion pyrite $(FeS_2)$, undergoes decomposition to $FeS$ and sulphur. The $FeS$ and elemental sulphur under the combustion conditions rapidly oxidises to $FeO$ with the release of $SO_2$.

The reactions involved are:

$$2\,FeS_2 \;\rightarrow\; 2\,FeS + S_2 \qquad\qquad (16.9)$$

$$2\,FeS + 3\,O_2 \;\rightarrow\; 2\,FeO + 2\,SO_2 \qquad\qquad (16.10)$$

and

$$5\,FeO + O_2 \;\rightarrow\; Fe_2O_3 + Fe_3O_4 \qquad\qquad (16.11)$$

The evolution of the reactive acid gas $SO_2$ from the sequence of reactions, and the oxidation of organically bound sulphur, has implications for further reactions, for example to produce $SO_3$ and subsequent corrosion of metal surfaces.

Minerals containing carbonate, for example calcite, dolomite and siderite are likely to fracture and decompose under the effects of temperature, giving rise to oxides such as $CaO$ and $MgO$ that are themselves susceptible to further chemical reactions, particularly from acidic reagents such as $SO_2$ and $SO_3$, to form sulphates.

The sodium and chloride ions will volatilise under the influence of high flame temperatures providing an opportunity to react with $SO_2$ and water vapour to produce sulphate. The reaction will be:

$$2\,NaCl + O_2 + SO_2 + H_2O \rightarrow Na_2SO_4 + 2\,HCl \qquad\qquad (16.12)$$

The particles, vapours and gases leaving the region of combustion (in stoker and pulverised coal firing but to a lesser extent in fluidised bed combustors) is a complex mixture. It will consist of melting silicates and sulphides, oxidising sulphides, vaporising sodium chloride, fragmenting and decomposing carbonates, and an oxide "fume" all undergoing reaction with sulphur dioxide and, in various combinations, with burning carbonaceous particles. The mineral particles embedded in the carbon will experience extensive coalescence and chemical interaction. It is the fate of this high temperature cloud of particles that will decide the extent and nature of the deposits on heat exchange surfaces and on the equipment. During the passage through the plant, extensive fusion and coalescence of the flame borne alumino-silicates, and their reaction with other minerals, produce particles that are responsible for deposition problems.

Because of its universal application to industrial processes there is a considerable background to attempts to characterise different types of coal in respect of their propensity to fouling and slagging. In general these are empirical assessments often involving the composition in terms of the major constituents of the coal.

16.3.2.1 Slagging assessment

The following is a brief survey of the methods adopted for slagging assessment.

*Temperature indicators*

A number of indicators have been devised, based on the effects of temperature on the ash produced from the coal under carefully controlled conditions. The temperature indices are an attempt to assess, under standard conditions, the temperature at which the solid ash becomes soft and fluid thereby giving an indication of the temperature at which slagging is likely to occur. The major standards that are used for these tests are, in no particular order:

1.  International Standard ISO 540-1981. Determination of Fusibility of Ash.

2.  British Standard BS 1016, Part 15-1970. Fusibility of Coal Ash and Coke Ash.

3.  American Society for Testing and Materials D1857-87. Fusibility of Coal and Coke Ash.

4.  German Standard DIN 51730-1984. Determination of Fusibility of Fuel Ash.

5.  State Standard of the UUS GOST 2057-82. Methods of Determining Ash Fusibility.

6.  Australian Standard AS 1038, Part 15-1972. Fusibility of Coal Ash and Coke Ash.

7.  South African Bureau of Standards SABS Method 932. Fusibility of Coal Ash.

8.  Peoples Republic of China National Standard GE 219-74. Coal Ash Fusibility Determination Method.

A much more detailed review is available [ESDU 1992].
The tests are based on the deformation of a particular shape (e.g. a pyramid) formed from a sample of the ash. An example is given in Fig. 16.17 which follows the change in shape of a pyramid of ash as the temperature is raised. The significance of the different temperature assessment points are as follows:

1.  Unheated specimen conforms to the original specimen shape and is the basis for comparison for assessing the extent of any deformation due to temperature.

2.  The initial deformation temperature (IDT) is the temperature at which the standard shape of ash just begins to show signs of deformation at its top as the temperature is raised.

The initial deformation temperature corresponds to the temperature in an operating coal combustor at which the particles of ash passing through the furnace, have cooled so far that they have only a slight tendency to stick together and form a deposit slowly on heat transfer surfaces. When the ash

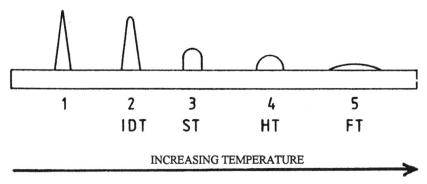

Temperature assessment points:
1. Unheated specimen
2. Initial deformation temperature, IDT
3. Softening temperature, ST
4. Hemispherical temperature, HT
5. Fluid temperature, FT

FIGURE 16.17. Specimen profiles for ASTM standard for fusibility of coal and coke ash

particles carried in the flue gas stream have cooled to below the initial deformation temperature, the deposit on cooled tubes tends to accumulate a "dry" product.

The implications for design are that the furnace exit gas temperature (that is the flue gas temperature leaving the radiant section) should be below the initial deformation temperature by about 50°C to avoid viscous flow sintering and potential slagging in the convection section.

Sanyal and Cumming [1985] proposed an alternative method of determining the initial deformation temperature based on changes in the electrical resistivity of the ash as the temperature is raised. At a temperature where melting commences, ionic conduction markedly reduces ash resistivity. A plot of the log of resistivity against the reciprocal of the absolute temperature will show a discontinuity where melting commences.

3. The softening temperature is the temperature at which the specimen has deformed to the extent that its height now equals its width. At temperatures in the coal combustor near the softening point there is a tendency for ash particles to stick together and to heat transfer surfaces.

4. The hemisphere temperature is the temperature above the softening temperature when the specimen becomes sufficiently fluid that it can assume a hemispherical shape, i.e. its height equals half its width.

5. The fluid temperature is the temperature at which the specimen assumes a flat "pancake" shape. Under plant conditions it is the temperature at which ash deposits are likely to flow and drip from the surfaces where they have accumulated.

Gray and Moore [1974] proposed an empirical formula for the so-called Slagging Index (*FS*) considered to be a measure of the slagging propensity of the coal. It is based on the initial deformation temperature (*IDT*) and the hemisphere temperature *HT*. Jones [1978] demonstrated that the Slagging Index is a useful measure of slagging potential in boiler equipment. For lignitic ash the index is given by:

$$FS = \frac{4(IDT) + HT}{5}$$
(16.13)

It will be seen that the Slagging Index has the units of temperature. In general the higher the index the lower the likelihood that slagging will occur. The range of Slagging Index and potential slagging propensity are given in Table 16.5.

TABLE 16.5

Slagging Index and potential for slagging propensity

| Slagging Index $K$ | Potential slagging |
|---|---|
| 1505 - 1615 | Medium |
| 1325 - 1504 | High |
| < 1325 | Severe |

Limitations of temperature indicators include:

1. The preparation of the ash for the test, although carried out under standard conditions, is unlikely to be representative of the burning process in industrial equipment. Under these different conditions the chemical reactions and equilibria within the ash may give rise to a different spectrum of constituents likely to influence the deformation characteristics.

2. Although the rate of rise of temperature in the test is carefully controlled, the rate of heating in industrial equipment will be quite different, and will vary between different combustion techniques, e.g. chain grate, pulverised coal and fluidised bed combustion. Furthermore in an industrial process the ash is

produced during combustion whereas in the test, combustion and ash deformation are of necessity, kept separate.

3.  There are difficulties in judging when a particular sample condition (and the associated temperature) has been reached.

Despite these restrictions however, ash fusion tests are a useful way of assessing the slagging propensity associated with different coal samples.

*Indices based on chemical composition*

Since the characteristics of coal ash will depend on its elemental components and compounds, it is not unreasonable to attempt to assess fouling and slagging potential in terms of the chemical constituents of the coal in question. In general these indices are largely based on the common minerals found in coal.

An early index is the so-called Silica ratio, *SR*, [Reid and Cohen 1944] defined on a weight basis as:

$$SR = \frac{SiO_2}{SiO_2 + Fe_2O_3 + CaO + MgO} \tag{16.14}$$

Although in principle the lower the values of *SR* (over the range 0.5 - 0.8) see Table 16.6, the greater is the tendency for slagging to occur, the ratio is not considered to be a particularly reliable method for slagging assessment. Raask [1985] suggested that the iron oxide ($Fe_2O_3$) content of ash is a better indicator of slagging propensity.

A comparison of Silica ratio and iron oxide content as indicators of potential slagging are presented in Table 16.6.

TABLE 16.6

Slagging characteristics in terms of Silica ratio and iron oxide content of coal ash

| SR | $Fe_2O_3$ wt% | Slagging potential |
|---|---|---|
| 0.5 - 0.65 | 15 - 23 | High |
| 0.65 - 0.72 | 8 - 15 | Some |
| 0.72 - 0.80 | 3 - 0 | None |

A more elaborate index, but still based on chemical composition, is the base to acid ratio $R_{b/a}$ [Raask 1985] defined on a weight basis as:

$$R_{b/a} = \frac{Fe_2O_3 + CaO + MgO + K_2O + Na_2O}{SiO_2 + Al_2O_3 + TiO_2} \tag{16.15}$$

*Fouling of Heat Exchangers*

where the ash composition is normalised to 100% on a sulphate free basis. The higher the value of $R_{b/a}$ the greater will be the tendency for slagging to occur.

Duzy [1965] and Duzy and Walker [1965] demonstrated that for US coals the lowest ash fusion temperature occurred when $R_{b/a}$ was 0.55, although in general most bituminous coals have their $R_{b/a}$ ratio between 0.2 and 0.4. Higher basic oxide contents, for example, increased $Fe_2O_3$, $CaO$ and $Na_2O$, would increase the ratio and the deposit forming propensity. A maximum value of $R_{b/a}$ of 0.5 is sometimes quoted as being acceptable for pulverised coal boilers [Raask 1985]. On the other hand, $R_{b/a}$ ratios below about 0.27 usually result in a slag that is too viscous under the temperature of operation, to flow readily from cyclone combustors [Winegartner 1974].

The base to acid ratio may be modified to improve its utility by multiplying by the sulphur content of the coal [Attig and Duzy 1969] to give the so-called slagging propensity, i.e:

$$R_s = \frac{Fe_2O_3 + CaO + MgO + K_2O + Na_2O}{SiO_2 + Al_2O_3 + TiO_2} \text{ x (total sulphur)} \qquad (16.16)$$

where the total sulphur is expressed as weight % in the dry coal.

The incorporation of the sulphur level in the formula recognises the importance of the $FeS_2$ in relation to slagging. Since the organic sulphur content is reasonably constant, the modified ratio allows for the level of pyrite. High pyrite contents in coal are known to promote slagging. For coals where pyrite is the major basic oxide, $R_s$ is effectively proportional to the proportion of ferrous iron. The ratio $R_s$ is widely used as a slagging indicator. Table 16.7 relates slagging propensity to $R_s$ but the observations do not apply to coal ashes with high calcium and low silica contents.

TABLE 16.7

$R_s$ and slagging propensity

| $R_s$ | Slagging propensity |
|---|---|
| < 0.5 | Low risk of slagging |
| 0.5 - 0.7 | Start encountering slagging problems |
| 0.7 - 1.0 | Likely to encounter severe slagging problems |

Calcium compounds are strong fluxing agents so that for coals where the concentation of this oxide is relatively high, even though the sulphur content may be relatively low, $R_s$ will tend to given an underestimate of the likely slagging and

$R_{b/a}$ is a more useful guide. Conversely if the iron content is relatively high an overestimate of the slagging propensity will be made.

Boiler design will also influence the slagging experienced in relation to the slagging index $R_s$, that is, for a given slagging index one design may be more susceptible to slagging than another. The separation between low and high risk of slagging as suggested by Table 16.7, is in the range $R_s = 0.5$ to 0.7. A great deal depends on the furnace volume per *MW* output. If the furnace volume is relatively large for a given output, a higher value of $R_s$ can be tolerated compared with a smaller furnace volume for the same output.

*Viscosity indices*

Sanyal and Cumming [1985] point out that it is the viscosity of the ash that determines its slagging propensity. The indices based on chemical analysis are really measures of viscosity since viscosity depends on the chemical composition and the ratio of acidic to basic oxides. Measurements of viscosity change with temperature and show a rapid rise in viscosity (a departure from Newtonian behaviour) when crystallisation occurs. The temperature at which the viscosity rises steeply has been called the temperature of critical viscosity, $T_{cv}$. The value of $T_{cv}$ may be used as an indication of slagging propensity: low values of $T_{cv}$ suggests a slagging problem. However, it is difficult experimentally to measure $T_{cv}$. Furthermore Gibb [1981] has drawn attention to the differences in the experimental values of the viscosity depending on whether the measurements were made as the molten slag is cooled or heated. The differences are thought to depend upon the presence of crystals within the melt that would undoubtedly affect its flow properties.

The elaborate equipment and the difficult experimental technique required coupled with the fact that ash used in the assessment may not be representative of that found during combustion (as discussed earlier), make the test unlikely to be universally acceptable as a method of assessing slagging potential.

*Phase equilibria at high temperature*

As a molten ash cools crystals will be formed. The nature of the crystalline phases formed during cooling are a function of the chemical composition of the ash, and may be made the basis of a predictive technique, as described by Kalmanovitch *et al* [1986]. The technique involves the use of high temperature phase diagrams for the system $CaO - FeO - Al_2O_3 - SiO_2$ to provide information on the slagging propensity. Good agreement between predicitons and actual boiler performance is reported.

The predictive technique involves a combination of the following steps:
1. An estimate is made of the liquidus temperature of the coal ash when the composition of the ash is normalised to the four major oxides, *CaO*, *FeO*, $Al_2O_3$ and $SiO_2$,

that is when $CaO + FeO + Al_2O_3 + SiO_2 = 100$                          (16.17)

2. The primary crystalline phase is determined from an appropriate section through the system.

3. The crystallisation path is determined and the change in composition of the residual liquid phase and the effect it has, in turn, on the viscosity of the ash is established.

The experimental technique involved with phase equilibria is of necessity, elaborate. For this reason, as with viscosity measurement, the method is unlikely to be adopted as a method for the assessment of slagging potential.

*Recommendation regarding slagging potential*

Temperature indices are useful but limited in application as discussed earlier in this section. Until improved methods of assessment become available predictions based on chemical composition are likely to yield the most useful data. Tables 16.4 and 16.5 probably provide the most reliable indication of slagging propensity along with the base to acid ratio $R_{b/a}$.

16.3.2.2 Fouling in coal combustion

In the same way that the chemical composition of the mineral matter in coal influences the slagging potential, the chemical content of the ash will also affect the potential for fouling (in the cooler regions of the combustion plant).

*Alkali metals*

Bryers [1987] has drawn attention to the importance of sodium in fouling during coal combustion, so that any attempt at predicting fouling potential should take the sodium content of the ash associated with the particular coal, into account. The concentration of $Na_2O$ in the ash may be used as an indication of the fouling potential. At the high temperatures resulting from combustion, the sodium volatilises and condenses in the cooler parts of the plant, e.g. in the convection sections of boiler equipment. Experience has shown that the element potassium can behave in a similar way to sodium and should also be considered in a fouling assessment of a coal.

An assessment of fouling potential based on $Na_2O$ is given in Table 16.8 [ESDU 1992].

Using the base to acid ratio in conjunction with the $Na_2O$ content in the ash on a weight basis, provides an alternative method of assessing fouling potential, i.e.

$R_f = R_{b/a} \times Na_2O$ in ash (weight basis)                        (16.18)

Equation 16.18 however, is considered to be less reliable than the knowledge of the $Na_2O$ content of the ash alone.

TABLE 16.8

Fouling potential and $Na_2O$ content of ash for bituminous coals

| $Na_2O$ content of ash, % weight basis | Potential fouling |
|---|---|
| < 1.5 | Negligible fouling |
| 1.5 - 2.0 | Fouling likley to occur |
| > 2.0 | Fouling inevitable |

In order to take into account the potassium content in addition to sodium a total alkalis index may be used. An empirical relationship for the total alkali content is given by:

$$\text{Total alkali content} = \% \, Na_2O + 0.6509 \, (\% \, K_2O) \qquad (16.19)$$

Table 16.9 [ESDU 1992] provides a guide to fouling potential of bituminous coal based on the total alkali of the ash.

TABLE 16.9

Total alkali of ash and fouling potential

| Total alkali content of ash wt % | Potential fouling |
|---|---|
| < 0.5 | Low |
| 0.5 - 1.0 | Medium |
| 1.0 - 2.5 | High |
| > 2.5 | Severe |

*Chlorine*

Earlier in this chapter the relationship between sodium and chlorine in coal was discussed. Chlorine is distributed throughout the coal including the mineral matter as well as the combustibles. It is logical therefore, to consider the potential for fouling in terms of chlorine as well as sodium (and potassium). The chlorine content of coal is relatively easy to determine.

An indication of the relationship between chlorine content and fouling potential is given in Table 16.10 [ESDU 1992].

TABLE 6.10

Chlorine in coal and fouling potential

| Chlorine content in coal wt % | Fouling potential |
|---|---|
| < 0.2 | Low |
| 0.2 - 0.3 | Medium |
| 0.3 - 0.5 | High |
| > 0.5 | Severe |

Although the chlorine content of coal may be used as an assessment of fouling, deposits are generally, not high in chlorine. The reason for this is probably due to sulphation of the chloride salts in the deposits. The reaction involved can be illustrated as:

$$CaCl_2 + SO_2 + O_2 \rightarrow CaSO_4 + Cl_2 \qquad (16.20)$$

With the passage of time, although the reaction may be relatively slow, the conversion to sulphate will be almost complete but with depositing particles consisting principally of chlorides [Osborn and Bott 1994].

16.3.3 Mineral oil combustion

In general deposit formation during the combustion of oils, particularly heavy fuel oils, is referred to as "fouling". Light oils may present fouling problems but in general, it is the heavier oils (i.e. viscous) that give rise to deposition problems on heat transfer surfaces. Heavy fuel oil or residual fuel oil (i.e. residues from the refining process) are very viscous at ambient temperatures, but these fuels are often used for steam raising purposes because of their relative cheapness.

Particulate matter, the source of deposit accumulation when fuel oils are burnt, may be of two kinds:
1. Carbon particles resulting from incomplete combustion of the fuel.

2. Inorganic material that may be present in the oil from the refining process or inherent in the original crude oil from which the fuel has been derived.

There is little resemblance between the mineral content of coal and fuel oil so that the fouling in each combustion process will be very different. Much higher proportions of sodium are encountered in fuel-oils than coal with relatively high vanadium contents (vanadium is virtually never present in coal). The total mineral content of oil is likely to be as low as 0.1% of the oil substance whereas the mineral content of some poor quality coal can be as high as 50%. Table 16.11

gives an indication of cationic elements that may be present in UK fuel oils [Lawn 1987].

TABLE 16.11

Major cationic components in UK residual fuel oils

| Metal | Content in *mg/kg* |
|-------|-------------------|
| Vanadium | 10 - 500 |
| Sodium | 10 - 150 |
| Nickel | 5 - 100 |
| Iron | 5 - 60 |
| Silicon | 10 - 80 |
| Aluminium | 5 - 50 |
| Calcium | 5 - 50 |

The vanadium generally associated with ashphaltenes, nickel and iron are constituents of the crude oil from which the fuel oil is derived, whereas the sodium occurs in a brine phase. The silicon and aluminium are present as a result of the refining process, generally derived from inert catalyst carriers. Sulphur is usually present in the form of organic sulphur compounds, the amount depending on the origin of the crude oil from which the fuel oil is obtained. The presence of these "impurities" is the source of fouling from oil combustion.

Whitehead *et al* [1983] have analysed the sequence of events that occur during combustion and likely to have a bearing on deposit formation. Following atomisation in the burner equipment a fine spray of droplets enters the spray diffusion flame and becomes ignited. In outline the following processes will be involved:

1. The volatile content of the fuel is lost from the droplets probably burning outside the liquid phase, and resulting in a generally "heavier" fuel within the remainder of the droplet.

2. Pyrolysis reactions occur within the droplet, giving rise to cracked and gaseous products that burn outside the fuel particles and residual porous coke particles.

3. Combustion of coke particles will usually occur near the flame region. The combustion process depends on the size and porosity of the particles which affect reactivity with the oxidising species present.

The chemical reactions occurring within an oil droplet subject to oxidising conditions at high temperatures are extremely complex. The extent of the reactions will depend not only on the chemical constituents of the fuel, but also on the droplet size as it affects the rate of diffusion of volatile components through the droplet and out of the surface. The amount of excess air present in the combustion space will also affect the rate and extent, of some of these complex chemical reactions. In addition, the metals present in the oil may serve to catalyse certain reactions.

It is likely that the large drops of oil contain the mineral matter since the volatiles will burn off rapidly. It is also likely that the burning process will not be complete if burner atomisation is poor, so that particles of carbon (coke) containing the mineral matter originally contained in the fuel, will go forward into the convection section of the combustion equipment. The porosity and pore structure of the coke particles is likely to be influenced by the initial evolution of volatile material and cracked products. The particulate matter going forward towards heat transfer surfaces is likely to consist of reacting and burning droplets that gradually solidify, but continuing to burn. The particulate matter that can accumulate on the heat transfer surfaces will consist of the inorganic and metallic residues together possibly with unburnt carbon.

A recent development in oil combustion technology is to use a stable emulsion of oil droplets (5 - 10 $\mu m$ diameter) in water using an emulsifying agent, that reduces the fouling potential by restricting the production of particulate material.

Table 16.12 provides an indication of some of the constituents of deposits found during fuel oil combustion [Tipler1964] together with their melting points. Whereas some compounds have high melting points (e.g. those of aluminium, calcium, magnesium and iron), others, particularly the sulphates of sodium, have quite low melting points.

It will be seen that many of the compounds contain sulphur (generally as the sulphate) or vanadium. Sodium is also present in different combinations.

It is the sodium-vanadium-sulphur reactions that will contribute substantially to the deposition problems in heavy fuel oil combustion. The vanadium in the fuel oil is generally associated with the combustible organic material so that, as the combustion process proceeds, the vanadium is subject to extremely high temperatures in the flame. Under these extreme conditions the vanadium will be converted to vanadium pentoxide ($V_2O_5$). At least some of the sodium present will be as sodium sulphate ($Na_2SO_4$) and Niles and Sanders [1962] suggest that only three sodium-vanadium compounds can form in reactions of $Na_2SO_4$ and $V_2O_5$:

$$Na_2SO_4 + V_2O_5 \rightarrow 2NaVO_3 + SO_3 \qquad (16.28)$$

$$Na_2SO_4 + 3V_2O_5 \rightarrow Na_2O.3V_2O_5 + SO_3 \qquad (16.29)$$

$$\text{and } Na_2SO_4 + 6V_2O_5 \rightarrow Na_2O.6V_2O_5 + SO_3 \qquad (16.30)$$

The extent of these reactions will depend upon the relative amounts of the $Na_2SO_4$ and $V_2O_5$ present in the combustion gases. Niles and Sanders [1962]

TABLE 16.12

Some compounds (and their melting points) present in deposits from fuel oil combustion

| Compound | Chemical formula | Melting point °C |
|---|---|---|
| Aluminium oxide | $Al_2O_3$ | 2049 |
| Calcium sulphate | $CaSO_4$ | 1450 |
| Ferric oxide | $Fe_2O_3$ | 1566 |
| Magnesium sulphate | $MgSO_4$ | $MgO$ at 1124 |
| Nickel sulphate | $NiSO_4$ | $NiO$ at 840 |
| Potassium sulphate | $K_2SO_4$ | 1069 |
| Potassium pyrosulphate | $K_2S_2O_7$ | 300 |
| Silicon oxide | $SiO_2$ | 1721 |
| Sodium sulphate | $Na_2SO_4$ | 882 |
| Sodium pyrosulphate | $Na_2S_2O_7$ | 401 |
| Sodium ferric sulphate | $Na_3Fe\ (SO_4)_3$ | 540 |
| Sodium metavanadate | $NaVO_3$ | 630 |
| Sodium orthovanadate | $Na_3VO_4$ | 856 |
| Sodium pyrovanadate | $Na_4V_2O_7$ | 632 - 654 |
| Sodium vanadyl vanadates | various | 535 - 625 |
| Vanadium pentoxide | $V_2O_5$ | 690 |

suggested that mixtures of vanadates are produced according to the molar ratio $V_2O_5/Na_2SO_4$, but it is recognised that these basic chemical reactions may only exist when the ash contains relatively large proportions of sodium and vanadium along with the sulphur.

Some heavy fuel oils cause severe corrosion that is attributed to the presence of vanadium compounds. In most examples of high corrosion rates these have been considered to be due to the ash composition or to a less favourable ratio of vanadium to sodium in the ash. Corrosion is very much a function of temperature (see Chapter 10). Naturally occurring contaminants other than sodium and vanadium present in the ash of heavy fuel oil tend to act as corrosion inhibitors by shifting the reaction toward the formation of the relatively non-corrosive metavanadate [Niles and Sanders 1962]. The relatively low melting points of the vanadium compounds enhance the corrosion effects.

The presence of $V_2O_2$ in the outer layers of a deposit can catalyse the oxidation of $SO_2$ to $SO_3$. This reaction has implications for the allowable flue gas exit temperatures as it affects the acid dew point in the cooler regions of the plant.

*Catalyst attrition*

The spectrum of petroleum products from straight distillation of crude oil rarely matches the market demand and there is a surplus of the "heavier" compounds. As a result it is usual to employ catalysts to assist cracking reactions to produce "lighter" products. The catalyst is often supported on alumino-silicate refractory type material. During the cracking operation a certain amount of spalling and attrition occurs giving rise to particles that accumulate in the residue. Although a certain amount of blending is practised to improve the quality of the fuel oil, the particles remain.

The quantity of particulate matter in the fuel oil will be dependent on a number of factors including the severity of the cracking process (that is, depending on the quality of the crude) and the age of the catalyst. The concentration of particles, may be up to 50 *mg/l*. The size of these particles (often referred to as "cat fines") is up to 25 *μm* with an average of about 15 *μm*.

Apart from the transport of these solid particles forward with the flue gas stream, their presence can erode burner tips. Erosion may ultimately cause malfunction and poor atomisation with consequent reduced combustion efficiency that may also accelerate the fouling process.

16.3.3.1 Assessing fouling potential

Tests based on the "evaporation" of an oil under carefully controlled laboratory conditions have been used to assess fouling potential. In general the test leaves a residue, the extent and quality of which gives some indication of the likely risk from fouling during combustion. The results of these laboratory assessments cannot be regarded as providing anything but a guide to fouling propensity and for comparison purposes. Two standard tests are the Conradson [Inst. Pet. 1986] and Ramsbottom [Inst. Pet. 1986] carbon determinations.

Whitehead *et al* [1983] point out that the Conradson carbon residue test has a poor correlation with measured particles in the combustion gases. They suggest that the differences are probably due to the very different conditions in the laboratory bench test compared with the practical combustion conditions. In an industrial furnace the oil droplets undergo a high heating rate and under these conditions the level of pyrolysis residue (coke) formed can be considerably reduced. Droplet size (a function of atomiser design and performance) is likely to be an important factor that affects the droplet heating rate as described earlier.

An additional factor affecting the extent of particulate formation in flames is the structure of the particles themselves, and their susceptibility to oxidation. The following could affect the particulate burnout efficiency:

1. The naturally occurring metals in the oil referred to earlier in this section can catalyse the post-burnout process.

2. The pore structure could be influenced by the evolution of volatiles and cracked products as discussed earlier in this section.

The high heating rates and the short reaction times present in industrial combustion equipment make experimentation and the determination of reactivity data difficult.

It may be concluded [Whitehead *et al* 1983] that the composition of a residual fuel can affect the combustion process in a variety of ways, either in the initial stages by influencing the pyrolysis reactions, or in the final stages by controlling coke reactivity and therefore the rate of burnout. A delay in the combustion process could cause a coke particle to reach a region in the flame where the lower temperatures can no longer maintain the combustion reactions. Under such conditions the particles burnout process will be quenched and unburnt coke particles will go forward through the equipment with the risk of fouling heat transfer equipment. A simple test or formula for fouling is unlikely to be satisfactory, since it will not represent combustion conditions.

### 16.3.4 Gas combustion

In general the combustion of natural gas does not give rise to problems of fouling provided that the equipment, particularly the burners, is well maintained and satisfactorily operated. The fuel is purchased as a relatively high priced commodity, the supplier having carried out any treatment that may be necessary, e.g. the removal of water and $H_2S$. The result is a fuel of consistent properties that makes combustion management a straightforward matter, in contrast to problems that may accompany the use of solid fuels such as coal or waste material, where consistency may be variable.

The generally high cost of purchasing gas is offset by the improved operations that are possible, not least the elimination of fouling problems that improves the unit cost of production. It is for this reason that there is a trend in the UK to utilise natural gas for the generation of electricity at the expense of coal fired stations.

### 16.3.5 Waste combustion

There is an increasing public awareness of problems associated with the disposal of waste. Indiscriminate dumping is no longer tolerated and consignment to landfill sites must be carefully monitored. Where combustible waste is concerned (liquid or solid) an alternative is to burn the material and this has been practised, particularly in respect of municipal waste, for many years. Often the resulting heat energy has been removed from the combustion gases by water quenching, since the emphasis has been on the disposal of the waste. In more recent times consideration has been given to utilising the heat released by combustion, e.g. to raise steam for industrial use or district heating. Power stations for the safe disposal of straw and other waste materials are being developed in the UK. The use of waste as a fuel is an extensive topic and therefore a detailed review cannot be included in this book, but a general discussion is provided as background to the problem of heat exchanger fouling.

Examples of wastes that may be used as fuel include:

1. *Agricultural wastes*
   a) Crop wastes: straw, pea vines, seed husks, nut shells, sugar cane residues, prunings.
   b) Manure: pig and cowshed slurries or solid manure.
   c) Logging wastes: residues from timber cutting.

2. *Industrial wastes*
   a) Residues: from chemical processing, off-specification products.
   b) Trimmings: textiles, plastics, furniture, wood manufacturing.
   c) Biological: fermentation residues, slaughterhouse wastes, meat processing.

3. *Urban wastes*
   a) Municipal refuse.
   b) Sewage.
   c) Hospital wastes.

Because of its widespread availability and the need for safe disposal, municipal refuse has received considerable attention as a potential fuel. One of the difficulties of burning for example municipal waste, is the relatively low calorific value and in some instances the heating value of the fuel has been improved by blending with a conventional fuel, e.g. coal. The major problems that attend the utilisation of combustible waste for heat generation are the problem of fouling of the heat transfer surfaces, and the stack emissions that may be the source of toxic gases and particulate matter. The problem of emissions (generally outside the scope of this book), may be actively addressed by attention to combustor operation, some form of filtration, e.g. electrostatic precipitation or bag filtration, to remove particles and the use of gas cleaning technology, e.g. gas scrubbing to reduce or eliminate the discharge of noxious gases.

There are alternatives to combustion that may release and utilise the heat energy contained in waste material. Municipal waste consigned to landfill for instance, degrades due to the activity of micro-organisms and other chemical reactions. One of the products is methane that may be gathered and used as a fuel.

It is possible to reduce the waste by pyrolysis and gasification techniques and to utilise the resulting gas as a fuel. Although these techniques may to some extent, overcome the problems of deposition on heat transfer surfaces, they introduce problems of their own. During the gasification process the sulphur contained in the waste material will also be volatilised, either in the form of mercaptans or hydrogen sulphide. These sulphur compounds are highly corrosive as described elsewhere in this chapter and this drawback will need to be taken into account in the design of the combustion equipment. An alternative, but more costly, method of dealing with the corrosion problem would be to remove the sulphur compounds by suitable gas cleaning methods.

Domestic refuse is a mixture of many different materials including non-combustible material, e.g. glass and metal, in addition to substances that will burn, e.g. paper, plastics and other materials with an organic base. For this reason urban waste may be regarded as non-specific. On the other hand some wastes are specific, i.e. being produced from a single original material, i.e. plastic sheet trimmings. Variations in quality in non-specific waste are almost inevitable, making it difficult in general, to assess potential fouling. On the other hand the constant quality of specific waste makes the assessment of fouling potential more straightforward.

A useful classification of waste has been provided by Antonini and Francois [1988] based on the specificity. Table 16.13 provides some information on properties of specific and non-specific wastes. The variability of the non-specific waste particularly municipal refuse is evident from Table 16.13.

TABLE 16.13

Properties of specific and non-specific wastes

|  | Specific | Non-specific |
|---|---|---|
| Origin of waste | Industrial or agri-processes | domestic or industrial refuse Hospital waste |
| Composition | Usually fixed | Variable |
| Calorific value | Constant | Likely to fluctuate |
| Combustion gas quality in terms of depositing species and pollution | Generally constant | Likely to fluctuate |
| Prime reason for combustion | Energy recovery | Waste reduction but can be associated with heat recovery |

It is not possible to provide a comprehensive discussion of the utilisation of all waste materials but, because of its widespread availability, the use of municipal waste as a fuel will be discussed in more detail. Mass burn technique generally based on the chain grate stoker technology developed for coal combustion (see Fig. 16.15), is the conventional method for disposing of large volumes of industrial solid waste and municipal waste. The same limitations and qualifications that apply to coal burning in the equipment, apply to waste combustion with problems associated with the fouling of heat exchange surfaces. The particulate matter content of the flue gases from the combustion of waste materials is likely to be much greater than for fossil fuels, a very large fraction of which (say > 80%) will be smaller than 1 $\mu m$ [Gunn 1991]. It is fouling rather than slagging that is predominant in domestic refuse incineration, although some sintering of deposits is possible as their thickness increases and the outer surfaces of the deposit approach the high temperatures of the flue gas stream. Agents leading to fouling are generated either from the non-combustible material, for example fillers from paper, or from entrained particulate matter and volatile molecular species. The fly ash associated with the combustion process contains calcium, aluminium and silica. Smaller amounts of other metals are also likely to be present in the particulate matter. Chlorine levels in deposits are significantly higher than those found in coal combustor deposits, and are considered to be derived from PVC polymer packaging contained in the refuse. In general, the flue gas atmospheres are more corrosive than those encountered from coal, oil or gas combustion. As a result, although deposit formation on heat transfer surfaces may be particulate or vapour phase condensation, corrosion and wastage of the metal surfaces can be a considerable problem.

Table 16.14 provides an analysis of some of the elements present in deposits found on surfaces associated with the incineration of a mixture of municipal waste and sewage sludge [Tsados 1986]. Some of the metals may be present in the deposit as a result of corrosion, but the high level of chlorine can only have originated from the waste (i.e. from the PVC). The $SO_4^{2-}$ content of this deposit was found to be 2.2% by weight.

TABLE 16.14

Analysis of some of the components of an incinerator deposit

| Element | % by weight |
|---------|-------------|
| *Si* | 7.3 |
| *Al* | 4.4 |
| *Fe* | 7.0 |
| *K* | 0.6 |
| *Na* | 1.5 |
| *Co* | 3.5 |
| *Ni* | 1.3 |
| *Cr* | 0.9 |
| *Mg* | 0.8 |
| *Cl* | 18.0 |

Equation 16.14 illustrates the possibility of reactions between $SO_2$ and *NaCl*. Similar reactions are possible in municipal waste combustion. Work by Osborn and Bott [1994] confirms that over a period of time chlorides in deposits are likely to be converted to sulphates. Rampling and Hickey [1988] have also presented evidence that deposits formed on heat exchanger surfaces during the combustion of

refuse derived fuel (RDF), consist largely of water soluble alkali sulphates (i.e. $Na_2$, $SO_4$ and $K_2SO_4$). There was a slight change of sulphate content from 80% in the layers close to the heat transfer surface, to around 70% at the outer deposit surface. Alkali metal salts in general have relatively low melting points. Changes in other constituents were noted between the inner and outer regions of the deposit. Rampling and Hickey [1988] attribute the changes to the increase in surface temperature as the deposit thickens making the higher melting material softer and sticky. Alkali silicates are formed in preference to alkali sulphates, resulting in a strong fused bond in the outer layer. Reported changes on a weight basis across the deposit include:

$SiO_2$ 6.2 to 9.1%
*Al*  1.6 to 2.4%
*Mg*  0.2 to 0.4%
*Ca*  1.4 to 3.1%

It may be concluded from the data that the severe fouling experienced in RDF fuelled boiler plant, is due to the volatilisation of alkali metals from the fuel bed. Oxides in reaction with acidic oxides of sulphur, or sulphates, are extremely sticky at the temperatures encountered in the equipment and tend to adhere to the cooler heat transfer surfaces. In coal burning, the alkali metal compounds constitute a sticky matrix which retains the impinging fly-ash, whereas in RDF combustion the concentration of alkali salts in the flue gas is so high that the rate of condensation exceeds the rate of fly ash retention.

Bott [1990] reports fouling problems in a boiler burning domestic refuse. In the particular case the flue gases from the moving grate mass burn furnace passed through the inside of vertical tubes of a shell and tube heat exchanger. Steam was raised in the shell. Due to fouling on the water side, fouling problems on the gas side were accentuated with undesirable consequences, particularly for the bottom tube plate in contact with the hot gases leaving the combustor and included:

1. Distortion of the tube plate.

2. Failure of the tube end/tube plate welds.

3. Fusion of deposits on the tube plate that eventually led to excessive pressure drop by virtue of the restrictions imposed at the entries of the tubes.

4. Corrosion of the tube plate.

Redesign of the boiler using conventional water tubes virtually eliminated the fouling problem.

An alternative technology to mass burn that overcomes some of the problems of waste combustion is the use of a fluidised bed. In this method of waste reduction the waste (solid or liquid) is fed directly into a bed of inert particles (e.g. sand) suspended in a gas stream. The advantages claimed for the technique [Institute Waste Management 1995] include:

1. Relatively low oxygen requirements result in lower gas volumes and as a consequence the plant size and the fan power requirements are minimised.

2. Relatively low uniform temperature allows good control of bed temperature to be achieved.

3. Due to the presence of the particles in the bed there is good "clean up" of the gas by the retention of partially burnt materials and noxious products till they have been completely oxidised.

4. Due to the lower volumes of oxygen present coupled with the relatively low combustion temperature, there are low $NO_x$ emissions.

5. The production of $CO$ and $CO_2$ is generally better controlled than in "mass burn" technology.

6. As particle size has to be controlled good "burnout" of the combustible material is achieved. With good control it is possible to obtain carbon retention in the ash as low as 0.5%.

7. The rapid mixing achieved within fluidised beds facilitates the introduction and dispersal of additives that may be required to improve operation and reduce fouling and slagging problems (e.g. to facilitate sulphur retention).

8. Furan and dioxin emissions are not a problem provided the flue gases are cooled rapidly to 260°C. Further research is being carried out to establish conditions to reduce the production of these noxious compounds.

9. Small sized units are economical for the disposal of household waste from small communities, and for the safe disposal of relatively small quantities of hazardous waste.

10. Reduced polluting emissions to the atmosphere occur during start-up, due to the rapid attainment of operating conditions and during shut down as a result of the relatively small amount of waste in the bed.

11. The amount of waste contained in the fluidised bed at any one time is relatively low.

These advantages have a direct bearing on the potential for fouling.

For successful operation it is usually necessary to treat solid waste prior to injection into the combustor unless it is of consistent quality. The purpose is to remove as far as possible, incombustible material, e.g. metals and glass from municipal waste, and to reduce the particle size of the feed.

In the past, waste e.g. municipal waste, has been incinerated in the "as received" condition in mass burn equipment, but modern methods involve the pretreatment of the waste before firing as for fluidised bed operation. The higher costs that this entails are generally offset by improvements in operation not least in respect of fouling of surfaces, due to the opportunity to maintain consistent conditions in the equipment.

In respect of operation of mass burn equipment for the incineration of municipal waste, Gunn [1990] has made some general comments in respect of fouling. He makes a comparison between refuse derived fuel (RDF) that has been densified (dRDF) by compression and a premium grade coal. His data are reproduced in Table 16.15.

TABLE 16.15

A comparison between densified refuse derived fuel (dRDF) and a premium grade coal

| Property | dRDF % by weight | Coal % by weight |
|---|---|---|
| Volatile matter | 67.5 | 25.9 |
| Fixed carbon | 10.2 | 55.5 |
| Ash | 15.0 | 10.2 |
| Moisture | 7.3 | 8.4 |
| Ash fusion temperature °C (IDT) | 1050 | 1120 |
| Bulk density, $kg/m^3$ | 500 - 600 | 650 |
| Gross calorific value, $kJ/kg$ | 18,700 | 27,200 |

Although both are graded solid fuels of broadly similar bulk densities there are significant differences in respect of other properties. The densified refuse derived fuel has a lower calorific value, higher ash content and lower ash fusion temperature. The major differences are the much higher proportion of volatile matter in the dRDF coupled with a very much lower value for the fixed carbon. These differences are likely to be sufficient to render not only the combustion properties different, but also the deposition from the respective flue gases.

For combustion of most solid fuels on a grate, the fixed carbon is either fully consumed on the grate by the primary air or partially converted to carbon monoxide which burns over the fire with the secondary air. The secondary air may be leakage air from the rear of the grate and that is common in coal firing, or it is introduced through ports or louvres. In contrast however, for refuse derived fuel (RDF) the fixed carbon content is generally lower than conventional fuels (see Table 16.15) such as coal, so that less primary air is required. Furthermore a large proportion of the relatively high level of volatile matter (in the waste) will vaporise out of the bed during the combustion. For this reason much more secondary air is required for satisfactory combustion. Gunn [1990] gives the respective ratios of secondary to primary air as:

for coal         0.48

for RDF         4.44

These ratios indicate that there is a need for a very different secondary air supply in the combustion of urban waste compared with coal. In addition, because the deposits associated with domestic refuse combustion are generally alkali metal salts of relatively low melting point, it may be necessary to consider diluting the gases leaving the combustion space to reduce their temperature to below 650°C, the sintering temperature of the salts. The dilution may be achieved by the use of tertiary air but this will, of course, reduce the thermal efficiency of the equipment.

16.3.5.1  Fouling assessment

Because of the broad similarity to coal, tests to assess the potential for fouling (and/or slagging) have been based on those developed for coal, for instance the temperature indicators described in Section 16.3.2.1 Dabron and Rampling [1988] have given some ash fusion data for refuse derived fuel from three different UK incinerators and compared them with a coal as presented in Table 16.16.

TABLE 16.16

Ash fusion temperature °C assessment of RDF from three UK incinerators compared with a standard coal

| Temperature indicator | Incinerator location | | | Standard British Coal (Tank 700) |
|---|---|---|---|---|
| | Byker | Govan | Doncaster | |
| *IDT*°C | 1050 | 1085 | 1125 | 1120 |
| *HT*°C | 1180 | 1155 | 1270 | 1390 |
| *FT*°C | 1210 | 1185 | 1290 | 1410 |

All the temperatures for the coal ash are significantly higher than the corresponding temperatures for the refuse derived fuel. The reason for these differences is that the ash produced by the combustion of domestic refuse, is different in chemical composition from the ash associated with coal combustion.

The combustion conditions, i.e. oxidising or reducing, are likely to affect the fusion temperatures [Dabron and Rampling 1988]. Work done in the United

States of America by Alter [1980] supports this view and is summarised in Table 16.17.

TABLE 16.17

Average ash fusion temperatures for RDF manufactured in the USA

| Temperature indicator | Combustion conditions | |
|---|---|---|
| | Reducing | Oxidising |
| $IDT°C$ | 1024 | 1065 |
| $ST°C$ | 1063 | 1092 |
| $HT°C$ | 1097 | 1131 |
| $FT°C$ | 1182 | 1193 |

The differences are greater for ash with a relatively high iron content due to the fluxing action of ferrous iron consistent with reducing conditions. Ferric iron (likely to be produced under oxidising conditions) will not have the same effect. Dabron and Rampling [1988] conclude that the reducing atmosphere gives a more dependable basis for standardisation and is more characteristic of the fuel bed where these conditions are more likely to exist. On the other hand oxidising conditions may alter the ash characteristics in local areas.

Rampling and Hickey [1988] suggest (based on industrial trials) that the deposits from refuse firing consist largely of soluble alkali metal sulphates (up to 80%) in contrast to coal ash which is largely insoluble. The melting point of these water soluble alkali metal sulphates is in the region of 650°C. Since the base:acid ratio $R_{b/a}$ discussed in Section 16.3.2.1, contains the weight % of $K_2O$ and $Na_2O$ it is possible that $R_{b/a}$ could be of relevance to deposition during domestic refuse incineration although there is no evidence in support of this suggestion.

In general, the empirical test methods developed over the years for traditional fossil fuels (coal, oil and natural gas) are not applicable to waste materials because they have been developed to match the expected impurities in the fossil fuels rather than those that may be present in the waste. Assessment is further complicated by the difficulty of obtaining a representative and reliable sample of the waste for testing. The problem is acute in solid wastes such as domestic refuse, where considerable variations in consistency can occur. Sampling of other solid wastes such as plastic offcuts (from a particular source) or straw, are likely to be more reliable. Ultimately it will be necessary for the operator to determine optimum operating conditions in the light of empirical experience as described in Chapter 17.

16.3.6 Mitigation of fouling and slagging in combustion systems

In general the techniques described in Chapters 14 and 15 for the mitigation of gas side fouling are employed in combustion systems to reduce or eliminate the accumulation of deposits on heat transfer surfaces. Soot blowing and sonic techniques are commonly employed, but the use of additives is of gaining importance.

## 16.4    FOOD PROCESSING

The fouling of equipment, particularly heat exchangers, associated with the processing of food, is closely related to the quality of the product in terms of its microbial content (i.e. maintenance of hygienic conditions) and the colour, texture and taste of the food. The presence of fouling deposits may encourage the growth of bacteria and fungi. Since in general food substances are usually heat sensitive, the colour, taste and texture may be affected by food damaged during processing. Often heat exchangers are employed to heat the food either to "cook" it, reduce its bulk volume by evaporation (e.g. concentration of fruit juices) or to sterilise by killing micro-organisms to improve the shelf life of the food and to render the food acceptably safe for consumption. During sterilisation for instance, the food (liquid or paste) in contact with the hot heat exchanger surface may become damaged, stick to the surface and be released subsequently as particles to affect the taste and texture of the product. The fouling process itself (i.e. the denaturation of the food) may also release breakdown products that detract from the acceptable product quality. It is for these reasons that food processing plant is frequently cleaned to ensure that the product meets stringent quality requirements. The extent and tenacity of fouling deposits will have a profound effect on the frequency of cleaning and the time necessary to return the plant to a satisfactory condition for further operation.

Where steam is used for cleaning, to sterilise either plant or product by direct injection, then it can become an integral part of the food. The composition of steam must therefore be controlled with respect to the additives employed for treatment. Since it is possible for chemicals added to boiler feed water to be carried with the steam to the point of use. No boiler feed water treatment chemical should be employed till it has been shown to be safe in respect of food product quality. In the UK permitted chemicals include [HMSO 1983]:

Sodium and potassium alginates
Unmodified starch
Polyethylene glycol with a molecular weight > 1000
Sodium and potassium carbonates
Magnesium and sodium sulphates
Monosodium dihydrogen orthophosphate
Disodium monohydrogen orthophosphate
Trisodium orthophosphate
Sodium tripolyphosphate
Tetrasodium pyrophosphate
Sodium silicate
Sodium metasilicate
Sodium aluminate

Sodium hydroxide
Sodium hexametaphosphate
Sodium sulphite (in neutral or alkaline solution)

Jones [1994] has drawn attention to the problems of biofilm formation in food processing plant. Hygiene and fouling are associated; surfaces that become fouled with food residues are likely to represent potential sources of bacterial contamination. Modern food processing plant is designed to reduce or limit, opportunities for food soil to accumulate where bacteria may thrive to the detriment of food safety and product quality. In general hygienic design avoids recesses, crevices and roughness elements where the access of deposit removal forces is severely limited or even non-existent. Food processing plant is generally made of a good quality stainless steel with highly polished surfaces to restrict the opportunity for accumulation of food deposits. Hauser *et al* [1989] have discussed the design and construction of food processing equipment with particular regard to hygiene.

Heat exchangers commonly used in the food industry include [Rose 1985]:
1.  Plate heat exchangers

2.  Concentric tube

3.  Spiral tube

4.  Shell and tube

5.  Scraped surface

Of these, the plate exchanger is extensively employed since its design may be made hygienic and it is very suitable to rapid and effective cleaning. Care in reassembling a plate heat exchanger after cleaning is required, to ensure that the tightening process is even. Unless particular attention is paid to this detail, it is possible to get variations in the spacing between plates with subsequent effects on flow distribution. Velocity variations can give rise to differences in deposit formation and possibly rapid deposit accumulation where velocities are reduced. In food processing, cleaning not only of the heat exchangers but the whole process plant may be required on a frequent basis. For particularly heat sensitive materials it may be necessary to use a scraped surface heat exchanger that may be made to be self cleaning (see Chapter 15). Where high pressures are employed the use of tubular designs may be necessary to contain the pressure at reasonable cost, e.g. the concentric tube, spiral tube or shell and tube heat exchanger.

An alternative way of heating liquid food products that eliminates surface fouling is the technique known as ohmic heating. Heat is not transferred across a surface (hence there is no hot surface to promote fouling), but is generated *in situ* by the electrical resistance of an electrically conducting lqiuid under the effect of an applied voltage [Skudder and Bliss 1987]. The process is particularly suitable for food materials containing solid pieces up to 25 *mm* cube, e.g. meat or diced vegetables.

### 16.4.1 Food fouling mechanism

Bott [1989] has described a process in the food industry, where a number of heat transfer surfaces are subject to different fouling mechanisms. Fig. 16.18 illustrates a potato crisp production unit.

The fryer temperature is maintained by passing the hot oil through a heat exchanger located in a fired heater. On the hot oil side of the exchanger, particles of say partially denatured potato from the fryer, may become attached to the heat exchanger surface, possibly with products from the decomposition of the vegetable oil used for frying. Further decomposition may take place at the hot surface that may give rise to severe fouling problems. The breakdown products could be carried forward to the fryer to the detriment of product quality.

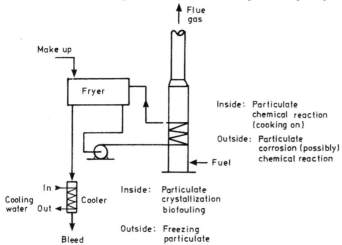

FIGURE 16.18. Schematic diagram of a potato crisp frying unit

On the flue gas side there could be problems of particulate deposition (see Section 16.3) and possibly corrosion that coupled with the deposits on the hot oil side, may lead to operating difficulties. The quality of the fuel and the maintenance of good combustion conditions will be factors in the fouling process.

In this example the heat exchanger fouling process is seen to be complex and it is unlikely that any fundamental knowledge regarding the fouling mechanism will be available, except that high surface temperatures in the heat exchanger will adversely affect deposit formation.

Problems of milk fouling in sterilisation equipment is better although not completely, understood because of the large volumes of milk and dairy products that are processed world-wide. Fouling of heat exchangers and sterilisation equipment in contact with milk, has attracted more attention than other food products. It would appear that two concurrent mechanisms are responsible for fouling from milk [Lalande and Rene 1987], i.e. mineral deposition and protein denaturation. At temperatures below 110°C the deposit is largely the so-called "milk film" or "Type A" with something like 50 - 60% protein and 30 - 35% mineral material [Burton 1968], i.e. much higher proportions than association with raw milk. It is well established that the mineral deposition is due to the complex inverse solubility of calcium phosphate [Burton 1968, Brulé 1981, Lalande *et al* 1984].

According to Lalande and Rene [1987], in an analysis of milk deposits the calcium deposit appears to be calcium phosphate dihydrates $(CaHPO_4.2H_2O)$, octocalcium phosphates $(Ca_8H_2(PO_4)_6.5H_2O)$ and hydroxyopatite $(Ca_5(PO_4)_3 OH)$ which is the least soluble form [Sandu and Lund 1985].

Milk proteins that are heat sensitive include immunoglobulins, $\beta$-lactoglobulin and $\alpha$-lactalbumine, but it would appear that it is the $\beta$-lactoglobulin that is at least, partially responsible for deposit formation from milk.

Lalande and Rene [1988] have given some data on the composition of whole milk and associated deposits. These data collected from fouling resulting from milk pasteurisation in a plate heat exchanger, are reproduced in Table 16.18 and are based on Lyster [1979] and Lalande *et al* [1984, 1985].

TABLE 16.18

Composition of whole milk and associated deposits

| Component | Milk | Composition g/100 g of dry material | |
|---|---|---|---|
| | | Heating section | Holding section |
| Lactose | 37.6 | - | - |
| Fat | 30.4 | 15.8 | 25.2 |
| Protein | 26.4 | 53.8 | 58.8 |
| Minerals | 5.6 | 26.8 | 10.0 |

Despite the relatively low level of $\beta$-lactoglobulin in the whole milk (2.5% based on dry weight of milk solids) the protein in the deposits contained almost 50% of the solids dry weight. A large proportion of the mineral deposit is calcium phosphate. Caseins in deposits were reported as 18% [Tissier *et al* 1984].

Lalande and Rene [1987] state that the protein fouling is dependent on the reactions involving -*SH* and *S-S* groupings in the protein. Fryer and co-workers [Fryer and Gotham 1988, Fryer *et al* 1989, Gotham *et al* 1989] and Schramel and Kessler [1994] have described how they consider the denaturisation mechanism responsible for milk fouling, occurs. The effect of heat on proteins is a two-stage process. Initially the effects of heat on the complex three dimensional structure of a protein, are to distort the structure, normally held together by intramolecular disulphide bridges, ion pair interactions and van der Waals forces [Fryer *et al* 1989]. Denaturation, which may be reversible when the protein is cooled, follows with an unfolding of the complex structure.  As the secondary and tertiary structure of the protein unfolds, more of the inner structure of the protein unfolds, so that more of its inner structure is able to react with other molecules.  The production of insoluble lumps of protein may be formed by intermolecular bridging reactions which are irreversible.  According to Fryer *et al* [1989] these reactions become significant at temperatures above 65°C.  A schematic representation of the denaturation and aggregation mechanisms is given in Fig. 16.8.

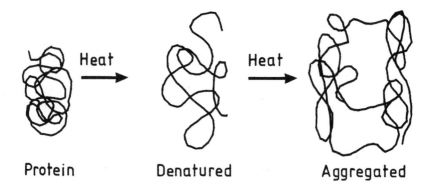

Figure 16.19. Protein denaturation and aggregation mechanisms [Fryer *et al* 1989]

Because these processes are temperature driven, the most likely place for aggregation to occur will be a hot exchanger surface with the strong likelihood that the surface will become fouled.  Temperature distribution will affect the uniformity of deposit accumulation, where the surface temperature is highest it is expected that the deposit will be more pronounced.  Many plate heat exchangers do not operate in a true countercurrent array with the possibility that the heating medium inlet temperature may coincide with the highest product temperature.  Under these circumstances the deposit formation may be more extensive than for true countercurrent operation where the metal surface temperature is lower.  On the other hand Belmar-Beiny *et al* [1993] suggest that protein denatured in the bulk gives more rapid fouling, the temperature having an effect on the total deposit.

Belmar-Beiny *et al* [1994] have also demonstrated differences in deposit morphology dependent on operating conditions.  The deposit may vary from a smooth thin film to a dense aggregate structure.  The addition of reagents which inhibit the formation of the *S-S* bonds has been shown to decrease the amount of deposit formed in UHT treatment of milk [Skudder *et al* 1981].  The addition of chemicals to limit the incidence of fouling in food processing however, would require careful consideration and would require authorisation from the appropriate regulating authority.

In connection with the fouling from milk proteins and salts, Kessler [1989] makes the following recommendations that are based on extensive studies, for heat exchanger (evaporator) operation.

1. To reduce the formation of deposits of salts the *pH* values should be as low as possible.

2. Supersaturation should be reduced by appropriate treatment of the product to be evaporated.

3. The heat flux should be no greater than 18 *kW/m²*.

4. The temperature (for evaporation) should be kept below 60°C to prevent denaturation of proteins and associated deposit formation.

In many examples of milk processing however, it is not possible to operate at such low temperatures. For instance pasteurisation requires a temperature of 70°C, the cooking of milk products may require temperatures up to 130°C and the production of ultra heat treated (UHT) milk may involve sterilisation temperatures up to 140°C. Furthermore the time the food product is held at a particular temperature will also offset the extent of the fouling.

Although Kessler [1989] suggests lowering the *pH* to reduce salt deposition in milk evaporation, extensive work by Skudder *et al* [1986] shows that lowering the *pH* can have an adverse effect on deposition of protein and fat. During UHT of milk in a plate-type plant deposit formation was greatly increased where the *pH* of the whole milk was reduced to 6.54, irrespective of whether the adjustment was made through the addition of mineral acid (*HCl*) or lactic acid. Most of the increase in deposition took place in the higher temperature sections of the plant. Conversely, an increase in *pH* by the addition of caustic soda to 6.8 resulted in less deposit being formed during heat treatment. Deposition of skim milk also increased when the *pH* was reduced to 6.51 before processing. Skudder *et al* [1986] confirmed that lowering the *pH* reduced the mineral deposition.

16.4.2 Fouling assessment

The fouling potential of food and food products is in general, assessed in the light of experience. Unlike other industrial processes there is little that can be done in terms of using additives to reduct or eliminate the problem, and much of the equipment is designed from a hygienic point of view rather than from the aspect of fouling, although they are clearly linked. Cleaning may only be undertaken when some item of equipment is unable to function in a satisfactory way, i.e. the required temperatures, or the desired throughput cannot be achieved. On the other hand cleaning may be carried out on a regular cycle irrespective of the condition of the surface in order to maintain hygienic conditions. Grasshoff [1994] states that plate heat exchangers have to be cleaned regularly at least once per day. Neither of these alternatives is entirely satisfactory since the costs involved in the cleaning process could be reduced by a more quantitative assessment of when fouling had reached a stage when cleaning would be appropriate, with minimum cost. A major problem in the assessment of fouling potential, is the variability of food composition even food or products that are considered to be identical in other respects.

Fryer and Pritchard [1989] emphasise the need to monitor fouling in a food process stream in order that the optimum time for cleaning may be determined so that for instance, a soft otherwise easy to remove deposit, is not converted to a hard intractable form by the action of heat. These authors describe two possible fouling monitors that may be used in a "side stream" from the main process stream. A more detailed discussion of fouling monitors is to be found in Chapter 17.

### 16.4.3  Cleaning of food fouled surfaces

Due to the complex nature of food products, the methods adopted for cleaning surfaces are very empirical. In general however, the so-called clean in place (CIP) is used wherever possible, to eliminate the need to dismantle the equipment. Cleaning in place usually involves several washing liquids, pumped through or recycled round the equipment and may include; an initial wash, the use of a dilute caustic solution, a biocide and a final rinsing with high quality water. Steam sterilisation may also be employed. The length of time involved in each stage of the CIP process will depend on the penetration of the deposits by the cleaning agents, and the removal forces that can be applied to the deposits. Cleaning a plate heat exchanger contaminated by tomato paste was discussed in Chapter 15. The example demonstrated that the temperature of cleaning and the concentration of the cleaning agent, may also influence the rate of deposit removal and hence the time required for effective cleaning of the surfaces. In addition to the cost of the cleaning agents, the downtime of the equipment will add to the overall cost of cleaning.

In the testing of a sensor to investigate fouling and cleaning of a UHT milk steriliser Corieu *et al* [1988] gave some information on the CIP procedures employed in the factory. The unit worked continuously during the week but it was not operated for 30 hours over the weekend. The steriliser (essentially a plate heat exchanger) was operated for two or three weeks depending on the quality of product and operating conditions, before being shut down and dismantled for inspection and maintenance.

During the operating periods regular cleaning procedures on a 24 hour cycle were adopted. At the 7th and 15th hours in the cycle a 35 minutes clean was employed under the conditions set out in Table 16.19.

TABLE 16.19

Specification of 7th and 15th hour CIP procedure (system under pressure)

| Rinse | Liquid | Temperature °C | Time s |
|---|---|---|---|
| Initial | Water | 140 | 300 |
| | followed by | | |
| | 2% *NaOH* | 140 | 600 |
| | (pumped around) | | |
| Intermediate | Water | 140 | 300 |
| | 1.5% *HNO₃* | 140 | 600 |
| | (pumped around) | | |
| Final | Water | 140 | 300 |
| | | | 2100 |
| | | | i.e. 35 minutes |

The maintenance of high temperatures during the cleaning procedure has two advantages; it preserves the sterility of the system and maintains the process operating temeprature for a rapid resumption of production.

At the end of the 24 hour cycle a more elaborate cleaning operation was undertaken as shown in Table 16.20.

TABLE 16.20

Specification of final CIP procedure

| Rinse | Liquid | Temperature °C | Time s |
|---|---|---|---|
| Initial | Water | 70 | 600 |
| | 1.5% *HNO₃* | 80 | 300 |
| | (once through) | | |
| Intermediate | Water | 80 | 600 |
| | 2% *NaOH* | 80 | 1200 |
| | (pumped around) | | |
| | Water | 80 | 600 |
| | 1.5% *HNO₃* | 80 | 900 |
| | (pumped around) | | |
| Final | Water | Temperature rising | 1200 |
| | Water under pressure | 140 | 1800 |
| | (sterilisation) | | |
| | | | 7200 |
| | | | i.e. 120 minutes |

Tables 16.19 and 16.20 indicate the rather arbitrary nature of the cleaning procedure. The work carried out by Corrieu *et al* [1988] suggests that the time dedicated to cleaning was very generous as the removal rate as observed by the monitor, was extremely rapid. Corrieu *et al* [1988] conclude that because of the pump round, or closed loop, system commonly employed in industrial installations the removal process is concealed and subsequently, this has led to long cleaning cycles. The circulation of the cleaning and rinsing solutions throughout a factory is not conducive to a better appreciation of when cleaning has been accomplished. It is understandable that operators of food processing equipment will err on the "safe" side. Corrieu *et al* [1988] also make the observation that cleaning might be enhanced at temperatures above 80°C for the initial and intermediate cleaning procedures, listed in Table 16.20.

It would be expected that all detergents and chemical cleaners would not be universally effective against all food soils, so that trial and error methods will be necessary to obtain optimum conditions in terms of concentration, temperature and the extent of mechanical removal applied. It would appear however, that increasing the concentration of a cleaning agent above a certain threshold does not increase the rate of cleaning and indeed, an increase in concentration could lead to a decrease in cleaning rate [Schlussler 1970]. The optimum concentrations of strongly alkaline detergents are in the region of 1% but for more mildly alkaline detergents, the optimum concentration may exceed 3% [Lalande *et al* 1988]. Work by Goedern *et al* [1989] suggests an optimum detergent concentration of about 2.5% (by weight) to minimise the cleaning time of plate heat exchangers previously used for milk processing.

According to some authors the effect of temperature of the cleaning operation is not well understood and it will clearly depend on the nature of the deposit. The mean removal rate of buttermilk derived solids, is very sensitive to higher temperatures, whereas those associated with milk are much less affected. It may be noted however, that removal of milk solids is at a maximum at around 60°C, but for other products (e.g. buttermilk, evaporated milk, and cocoa) the maximum removal of asociated deposits is in the region of 85 - 90°C.

Flow velocity or turbulence will affect the removal of deposits, as discussed in Chapters 7 - 12 for the different deposition mechanisms. It is to be expected that the effect of increasing the velocity above a certain threshold will not show a corresponding increase in cleaning rate. A threshold Reynolds number of 25,000 was suggested by Jennings [1965]. Goedern *et al* [1989] report that cleaning times were greatly reduced by using flow velocities in excess of 0.5 *m/s* for milk pasteurisation and sterilisation plant.

Holah *et al* [1994] have drawn attention to the fact that disinfection or sanitation are an essential part of the cleaning requirements for food processing equipment to maintain product quality. Ozone has been suggested as a possible sterilising agent in the cleaning of heat exchangers [Bott 1991]. The ozone in solution reverts to oxygen in a relatively short time but after disinfection has been achieved so that the quantity of the high quality rinse water is reduced.

Flagg and Thompson [1989] have demonstrated that it is possible to improve the management of CIP systems by effective computer control. Microprocessors dedicated to the CIP system have a store of memory so that within the machine provision can be made for a series of operations such as burst rinses, rinse to drain, chemical addition, application of heat, post rinses, recirculation of detergent wash and whatever operation may be required. It is possible to combine these CIP operations in any sequence necessary to clean and sterilise the plant. The technique is particularly useful for multipurpose plant where different cleaning techniques may be necessary for different products.

## REFERENCES

Alter, H., 1980, The preparation and properties of densified refuse derived fuel, in: Jones, J.L. and Radding, S.B. eds. Thermal Conversion of Solid Wastes and Biomass. ACS Symposium Series, 130, Paper 10, Washington.

Antonini, G. and François, O., 1988, Wastes as Fuel, in: Collin, R., Leukel, W., Reis, A. and Wood, J. eds. Proc. 1st European Conf. on Industrial Furnaces and Boilers. Lisbon.

Attig, R.C. and Duzy, A.F., 1969, Coal ash deposition studies and application to boiler design. Proc. Am. Power Conf. 31, 290.

Belmar-Beiny, M.T., Gotham, S.M., Fryer, P.J. and Pritchard, A.M., 1993, The effect of Reynolds number and fluid temperature in whey fouling. J. Food Eng. 19, 119 - 139.

Belmar-Beiny, M.T., Toyoda, I. and Fryer, P.J., 1994, Initial stages of fouling to surfaces above 100°C. Proc. Conf. Fouling and Cleaning in Food Processing. University of Cambridge, 50 - 59.

Bott, T.R., 1989, Fouling of heat exchangers : a review, in: Field, R.W. and Howell, J.A. eds. Process Engineering in the Food Industry. Elsevier Applied Science, London.

Bott, T.R., 1990, Fouling Notebook. Instn. Chem. Engrs. Rugby.

Bott, T.R., 1991, Ozone as a disinfection in process plant. Food Control, 2, 1, 44 - 49.

Brulé, G., 1981, Les minéraux du lait. Rev. Lait, 400, 61 - 65.

Bryers, R.W., 1987, On-line measurements of fouling and slagging (full scale and pilot units) and correlation with predictive indices in conventionally fired steam generators. Proc. Workshop Low Rank Coal Technology, San Antonio, Texas.

Burton, H., 1968, Reviews of the progress of dairy science. J. Dairy Research. 35, 317 - 330.

Butcher, C., 1991, Controlling *Legionella*, Chem. Engr. 14, 11th April

Corrieu, G., Lalande, M. and Ferret, F., 1988, On-line measurement of fouling and cleaning of industrial UHT exchangers, in: Melo, L.F., Bott, T.R. and Bernardo, C.A. eds. Fouling - Science and Technology. Kluwer Academic Publishers, 575 - 587.

Dabron, G.R. and Rampling, T.W., 1988, A literature review of published research on refuse derived fuel and its components. Appendix A the laboratory characterisation of refuse derived fuel. Report LR643 (MR), Warren Spring Laboratory.

Duzy, A.F., 1965, Fusibility - viscosity of lignite type ash. ASME Paper 65 - WA/FU 7.

Duzy, A.F. and Walker, J.B., 1965, Utilization of solid fuels having lignite type ash. US Bureau of Mines, Inf. Circular 8304.

EPRI, 1982, Design and operating guidelines manual for cooling water treatment. CS2276, 1st Edition Research Project 1261 - 1, EPRI, Palo Alto.

EPRI, 1984, Dechlorination Technology Manual, EPRI, Palo Alto.

EPRI, 1984, Condenser Macrofouling Control Techniques, CS3550 Project 1689-9, EPRI, Palo Alto.

EPRI, 1985, Condenser-targetted Chlorination Design, CS4279 Project 2300, EPRI, Palo Alto.

EPRI, 1987, Guidelines on Macrofouling Control Technology, CS5271 Project 1689-9, EPRI, Palo Alto.

EPRI, 1985, Proceedings: Condenser Biofouling Control - State-of-the-Art Symposium, EPRI, Palo Alto.

ESDU, 1992, Fouling and slagging in combustion plant. Item 92012, ESDU Int. London.

ESDU, 1988, Fouling in Cooling Water Systems, Item 88024, ESDU Int., London.

Flagg, P.L. and Thompson, R.P., 1989, Management information, food processing and CIP systems, in: Field, R.W. and Howell, J.A. eds. Process Engineering in the Food Industry. Elsevier Applied Science, 159 - 166.

Fryer, P.J. and Gotham, S.M., 1988, Fouling and cleaning in food processing. Proc. Conf. Fouling in Process Plant. Inst. Cor. Sci. Tech. and Inst. Chem. Engrs. St. Catherine's College, Oxford, 167 - 181.

Fryer, P.J. and Pritchard, A.M., 1989, A comparison of two possible fouling monitors for the food processing industry, in: Field, R.W. and Howell, J.A. eds. Process Engineering in the Food Industry. Elsevier Applied Science, London.

Fryer, P.J., Gotham, S.M., Paterson, W.R., Slater, N.K.H. and Pritchard, A.M., 1989, A systematic approach to the study and modelling of food fouling, in: Field, R.W. and Howell, J.A. eds. Process Engineering in the Food Industry. Elsevier Applied Science, London, 111 - 129.

Gibb, W.H., 1981, The slagging and fouling characteristics of coals Part 1. Ash viscosity measurements for the determination of slagging propensity. Power Ind. Res. 1, 29.

Goedern, G. de., Pritchard, N.J. and Hasting, A.P.M., 1989, Improved cleaning processes for the food industry, in: Kessler, H.G. and Lund, D.B., eds. Fouling and Cleaning in Food Processing. Prien, 115 - 130.

Gotham, S.M., Fryer, P.J. and Pritchard, A.M., 1989, Model studies of food fouling, in: Kessler, H.G. and Lund, D. eds. Fouling and Cleaning in Food Processing. Prien Bavaria, Germany, 1 - 13.

Grasshoff, A., 1994, Efficiency assessment of a multiple stage CIP procedure for cleaning a dairy plate heat exchanger. Proc. Conf. Fouling and Cleaning in Food Processing. University of Cambridge, 103 - 110.

Gray, R.J. and Moore, G.F., 1974, Burning the sub bituminous coals of Montana and Wyoming in large utility boilers. ASME Paper 74 - WA/FU1.

Gunn, D., 1990, The combustion of densified refuse derived fuel in shell boilers. dRDF Workshop, Department of Energy (ETSU) and Glasgow District Council, Glasgow.

Gunn, D., 1991. Private communication.

Hauser, G., Michel, R. and Sommer, K., 1989, The design and construction of food processing equipment with particular regard to hygiene, in: Field, R.W. and Howell, J.A. eds. Process Engineering in the Food Industry. Elsevier Applied Science, 143 - 158.

Hewitt, G.F., Shires, G.L. and Bott, T.R., 1994, Process Heat Transfer, CRC Press Inc., Boca Raton.

Highley, J. and Kaye, W.G., 1983, Fluidised bed industrial boilers and furnaces, in: Howard, J.R. ed. Fluidised Beds Combustion and Applications. Applied Science Publishers, London.

HMSO, 1983, Statutory Instrument No. 1508. The milk-based drinks (hygiene and heat treatment) regulations HMSO.

Holah, J.T., Bloomfield, S.F., Walker, A.J. and Spenceley, H., 1994, Control of biofilms in the food industry, in: Wimpenny, J., Nichols, W., Stickler, D. and Lappin-Scott, H. eds. Bacterial Biofilms and their Control in Medicine and Industry. Bioline, School of Pure and Applied Biology, University of Wales College of Cardiff, 163 - 168.

Inst. Pet., 1986, Methods for the analysis and testing. Vol. 1, pages 13 - 14, London.

Institute of Waste Management, 1995, Fluidised Bed Incinerators. Institute of Waste Management.

Jacob, A., 1992, *Legionella* Control in Practice, Chem. Engnr. 27 - 28, 14th May.

Jennings, W.G., 1965, Theory and practice of hard surface cleaning. Advanced Food Research, 14, 325 - 458.

Jones, A.R., 1987, An appraisal of the applicability of common slagging indices to British coals. EPRI Conf. Effects of coal quality on power plant, Atlanta.

Jones, M., 1994, Biofilms in the food industry, in: Wimpenny, J., Nichols, W., Stickler, D. and Lappin-Scott, H. eds. Bacterial Biofilms and their Control in Medicine and Industry. Bioline, School of Pure and Applied Biology, University of Wales College of Cardiff, 113 - 116.

Lalande, M., Gallot-Lavallée, T. and Corrieu, G., 1984, Chemical reactions and mass transfer associated with cleaning of heat exchange surfaces fouled by milk deposits, in: McKenna. Engineering and Food, Vol. 1. Elsevier Applied Science, London, 59 - 68.

Lalande, M., Tissier, J.P. and Carrieu, G., 1985, Fouling of heat transfer surfaces related to $\beta$-lactoglobulin denaturation during heat processing of milk. Biotech. Prog. 1, 2, 131 - 139.

Lalande, M. and Rene, F., 1988, Fouling by milk and dairy products and cleaning of heat exchange surfaces, in: Melo, L.F., Bott, T.R. and Bernardo, C.A. eds. Fouling - Science and Technology. Kluwer Academic Publishers, Dordrecht.

Lawn, C.J., 1987, The combustion of heavy fuel oils, in: Lawn, C.J. ed. Principles of Combustion Engineering for Boilers. Academic Press.

Lyster, R.L.J., 1979, Milk and dairy products, in: Effects of Heating on Foodstuffs, 353 - 368, Applied Science Publishers, London.

Niles, W.D. and Sanders, H.R., 1962, Reactions of magnesium with inorganic constituents of heavy fuel oil and characteristics of compounds formed. Trans. ASME J. Eng. Power, 178.

Osborn, G.A. and Bott, T.R., 1994, Chemical reactions of fly ash from refuse incineration. 10th Int. Heat Trans. Conf. Brighton, Paper144.

Raask, E., 1985, Mineral impurities in coal combustion. Hemisphere Pub. Corp., Washington.

Rampling, T.W. and Hickey, T.J., 1988, The characterisation of refuse derived fuel. LR643 (MR). Warren Spring Laboratory.

Reid, W.T. and Cohen, P., 1944, The flow characteristics of coal ash slags in the solidification range. J. Eng. Power, Trans. ASME Series A, 66, 83.

Rose, D., 1985, Guidelines for the processing and aseptic packaging of low acid foods. Part 1. Principles of design installation and commissioning. Campden Food Preservation Research Association. Chipping Campden.

Sandu, C. and Lund, D. Fouling of heating surfaces. Chemical reaction fouling due to milk, in: Lund, D., Platt, E.A. and Sandu, C. eds. Fouling and Cleaning in Food Processing. University of Wisconsin, 122 - 167.

Sanyal, A. and Cumming, J.W., 1985, Steam coal quality. Proc. Coaltech '85, 1, 121, London.

Schlussler, H.J., 1970, Zur reinigung fester, oberflächen in der lebensmittel industrie, Milchwissenschaft, 25, 3, 133 - 149.

Schramel, J.E. and Kessler, H.G., 1994, Effects of concentration on fouling at hot surfaces. Proc. Conf. Fouling and Cleaning in Food Processing. University of Cambridge, 17 - 31.

Skudder, P.J., Thomas, E.L., Pavey, J.A. and Perkin, A.G., 1981, Effect of adding potassium iodate to milk before UHT treatment. J. Dairy Res. 48, 99.

Skudder, P.J. and Bliss, C., 1987, Aseptic processing of food products using ohmic heating. Chem. Engr. February, 26 - 28.

Skudder, P.J., Brooker, B.E., Bonsey, A.D. and Alvarez-Guerrero, N.R., 1986, Effect of *pH* on the formation of deposit from milk on heated surfaces during ultra high temperature processing. J. Dairy Res. 53, 75 - 87.

Tipler, W., 1964, Some results obtained from a fuel oil water washing plant in studies of the fouling of a marine superheater. Trans. Inst. Mar. Engrs. 76, 37.

Tissier, J.P., Lalande, M. and Corrieu, G., 1984, A study of milk deposit on heat exchanger surface during UHT treatment, in: McKenna, B. ed. Engineering and Food. Elsevier Applied Science, London, 49 - 58.

Tsados, A., 1986, Gas side fouling studies. PhD Thesis, University of Birmingham.

Whitehead, D.M., Finn, J.C. and Beadle, P.C., 1983, Studies related to residual fuel oil composition and the formation of burnout of particles during combustion. Erdöl und Kohle-Erdgas-Petrochemie Vereinight mit Brennstofe Chemie, 36, 12, 579.

Winegartner, E.C. Coal slagging and fouling parameters. ASME Special Publication.

# CHAPTER 17

## Obtaining Data

### 17.1  INTRODUCTION

To operate heat exchangers efficiently and perhaps more importantly in the design of heat exchangers, it is imperative that at least some fouling data is at hand to give guidance. Inaccurate fouling data as described in Chapter 2, can give rise to cost penalties, not only in terms of capital cost (i.e. overdesign) but in operating costs where the heat exchanger underperforms and the costs of cleaning may become excessive. There is some flexibility in operation to make allowance for limitations of design by the adjustment of flow streams through the exchanger and through any bypass, but the opportunity "to trim" the heat exchanger may be quite restricted however, due to the process requirements.

In general there are three ways in which data on fouling may be obtained, namely from experience, from laboratory studies and by the use of plant data.

### 17.2  EXPERIENCE

Many heat exchanger designs are based on experience, either of the client or the manufacturer of heat exchanger equipment. The experience may be derived from the operation of identical exchangers and processes in the past, or on processes that are reckoned to be similar. The earlier chapters in this book have stressed the importance of different variables, notably velocity, temperature and concentrations of the fouling species on the deposition process, and so unless the design is based on exactly the same conditions as the experience was obtained, reliance on experience alone could lead to error. Where relatively low fouling resistances apply, e.g. pure organic liquids at low temperature, the errors involved in a poor choice of fouling resistance are not great. On the other hand where heavy fouling is present, e.g. slagging in combustion systems, gross errors could result from incorrect choice of fouling resistance.

The TEMA recommendations reported by Chenoweth [1990], are essentially based on the common experience of the expert members of the working party, but in all but a few cases, the conditions to which they apply are not stated. Even where some conditions are given, in general they are far from complete. The experiences of different members of the team considered to concern similar problems, could quite easily, be affected by the presence of relatively small concentrations of an unknown impurity. The total temperature driving force may, in different applications, give rise to different heat exchanger surface temperatures which might result in very different fouling effects, particularly where chemical reactions are involved. Experience may have been gained with over or under designed heat exchangers, where the velocity effects are likely to have affected the fouling resistance making it atypical.

Although data based on experience have to be treated with caution because of the potential sources of error, they do offer a quantitative measure of the possible fouling likely to be encountered in particular situations. It has to be admitted also, that in the absence of real data (often the situation in which a designer is placed), these TEMA recommendations do give a basis for design, perhaps to be tempered by the designers or client's particular experience.

Bott [1990] gives an example of the incorrect choice of fouling resistance resulting from inappropriate experience. It concerns the design of a plate heat exchanger for the reduction of the temperature of a stream of demineralised water from 75 to 40°C, using site cooling water with a temperature rise of 25 to 45°C in the heat exchanger. Little or no fouling was anticipated on the demineralised water side of the heat exchanger. From experience the designer of the heat exchanger suggested that a fouling resistance of $6 \times 10^{-5}$ $m^2$ $K/W$ would allow for the cooling water fouling. The client however, whose experience was all gained with shell and tube heat exchangers, insisted on a fouling allowance of ten times the value, i.e. $6 \times 10^{-4}$ $m^2$ $K/W$. The pertinent calculated data are set out in Table 17.1.

TABLE 17.1

Comparison of plate heat exchanger area requirements with two different cooling water fouling resistances

| Fouling resistance $R_f$ $m^2$ $K/W$ | Clean overall heat transfer coefficient $U_c$ $W/m^2K$ | Overall heat transfer coefficient $U_D$ for fouled conditions $W/m^2K$ | Area required for heat transfer $m^2$ |
|---|---|---|---|
| $6 \times 10^{-5}$ | 4831 | 3745 | 54.3 |
| $6 \times 10^{-4}$ | 4831 | 1239 | 164.3 |

Using the client's experience the "overdesign" amounted to approximately 200% over the area obtained using the lower value of the fouling resistance. The result of installing the larger heat transfer area was a series of operating difficulties that may be summarised as follows:

1. The outlet temperature of the demineralised water was reduced to the unacceptable level of 26.5°C. In order to bring the demineralised water to the required working temperature of 40°C the cooling water flow rate was reduced.

2. The cooling water velocity reduction had two consequences

a) The cooling water outlet temperature rose to 72.5°C instead of the designed value of 45°C with a corresponding rise in wall temperature.

b) The rate of fouling increased dramatically.

3. The combined effects of reduced flow velocity and increased wall temperature resulted in an increase of the fouling rate to $10^{-4}$ $m^2$ $K/W$ per month.

4. Increasing the cooling water velocity again in an attempt to overcome the effects of fouling had little effect.

5. After six months it was apparent that the heat exchanger would have to be shut down for cleaning (12 months would have been considered acceptable). The poor heat transfer performance and the fact that it was no longer possible to maintain the cooling water velocity because of excessive back pressure in the heat exchanger, meant that cleaning was the only sensible course of action.

6. More plates were added to the heat exchanger in an attempt to improve performance but this only served to reduce the cooling water velocity still further, with an attendant increase in fouling rate.

Ultimately, the oversized heat exchanger was replaced by one having an area of 54 $m^2$ (i.e. the plate heat exchanger designer's original suggestion). The alternative design proved satisfactory in terms of heat transfer, pressure drop and cleaning frequency.

Reliance on "experience" either personal or from published fouling data, can present the operator of heat exchangers with serious problems.

## 17.3 LABORATORY METHODS

The aim of any laboratory technique is to simulate the conditions that are likely to occur in a heat exchanger, particularly in terms of velocity surface temperature and residence time. Wherever possible the fluid used in the laboratory investigation will be that used in the actual process, but this may not be possible due to the difficulties of maintaining chemical and physical conditions in the laboratory equipment the same as those present in the full scale plant. It is not within the scope of this book to provide a comprehensive list of the problems associated with laboratory evaluation because each system will have different characteristics. It is possible however, to provide some examples of where difficulties are likely to occur as a guide, and these may be summarised as follows:

1. The fluid may be chemically unstable in some way, so that during transfer to the laboratory and storage, it undergoes a chemical change so that it is no longer representative of the process fluid. The chemical instability of the fluid may be the cause of heat exchanger fouling so that the chemical changes that have occurred before the experiments have been carried out could have a

drastic effect on the fouling data obtained. Examples of these process fluids are food liquids and pastes, solutions of polymers and intermediates.

2. Storage and disposal of large volumes of liquid wastes that may be flammable or toxic, could be a basic difficulty. In attempting to assess fouling over long periods of time large volumes of fluid will be required. It is tempting to recycle or reuse the fluid, but passage across a simulated heat exchanger surface that is at a raised temperature, may bring about permanent chemical changes in the fluid so that the fluid is no longer representative of the process fluid. The problem is particularly acute where chemical reaction fouling is experienced (see Chapter 11), e.g. crude oil processing.

3. The test fluid may be subject to settlement or separation problems. If a contributory factor in the fouling process is particulate deposition, any tendency for settlement of the particles in the storage system will need to be addressed. More importantly there is a strong possibility that particles under storage conditions, will tend to agglomerate and since particle size has a profound effect on particulate fouling (see Chapter 7), the results using a stored solution may therefore be erroneous. For the same reason it may be unwise to recycle the fluid through the equipment, since the agglomeration mechanism may not be reversible simply by agitation of the stored suspension.

4. Where suspensions of living organisms are present, e.g. in a cooling water system it is extremely difficult if not impossible, to maintain the ecological conditions that exist in the industrial system. The changed scale in the laboratory, compared with the full scale plant, may affect the distribution of micro-organisms. Simulation of macro-organic fouling in the laboratory is impossible because of the large size of the organisms in comparison with the scale of operations that are feasible in the laboratory.

Despite these difficulties however, laboratory investigation can provide insights into fouling mechanisms and as such, provide valuable data that may be useful in formulating models of fouling. Furthermore it is possible through laboratory investigations, to provide data on mitigation techniques particularly in terms of additives, e.g. biocides, corrosion and scale inhibitors. It is also possible to study the effects of velocity and temperature on the fouling process that will be invaluable in the design and operation of full scale heat exchangers. Much of the work associated with fouling of heat exchangers reported in the literature, is based on laboratory work.

17.3.1 Experimental techniques

It is not possible within the scope of this work, to provide a comprehensive discussion of the laboratory equipment that has been used to study heat exchanger fouling, but in general the principle is to pass the fluid under test through some form of test section that is meant to simulate a heat exchanger surface. The fluid under test is as near to that encountered in full scale operations as possible and the conditions within the test section are those associated with heat transfer surfaces. It may be possible to simplify the test fluid to make the experimentation easier; for instance to use a single species of micro-organism in a simulated cooling water to test the efficacy of a biocide to prevent biofouling. The organism chosen may be an example of a slime former. *Pseudomonas fluorescens* has been used extensively for this purpose. Fig. 17.1 is a simplified flow diagram of a laboratory apparatus to study biofilm formation [Pujo 1993]. The fermenter provides a continuous supply of a microbial suspension, operated aseptically under carefully controlled conditions to maintain a consistent product. A mixing vessel is provided where the simulated cooling water is generated. Separate streams of nutrients (medium) iron, alkali for *pH* control, air (to provide dissolved oxygen), water and microbial suspension are thoroughly mixed before the simulated cooling water is pumped through the test sections at a constant velocity. In this particular example the test sections represented heat exchanger tubes of similar diameter to tubes found in a shell and tube heat exchanger. The specimen tubes were mounted vertically to avoid problems associated with the effects of gravity that might otherwise confuse the results. Sufficient inlet lengths are necessary in order to be sure that the velocity profile in the tubes is fully established. The simulated cooling water may be recycled back to the mixing vessel. The residence time of the water in the apparatus may be varied depending on the desired conditions for the establishment and growth of a biofilm. On the other hand for the testing of biocides it would be necessary to have a "once through" system, i.e. there was no recirculation, otherwise the effects of the biocide would be different from those in a full scale cooling water system and under recirculating conditions the true efficacy of the biocide would be difficult to establish. Furthermore the effects of the biocide would be apparent in the mixing vessel in addition to the simulated heat exchanger tube, i.e. not representative of full sized plant conditions.

Fig. 17.2 [Pujo 1993] is an example of the data on the development of biofilm thickness obtained using the equipment shown in Fig. 17.1.

A laboratory system to study the removal of particulate haematite deposits is shown on Fig. 17.3 [Williamson 1990]. The similarities to the equipment shown in Fig. 17.1 are apparent. In this particular experiment radioactivity is used to measure changes in the deposit thickness.

The apparatus consists of two 50 *l* stainless steel tanks each with a stainless steel recirculating pump and loop capable of a circulating rate of around 1.3 *l/s*. One tank contains a suspension in water of previously irradiated haematite particles (referred to as "active") and the other contained water (referred to as "inactive"). Due to the high cost and disposal problems freshly irradiated haematite is not used in each run. Haematite particle diameters are in the range 1 - 3 *μm*. Both tanks

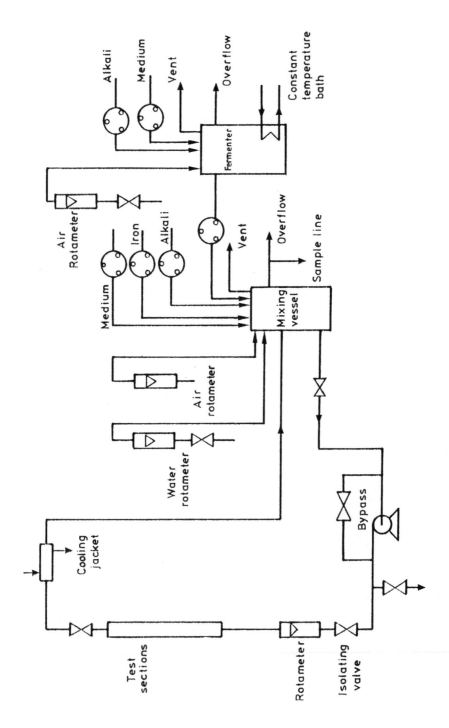

FIGURE 17.1.  Simplified flow diagram of a laboratory apparatus to study biofilm formation

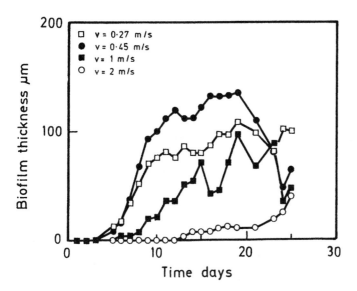

FIGURE 17.2. The change in thickness of a biofilm with time at two different velocities

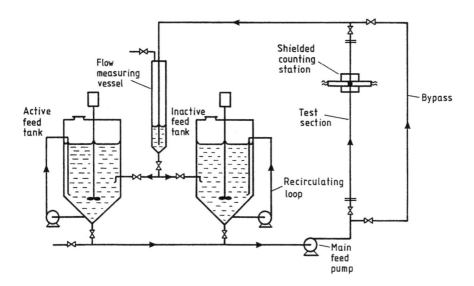

FIGURE 17.3. Simplified sketch of laboratory equipment to study the removal of particulate haematite particles from stainless steel heat exchanger tubes

are fitted with 3 *kW* electrical titanium plated heaters. Liquid from either tank can
be fed to a third stainless steel pump (main circulating pump), that passes liquid
through the test section or bypass. The test section consists of 17 *mm id* x 2 *m*
length of stainless steel tube held vertically and having the liquid inlet at the
bottom. The diameter of the tube is consistent with an actual heat exchanger tube.
A stainless steel test section was chosen to avoid problems of corrosion that could
otherwise mask the results. The research work is concerned with the deposition of
particles. The tests are isothermal.

The test sections are equipped with a radioactive counting station using sodium
iodide detectors about two thirds up from the bottom of the tube. It is considered
that after approximately this distance the velocity profile will have been established.
The length after the counting station is included to avoid any interference by the
exit of the liquid from the tube. After passing through the test section, liquid can
be directed to either of the feed tanks or into the glass vessel (for flow
measurement). Fig. 17.4 shows data obtained from the test apparatus [Williamson
1990].

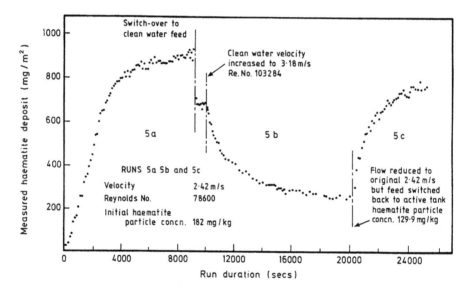

FIGURE 17.4. Data obtained on haematite particle deposition

Although the two examples of laboratory equipment shown in Figs. 17.1 and
17.3 are for heat exchanger tubes other test sections may be used depending on the
data desired. Square or rectangular flow sections may be employed where it is
desired to remove a flat plate or coupon, for visual or other examination. Fig. 17.5
is an example [Patel 199 ]. Suitable studs may be incorporated into the wall of a
tubular test section for the same purpose, but because of the curvature, the area of
stud surface exposed to the flow is of necessity quite small. Fig. 17.6 is an
example of a so-called Robbins device used in biofilm assessment.

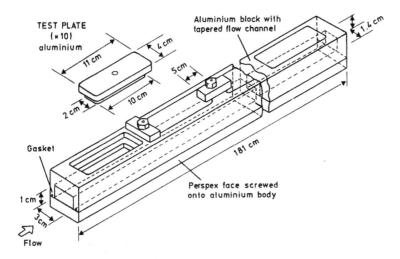

FIGURE 17.5. Test section with a removable test plate

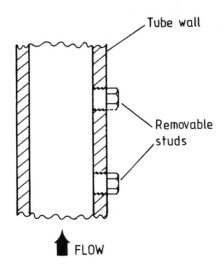

FIGURE 17.6. Section of a Robbins device with removable studs used for the study of biofilm formation.

In any test apparatus that uses variable plates or studs it is essential that the surfaces of the test section and the removable piece are flush. Irregularities in location will cause a flow disturbance that will affect the flow velocity near the wall, with consequences for deposit formation.

Because of the low productivity from laboratory equipment associated with the slow development of deposits (a single test may take several weeks) attempts have been made to provide a more rapid assessment, for instance of the effects of velocity. Fryer and Slater [1987] reported the use of a tapered tube. The taper on the tube means in effect, that the flow area continuously changes along the tube length. For a given volumetric flow rate therefore the velocity changes from inlet to outlet of the tube. If the tube is so arranged that the fluid enters at the largest cross section the velocity, and hence the shear stress, increases through the tube. Fig. 17.7 shows a typical fouling pattern although the boundaries between the regions will not be clearly defined, but it does reflect the effect of velocity changes.

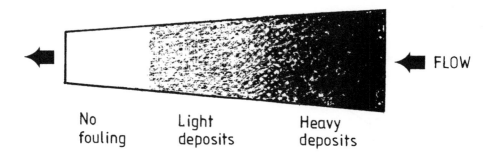

FIGURE 17.7. A typical fouling pattern in a tapered tube

Fryer and Slater [1987] used the device to study milk fouling and in particular the critical shear stress that was necessary to maintain a clean surface under the conditions of the experiment. Fig. 17.8 is taken from the work of Fryer and Slater [1987] and shows a plot of the variation of the logarithm of the critical shear stress with the reciprocal of the absolute surface temperature (the range of surface temperature being 95 - 105°C). These data clearly show the interaction between critical shear stress (velocity) and surface temperature, i.e. as temperature is lowered the shear stress necessary to maintain a clean surface is reduced.

A tapered duct with removable plates was used by Patel [1981] to study biofouling. A tapered device to study the attachment of cells is commercially available [Anon. 1991].

A difficulty of using a tube with a straight taper is that the change in shear stress with distance is not linear. In order to overcome this particular problem Reid [1992] developed a test section with a rectangular cross section with a taper such that the shear stress changed linearly with distance.

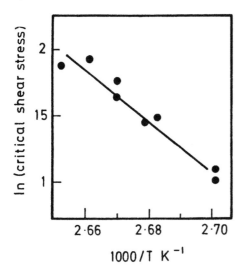

FIGURE 17.8. The relationship between critical shear stress and surface temperature

Changing velocity in a test apparatus in order to investigate the effects of shear stress is the basis of the so-called radial flow growth chamber. The device was developed to study biofilm adhesion. The principle of the apparatus is that contaminated water is passed between two parallel horizontal circular plates separated by a small distance (see Fig. 17.9). The fluid enters the chamber through the centre of the bottom plate, flows radially between the two plates and leaves through suitable ports at the periphery of the plates. Data on biofilm removal obtained using the radial flow growth chamber are given in Fig. 17.10 [Mott 1987].

Variations of the tubular geometry for test sections include:

1. An annular flow area contained between two concentric tubes. The centre tube, although not exclusively, may be used to provide the hot surface. The technique also has the advantage that is possible to have the outer tube fabricated from transparent material (e.g. glass or plastic) that allows visualisation of the deposit as it develops. The technique is suitable for all fouling mechanisms. It could be argued however, that the flow conditions in an annulus are different from those in a tube which, in turn, might affect the resulting data.

2. An electrically heated wire within the flowing stream. The wire provides the hot deposition surface. Changes in the resistance of the wire are a measure of its temperature and for constant experimental conditions therefore, a measure of the fouling on the wire surface. The method has been used for the assessment of potential fouling in crude oil heaters, for milk fouling and for measurement of *in situ* corrosion.

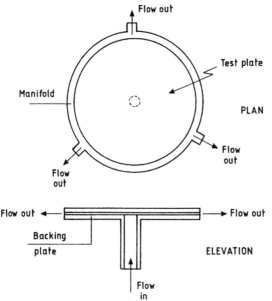

FIGURE 17.9. The radial flow growth chamber

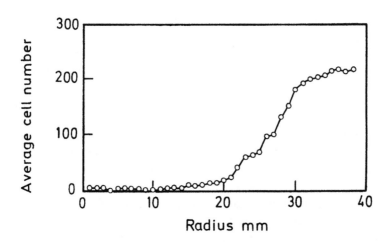

FIGURE 17.10. Data obtained using a radial flow growth chamber

3. The inclusion of an array of tubes so that the fouling fluid flows across the tubes. Plain tubes were studied in this way by Melo and Pinheiro [1983] for the deposition of kaolin from aqueous suspensions. Bemrose and Bott [1983] and Yeo *et al* [1994] investigated particulate deposition from air flowing over finned tubes.

The detailed techniques for investigating fouling under flowing conditions in the laboratory are virtually infinite, limited only by the inventiveness of the investigator and the requirements of the study. The basic principle however, is that a surface is held in contact with the flowing fluid under well defined and controlled conditions.

17.3.2 Estimating the extent of the fouling

Almost any physical or chemical measurement technique, provided it is appropriate, may be used to estimate the extent of fouling in laboratory apparatus. Melo and Pinheiro [1984] discussed some of the more common methods.

*Direct weighing*

The simplest method for assessing the extent of deposition on test surfaces in the laboratory is by direct weighing. The method requires an accurate balance so that relatively small changes in deposit mass may be detected. It may be necessary to use thin walled tube to reduce the tare mass so as to increase the accuracy of the method. Mott and Bott [1991] used the technique satisfactorily, to estimate the amount of biofouling that had accumulated on the inside of simulated heat exchanger tubes. The mean thickness of the biofilm over the whole inner surface of the tube may be made from a knowledge of the deposit density; for biofilms the density may be assumed to be that of water since the biofilm is substantially made up of water [Patel 1981]. The inside surface area of the tube $A_i$ is given by

$$A_i = \pi d l \qquad (17.1)$$

where $d$ and $l$ are the inside diameter and length of the tube respectively.

If the density of the deposit is $\rho_f$ then $M_f$ the mass of deposit for thin deposits, is given by

$$M_f = \rho_f \pi d l x_f \qquad (17.2)$$

where $x_f$ is the deposit thickness
hence $x_f = \dfrac{M_f}{\rho_f \pi d l}$ $\qquad (17.3)$

From a knowledge of the thermal conductivity of the deposit, for biofilms it may be assumed to be the same as water, it is possible to estimate directly the fouling resistance $R_f$ using Equation 2.1.

i.e. $R_f = \dfrac{x_f}{\lambda_f}$ (2.1)

where $\lambda_f$ is thermal conductivity of the deposit

*Thickness measurement*

In many examples of fouling the thickness of the deposit is relatively small, perhaps less than 50 $\mu m$, so that direct measurement is not easy to obtain.

A relatively simple technique provided there is reasonable access to the deposit, is to measure the thickness. Using a removable coupon or plate the thickness of a hard deposit such as a scale, may be made by the use of a micrometer or travelling microscope. For a deformable deposit containing a large proportion of water, e.g. a biofilm it is possible to use an electrical conductivity technique as described by Harty and Bott [1981]. Fig. 17.11 illustrates the principle. The instrument consists of a steel needle mounted on a micrometer which forms part of

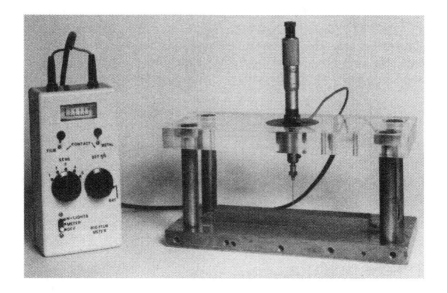

FIGURE 17.11. The principle of electrical conductivity for measurement of the thickness of soft wet deposits

an electrical circuit in conjunction with the metal plate on which the deposit resides, having been previously in contact with the fouling fluid. When the needle touches the top of the biofilm an ammeter shows that a current is flowing. The micrometer reading is noted. The needle is made to penetrate the deposit by screwing down the micrometer. When the needle touches the metal plate a second deflection is given by the ammeter. The micrometer reading is noted and the difference between the two readings gives the thickness of the deposit. If the deposit is uneven, i.e. the usual condition, it will be necessary to record several thicknesses to obtain a reliable mean. The fouling resistance may be estimated in the same way as for direct weighing using Equation 2.1 when the thickness has been determined.

### Heat transfer measurements

Since it is the fouling resistance that is under scrutiny an obvious choice of experimental method involves the determination of changes in heat transfer during the deposition process. The basis for subsequent operations will be Equations 2.4 and 2.5.

The amount of heat transferred $Q$ is given by:

$$Q = \frac{T_1 - T_2}{\left(\dfrac{1}{\alpha_1}\right)} = \frac{T_5 - T_6}{\left(\dfrac{1}{\alpha_2}\right)} \qquad (2.4)$$

where $\alpha_1$ and $\alpha_2$ are the heat transfer coefficients of the two fluids involved in the heat transfer process and $(T_1 - T_2)$ and $(T_5 - T_6)$ are the temperature driving forces across the individual thermal resistances of the two fluids respectively, represented by the inverse of the individual heat transfer coefficients. Equation 2.4 may be modified to give

$$Q = \frac{\Delta T_1}{\left(\dfrac{1}{\alpha_1}\right)} = \frac{\Delta T_2}{\left(\dfrac{1}{\alpha_2}\right)} \qquad (17.4)$$

where $\Delta T_1$ and $\Delta T_2$ are the temperature differences $(T_1 - T_2)$ and $(T_5 - T_6)$ respectively.

Equation 2.5 provides an estimate of the total thermal resistance $R_T$

$$R_T = \left(\frac{x_1}{\lambda_1}\right) + \left(\frac{x_2}{\lambda_2}\right) + \left(\frac{x_m}{\lambda_m}\right) + \frac{1}{\alpha_1} + \frac{1}{\alpha_2} \qquad (2.5)$$

so that for the fouled conditions

$$R_T = \frac{1}{U_D}$$
(2.10)

There are two ways of using these equations, but the principle is the same in each method and they represent the bulk of laboratory experimentation.

1.  The heat transferred is obtained from a heat balance between the two fluids flowing through the equipment. By estimating the temperature driving force between the two fluids, see Equations 2.11 and 2.12, it is possible using Equation 2.13, to estimate the overall fouling resistance.

$$Q = U_D A \Delta T_m$$
(2.13)

$$\text{or } U_D = \frac{Q}{A \Delta T_m}$$
(17.5)

where $A$ is the area for heat transfer.

If the initial overall heat transfer obtained from the experiment, i.e. when the surfaces are clean is assumed to represent $U_c$ (see Equation 2.9), then subtracting Equation 2.9 from Equation 2.10 yields:

$$\frac{1}{U_D} - \frac{1}{U_C} = \left(\frac{x_1}{\lambda_1}\right) + \left(\frac{x_2}{\lambda_2}\right)$$
(17.6)

i.e. $\left(\dfrac{1}{U_D} - \dfrac{1}{U_C}\right)$ represents the total fouling resistance to heat transfer at the time the heat balance was made. The data may be reported in this way, i.e. in terms of changes in overall heat transfer coefficient. Fig. 17.12 illustrates the general form of the change in overall coefficient with time (note the relationship to Fig. 1.2).

If the fouling on one side of the heat exchanger is negligible then the total change in fouling resistance may be attributed to the deposits on the side of the test surface where fouling occurs. The shape of the curve of the change in fouling resistance with time will also be asymptotic. For design purposes it is likely that the thermal resistance near the asymptote would be chosen provided of course, that the conditions in the full scale heat exchanger are similar to those in the experiment.

For the relatively small scale equipment used in the laboratory the temperature changes involved may be quite small so that an accurate heat balance is difficult to obtain.

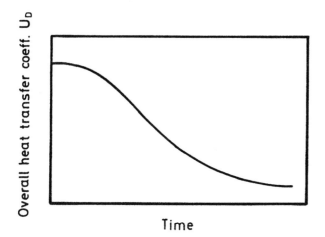

FIGURE 17.12. The change in the measured overall coefficient with time

The technique may be improved by inserting thermocouples in the wall separating the two fluids. The temperature driving force between the wall and the bulk fluid, whose fouling potential is being assessed, eliminates any difficulties introduced by fouling from the other fluid. Provided the location of the thermocouples is accurately known it is possible to make a more accurate calculation of the changes in thermal resistance of the deposit during the experiment.

A major assumption in the method is that the presence of the deposit does not affect the hydrodynamics of the flowing fluid. As mentioned in Chapter 5 however, the surface of a deposit is unlikely to be the same as that of the metal on which it resides. In general, the surface of the deposit is rough so that the turbulence within the fluid is greater than when it is flowing over a smooth surface. As a result the fouling resistance calculated from the data will be lower than if the increased level of turbulence had been taken into account. It is possible that the increased turbulence offsets the thermal resistance of the deposit and negative values of thermal resistance will be calculated. Fig. 17.13 illustrates the problem and involves scale formation from geothermal water [Bott and Gudmundsson 1976].

The method relies on constant inlet temperature conditions.

2. It is possible to use only the fluid whose fouling properties are required in the experiments. For these experiments it is usual to employ electrical heating so that fouling data is only obtained during heating. The temperature driving force from the electrically heated surface to the bulk fluid, may be estimated as described above using thermocouples embedded in the wall. One of the difficulties of the method is to obtain a uniform temperature distribution along the heated surface. A major advantage of the technique however, is that the electrical energy input provides an accurate estimate of the heat transferred

provided the heat losses are low or they can be calculated. Because the electrical energy input is constant the technique involves constant heat flux so that the fouling resistance affects the temperature of the wall. The change in wall temperature is a measure of the fouling resistance. A constant heat flux would be difficult to obtain using two flowing fluids.

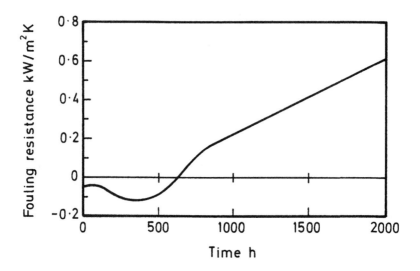

FIGURE 17.13. The change in fouling resistance with time as a result of scale formation from geothermal water

## Pressure drop

As an alternative to direct heat transfer measurements it is possible to use changes in pressure drop brought about by the presence of the deposit. The pressure drop is increased for a given flow rate by virtue of the reduced flow area in the fouled condition and the rough character of the deposit. The shape of the curve relating pressure drop with time will in general, follow an asymptotic shape so that the time to reach the asymptotic fouling resistance may be determined. The method is often combined with the direct measurement of thickness of the deposit layer.

Changes in friction factor (see Chapter 5) may also be used as an indication of fouling of a flow channel. The method was used by Bott and Bemrose [1983] for estimating the extent of particulate fouling on finned tubes. Fig. 17.14 is a schematic diagram of the apparatus used in this research.

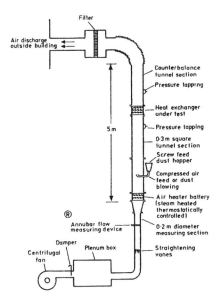

FIGURE 17.14. Equipment for fouling assessment on a bank of finned tubes

Fig. 17.15 shows how the friction factor increased with time in a "sample" finned tube heat exchanger exposed to an air flowing through laden with calcium carbonate particles the apparatus shown in Fig. 17.14. It may be regarded as a typical curve obtained in the laboratory apparatus. During the experiments water at temperature in the range 10 - 90°C was passed through the tubes. The equipment allowed thermal performance and pressure drop to be measured. From Fig. 17.15 it can be seen that the friction factor rises asymptotically to a level 50%

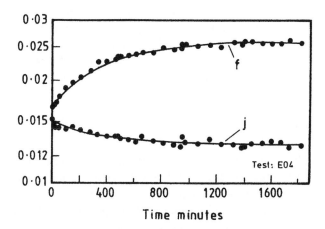

FIGURE 17.15. The effect of particulate fouling on the performance of finned tubes

above its initial value. The shape of the friction factor-time curve is similar to that expected for the development of fouling thermal resistance with time. The effect of the increasing friction factor $f$ is to increase the pressure drop across the experimental heat exchanger as demonstrated by Equation 5.4, i.e. $\Delta P \alpha f$. In an air cooled heat exchanger this would represent an increase in fan power.

The reduction in heat transfer is reflected in the other curve on Fig. 17.15 which is a plot of the change in the so-called Colburn $j$ factor with time as the deposition progresses.

$$j = \frac{Nu}{Re \ Pr^{\frac{1}{3}}} \qquad (17.7)$$

where $Nu$, $Re$, $Pr$ are the Nusselt, Reynolds and Prandtl numbers respectively

$$Nu = \frac{\alpha l_m}{\lambda} \qquad (17.8)$$

where $\alpha$ is the heat transfer coefficient

$l$ is characteristic length

$\lambda$ is the thermal conductivity of the flowing fluid (in this example the fluid is air)

The reduction in $j$ with time therefore represents an apparent reduction in $\alpha$ (by about 20% from the initial value), but in reality the relatively low value of $\alpha$ associated with gas streams will not be affected greatly by the presence of the deposit unless the deposit becomes thick enough to change substantially the flow pattern across the fins. The presence of the fouling layer however interposes a thermal resistance between the air flow and the metal fins and core tube of the model heat exchanger, thereby reducing the effective heat transfer and giving the impression that the heat transfer coefficient has diminished.

*Other techniques for fouling assessment*

In terms of their effect on heat exchanger performance the measurement of heat transfer reduction or increase in pressure drop provide a direct indication. The simple methods of measuring deposit thickness described earlier are useful, but in general they require that the experiment is terminated so as to provide access to the test sections. Ideally non-intrusive techniques would allow deposition to continue while the experimental conditions are maintained without disturbance. Such techniques include the use of radioactive tracers as already referred to earlier in this section and optical methods. McKee *et al* [1994] used a laser technique to investigate the conditions close to a solid surface that influence the removal of deposit from the surface. Moving pictures of the removal process were obtained using high speed ciné film and video recording. Time lapse photography may also be used for investigating the accumulation or removal of deposits. Bartlett *et al* [1995] used an infra red system to investigate the development and removal of biofilms from tubular test sections. The basic concept is that the absorbence of infra red radiation is a function of the "density" of the biofilm, i.e. related to its thickness and the mass per unit volume. In biofilms the correlation between infra red absorbence and thickness is not easily obtained because wide variations of biofilm structure are possible depending on the conditions under which the growth occurred. Furthermore optical methods require a transparent "window" to allow access of the appropriate beam of light or radiation. Unless the equipment is carefully designed and operated the "window" itself may affect the validity of the data obtained. The experimental method may in fact demand that deposition takes place on the window that is likely to have very different surface characteristics to the surfaces of metallic heat exchangers.

Microscopic examination of deposits may provide some further evidence of the mechanisms of fouling, but this is generally a "back up" system rather than to give quantitative data.

Tenacity of the deposit depends on its adhesion to a surface as discussed in Chapter 6. Specially designed laboratory equipment has been used to study adhesion. Visser [1970] for instance, used liquid flow between concentric cylinders, one of which was rotating to study the adhesion of particulate deposits.

17.3.3 Comments on laboratory techniques

The principle drawback of laboratory techniques is the difficulty of simulating exactly the conditions that apply to an actual heat exchanger, i.e. flow conditions, temperature, fluid quality, geometry and surface conditions. It is necessary therefore for the experimenter to determine the optimum conditions that will provide the most realistic and reliable data relating to the particular study. In more fundamental work, which seeks to elucidate the effects of certain variables on the mechanisms by which a deposit forms or is removed from a surface, it may not be necessary to investigate all these variables. For instance in the investigation of biofilm formation it may be acceptable to run the experiments under isothermal conditions using glass tubes (to provide access for optical measuring devices) so that comparisons may be made by changing a particular variable, e.g. velocity or nutrient concentration while all other variables are held constant. The technique is very useful for the comparison of the effects of different biocides and their concentrations on the prevention of biofilm growth or biofilm removal.

17.4    THE ACQUISITION OF PLANT DATA

There are in general, two ways in which plant data on fouling may be directly obtained. They involve an assessment of heat exchanger performance and the use of a side stream of process fluid. The use of probes may provide an indirect method of fouling evaluation.

17.4.1  Assessment of heat exchanger performance

The performance of a heat exchanger may be assessed by a number of different techniques, but they are all dependent on good instrumentation, particularly for the measurement of flow rate and temperatures. Without these measurements accurate data are impossible to obtain. The acquisition of plant data is facilitated if temperatures and flow rates are recorded as a matter of routine. A data logger may be used in conjunction with a computer and a suitable program, to provide values of fouling resistance and to give a graphic printout of the change in fouling resistance with time.

The technique requires sufficient temperature measurements so that the inlet and outlet temperatures of the fluid stream entering and leaving the heat exchanger can be measured, or calculated from a suitable heat balance.

Bott [1990] reports data obtained in this way for a a steam heated shell and tube exchanger raising the temperature of a mixture of $C_4/C_5$ hydrocarbons from 15 - 130°C. The steam was admitted to the shell side of the exchanger. As the heat exchanger became fouled it was necessary to raise the steam temperature (i.e. raise the steam pressure in the shell) to increase the temperature driving force to compensate for the effects of the deposit on heat flow. In turn this raises the tube wall temperature. The data obtained or calculated from the performance of the heat exchanger records are given in Figs. 17.16 - 17.18.

It is assumed that the thermal resistance on the steam side is negligible. for this particular example the assumption is justified since the heat transfer coefficient for condensing steam is high and the heat transfer surface is generally clean. The tube wall temperature is close to the steam temperature.

In other plant heat exchangers for example where the two fluids produce deposits on both sides of the heat exchanger, it may only be possible to estimate changes in overall fouling resistance. It is not possible to separate the individual components that make up the overall fouling resistance. These data are then only strictly applicable to conditions similar to those for which the data were obtained.

In order to reduce capital cost many plants are installed with the minimum of instrumentation. Under these circumstances it may be difficult or even impossible, to make a proper assessment of the fouling encountered in heat exchangers under operating conditions. Temperatures and flow rates may have to be inferred with the problems of accuracy that this raises.

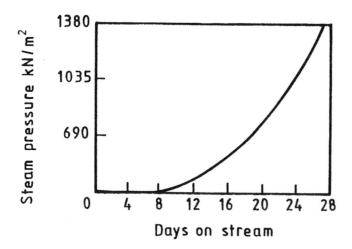

FIGURE 17.16. The change in steam pressure with time

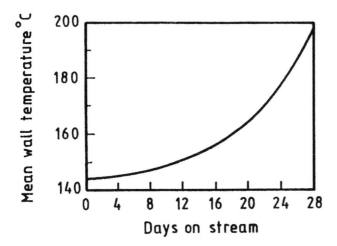

FIGURE 17.17. The change in tube wall temperature with time

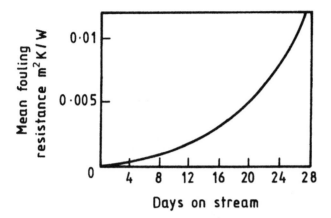

FIGURE 17.18. The increase of mean fouling resistance over 28 days operation

A further limitation of data obtained during plant operation is that brought about by changes in operating conditions. Many heat exchangers operate under very constant flow rates and temperatures, but there are also many heat exchangers in which conditions change with time in response to production requirements and problems in other parts of the plant. Temperatures may change and flow rates adjusted that will affect the rate of fouling. Under these circumstances it becomes impossible to relate the accumulation of deposits to a fixed set of conditions. Although the data might provide some guidance on the severity of the fouling problem, it would be difficult to use the data so obtained, with any kind of confidence for the design of heat exchangers for a similar duty.

Even heat exchangers that operate for long periods under steady conditions may also provide data of doubtful value. For instance, a reduction in fluid velocity, accompanied by a rapid rise in wall temperature, even for a short time, could result in a very high rate of fouling that is inconsistent with the rate of deposit formation under steady conditions.

Czolkoss [1991] reports a device for direct on-line monitoring of single tubes in steam condensers for the assessment of fouling. The technique involves closing off the flow to a single tube, allowing the water in the tube the rise in temperature, and then allowing the flow to recommence. The result is a transient exit temperature from which it is possible to assess accurately temperature differences that are directly related to the fouling resistance. It is claimed that biofouling layer thickness as low as 5 $\mu m$ can be detected by the device, so that early warning of potential problems may be obtained.

Corrieu *et al* [1988] have described a monitor for the on-line assessment of fouling during the production of ultra high temperature sterilisation of milk in a plate heat exchanger. The monitoring equipment consists of three pressure transducers, an electromagnetic flow meter and a microprocessor. The pressure transmitters are suitably installed so as to be hygienic, and measure the differential pressure in the two sections of the steriliser most likely to become fouled on account of the temperatures the milk encounters, i.e. the preheat zone where the temperature is raised from 80 to 120°C, and the heating zone where the temperature rises from 120 to 140°C. The flowmeter operates externally so that it does not present a problem of hygiene. The pre-programmed microprocessor receives data and converts it into an empirical degree of fouling (zero represents a clean heat exchanger). These data may be used to assess the effectiveness of cleaning (see Chapter 15) carried out three times per day.

Despite notable successes, some mentioned in this section, all plant derived data need careful scrutiny before they are accepted as being reliable in relation to their application to systems other than those for which the data were obtained.

### 17.4.2 Sidestream monitoring

An improvement on the assessment of individual heat exchanger performance is to pass a sidestream through some kind of monitor. In general monitors resemble the test equipment used in the laboratory. The use of monitors allows a more careful control of the fouling conditions to be exercised particularly in terms of temperature, temperature changes and flow rates, using actual plant streams. The data obtained are more reliable than those obtained directly from the operation of full scale plant.

The installation of the monitor of course, has to be consistent with the safety standards of the process with particular care in respect of flammable gases and liquids. It is imperative that the monitor and associated equipment should be robust to withstand the rigours of site conditions.

The ultimate technique in plant monitoring is to use a mobile test facility that can accept a sidestream from a plant and pass it through suitable equipment. Heating and cooling could be achieved by the use of site utilities. Electrical heating may be employed to give better control of experimental conditions than might be possible using process steam. Mobile test facilities were used with success by Mott [1991] for the assessment of fouling due to corrosion. The National Engineering Laboratory at East Kilbride has designed and developed a portable fouling assessment unit (PFAU) that has obtained data from a variety of locations. Fig. 17.19 is a photograph of the unit which, in addition to the test monitor, also has data logging and computer facilities. The mobile equipment is capable of simulating the conditions found in say a tube of a shell and tube heat exchanger in respect of temperature, velocity and geometry. It is possible to use a sample of heat exchanger tubing in the test. The microprocessor is programmed to provide time related heat transfer fouling resistances. Simple chemical analysis may also be carried out in the mobile laboratory as required by the process.

One of the main drawbacks to the use of portable equipment and mobile laboratories is one of cost. The initial capital cost may be high and operating costs can also be high where extended timescales are involved. Many fouling deposits

FIGURE 17.19. The portable fouling assessment unit developed by the National Engineering Laboratory

take a long time to establish so that the test facility will be required on site for relatively long periods. Although the monitor in the mobile laboratory can be automatically controlled it will require attention from skilled personnel for a considerable time while data are obtained. The total cost of operation will make the technique prohibitive unless the data obtained can be shown to offer opportunities for improved heat transfer efficiency and substantial financial savings.

Because of the widespread use of cooling water and the need to assess treatment programmes, monitors have been developed specifically for use in cooling water systems. Many of these devices are relatively simple and do not necessarily require the use of a mobile laboratory. Moore and Hays [1985] produced a survey of commercial fouling monitors that could be used in conjunction with cooling water and EPRI [1985] made a status survey of 18 biofouling detection monitoring devices. Chenoweth [1988] provides detailed information on ten such monitors. He observes that the choice of monitor depends on a number of questions, the answers to which will not only affect the cost but also the quality of the data obtained.

1. What, where, when and how long are the monitor activities to continue?

2. How much data is required and how accurate do the data need to be?

3. Can the data be intermittent or need they be taken continuously?

4. Will the monitor be used to reflect the operation of a particular plant heat exchanger?

5. Are the data obtained from the monitor easy to correlate to plant required?

6. Will the data be used to predict plant performance?

7. What personnel attention will be required?

8. What are the operational and maintenance requirements of the monitor?

The EPRI [1985] survey reveals that a wide range of techniques have been employed in commercially available biofouling monitors including annular flow devices, externally heat tubes, unheated surfaces, condenser simulators, electrically heated wires, biofilm sampling devices and acoustic detectors.

The importance of fouling measurement techniques using monitors is likely to increase as the competitive nature of the process and power industries intensifies. In cooling water applications more effective use of additives can be achieved through their use with reduced costs and less environmental impact, as a result of accurate assessment of the fouling problem.

## 17.4.3 The use of probes

As already mentioned in this chapter, interpretation of plant data is extremely difficult since they are likely to be averaged over a wide range of physical and chemical conditions. Laboratory studies are unlikely to reproduce the complex exhaust environments that are to be found in industrial combustion systems. The use of sidestreams may be costly in terms of equipment and labour. The use of a probe in industrial plant to determine fouling resistances (or data from which fouling resistance may be obtained) would seem to be the most appropriate route [Isdale 1988]. It has the added value that the data are obtained under plant conditions. The principle is simply to introduce a suitable surface into the fluid flow and to make some kind of measurement in respect of the deposit formed on the surface. The technique is in general, only suitable for large spaces such as the flue gas channels in combustion plant, where the probe is unlikely to disturb the flow conditions to any extent that could affect the data obtained. Because of the wide application of fouling probes to the investigation of fouling and slagging in combustion systems (see Chapter 16) the discussion will concentrate on this particular technology. Often probes are used in conjunction with large scale test facilities rather than an actual industrial plant.

Isdale [1988] in a report on discussion concerning fouling in gaseous systems, makes the following observations. The production and interpretation of data obtained by the use of probes is a key issue. Many of the significant and costly problems in gas side fouling often occur in extremely complex environments, particularly in terms of the composition of the gas stream. For instance in the combustion products from coal combustion there are many components ranging from simple gases, through liquid droplets to solid mineral compounds. Furthermore, chemical reactions will be occurring as the gas passes through the equipment (see Chapter 16).

A wide variety of devices has been used to obtain fouling data in gas streams and they range from simple impact discs to heat exchangers. Most fouling probes simulate a heat exchanger surface (i.e. a cylindrical shape to represent a tube) capable of maintaining a desired surface temperature by the use of air or water through inner ducts or passages in the probe. In general the probe surface is required to be below the temperature of the flue gases flowing past the outside of the tube.

Marner and Henslee [1984] reviewed the design of devices that have been used to study gas side fouling under plant conditions and they were able to identify five types, i.e:

1. Heat flux meters that measure changes in heat flux for given conditions. The reduction in heat flux is a measure of the thermal resistance of the accumulated deposit.

2. Deposition probes are simple ways of trapping a deposit so that its physical and chemical characteristics may be investigated. In general these devices do not yield quantitative data in terms of fouling rate or thermal resistance.

3. Mass accumulation probes as the name suggests, are used to obtain a deposit that can be weighed, i.e. a direct measurement of the fouling per unit area of probe for given conditions. The technique has also been used to assess corrosion under the conditions within a combustion space.

4. Acid deposition probes that are used generally to determine the acid dew point of combustion gases, i.e. the temperature at which acid condensation occurs, usually near the flue gas exit where the temperature is relatively low. Acid dew point temperatures are often in the region of 120°C for most flue gases.

5. Optical devices may be used to assess some aspect of fouling (e.g. particle size distribution in a flue gas). Although they have been primarily used in laboratory tests, they are increasingly being used for industrial measurements.

Other devices for special applications have been developed but in general, they are variations of the five types suggested by Marner and Henslee [1984].

Tsados [1988] has listed the requirements of probe design for satisfactory service and include:

1. The probe should be of robust construction in order to resist the arduous environment in combustion gases, i.e. high temperatures and corrosive conditions.

2. Combustion spaces are generally large so that the probe assembly must be capable of reaching the desired location without undue distortion or bending. The problem may be accentuated due to difficulties of accessibility as many combustion plant will have few suitable ports.

3. Minimum weight commensurate with (1) and (2) above to provide the operator with ease of handling.

4. Minimum service requirements, i.e. water, compressed air or power. There is the likelihood that these are not in general, readily available at the test location.

5. Automatic acquisition of data would be a distinct advantage, particularly for long term testing.

Some observations on the design of the test surface itself may be made:
1. The surface temperature should be controlled over a temperature range so that the effects of temperature may be studied.

2. The target surface should be large enough so that adequate amounts of deposit are collected to permit an accurate weight to be determined and any chemical analysis to be made.

3. The texture of the receiving surface should be reproducible since roughness or some other surface characteristic, is likely to affect the extent of deposit accumulation. It may be necessary to use a fresh sample surface for each test since cleaning may modify the surface and virginal surface may not be restored.

4. A facility to allow the concurrent testing of different surfaces (e.g. roughness or material) would be a distinct advantage although it may be difficult to design the probe head to accommodate several surfaces that are substantially subject to the same conditions of flow and temperature.

It is outside the scope of this book to describe in detail the various designs of probe that have been designed to provide data. It is useful however, to refer to examples of different probes to investigate fouling in combustion systems.

*Heat flux meters*

Neal and Northover [1980] describe three types of measuring devices able to handle heat fluxes up to 600 $kW/m^2$. Two of the meters have to be attached to the heat receiving working surface in the boiler, therefore good thermal contact is essential for reliable data to be obtained. The third device described, is portable, self contained and easily inserted into the combustion space through a 10 *mm* diameter port. The probe essentially a heat pipe, is able to monitor local values of heat flux from flames, but it is not suitable for long term measurements of heat flux. Comparisons between clean and fouled surfaces have been made by Chambers *et al* [1981] using heat flux meters commercially available.

Tsados [1988] reports the use of a heat flux meter to assess fouling in a pilot scale combustor. Fig. 17.20 provides some of the dimensions and details of his design. The measuring section of the water cooled probe consisted of a cylindrical block 50 *mm* long x 30 *mm* in diameter. The wall thickness was 13 *mm*. The measuring section consisted of eight equal segments separated from each other by insulating material. Thermocouples were used to measure the temperatures at the probe surface and close to the cooling water flow. The two ends of the measuring section were thermally insulated with fibreglass to provide conditions of near one dimensional heat flow and so improve the quality of the data. Fig. 17.21 provides an example of data on the combustion of domestic refuse obtained using the probe.

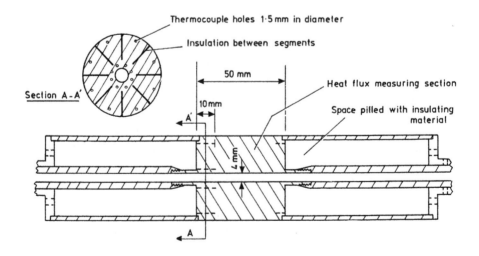

FIGURE 17.20. An experimental heat flux meter

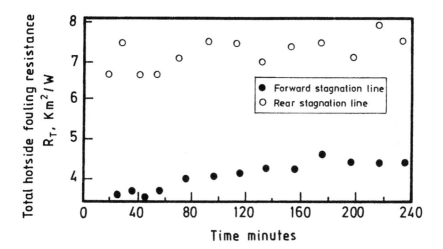

FIGURE 17.21. Data obtained using a heat flux probe

## Deposition and mass accumulation probes

In general deposition and mass accumulation probes are similar in principle but differ in sophistication. The deposition probe is simple whereas the mass accumulation probe is likely to be more elaborate.

Jackson and Raask [1961] and Jackson [1978] designed a probe for use in the high temperature regions of boilers. The probe was 4 *m* x 25 *mm* diameter. Thermocouples were located just below the target surface of the probe. The air used for cooling the probe was discharged into the combustion space.

A mass accumulation probe that has been used in a variety of test locations was first reported by Isdale *et al* [1984]. The essential design of the probe is shown on Fig. 17.22. The sample collection surfaces are in the form of a simple cylindrical shell split into two identical half cylinders as shown in Fig. 17.23. The two simple half cylinders are clamped together over the end of the probe, which may be cooled with air or water depending on conditions.

The support pipe is 38 *mm* in diameter and is inserted into the combustion space through a specially designed valve. The valve reduces the influx of air that could

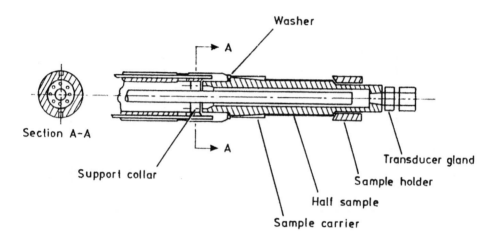

Washer

A

Section A-A

A

Support collar

Transducer gland

Sample holder

Half sample

Sample carrier

FIGURE 17.22. The National Engineering Laboratory mass accumulation probe

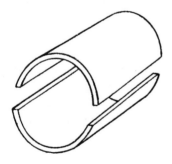

FIGURE 17.23. The cylindrical target shells of the NEL mass accumulation probe

affect the deposit, provides an enclosure where the probe surface temperature reaching the required level before final insertion into the high temperature region, and supports the probe while it is in the combustion place.

The temperature of the cylindrical sample surface is measured by means of thermocouples inserted in 0.5 *mm* diameter holes running parallel with the axis. The inlet and outlet temperatures of the coolant and its flow rate can be measured. A pitot tube located near the test surface measures the local gas velocity. Dew point measurements and particle size analysis, are carried out at the same time as the mass accumulation tests to provide additional relevant data.

Tsados [1988] lists some of the advantages of the probe described by Isdale *et al* [1984] and include:

1.  The total amount of accumulated material can be quickly and easily assessed by direct weighing before and after the test.

2.  The amount of deposit formed on the front and rear surfaces of the probe can be separately weighed thereby providing an insight into the mechanisms of deposition.

3.  Rapid changes of the test surfaces is possible so that the productivity of test data is good.

4.  Removal of the cylindrical shells together with the associated deposits make chemical analysis a straightforward matter.

5.  It is possible to use different materials for the shells so that the effects of surface on deposit accumulation can be determined.

Fig. 17.24 presents some data obtained using the probe during the combustion of simulated domestic refuse.

*Acid dew point and salt condensation measurement*

The essential feature of an acid dew point probe is that there is a known temperature gradient along its "active" length or the temperature of a probe surface can be progressively reduced. The temperature at which condensate appears is considered as the acid dew point of that particular system of combustion gases.

An early example of a dew point meter was developed by the British Coal Research Association (BCURA) reported by Flint [1948]. The dew point meter consists of a glass thimble located at the end of the probe inserted into the flue gas. The thimble is progressively cooled by an air stream and the surface temperature measured by means of a suitable thermocouple. The junction of the thermocouple is fused into the glass and forms one electrode of an electrical conductivity cell. When condensation occurs the aqueous acid film produced on the surface of the

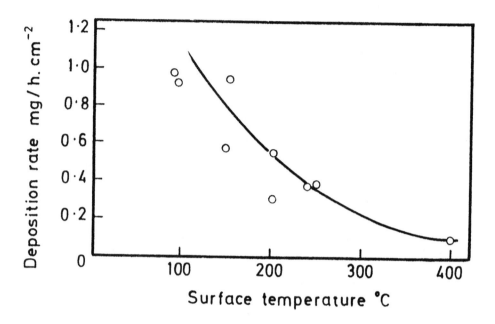

FIGURE 17.24. Mass accumulation data obtained during the combustion of simulated domestic refuse

probe becomes a conductor and the temperature recorded by the thermocouple is considered to be the acid dew point temperature. A problem associated with the technique is that it relies on the continuity of the film to complete the electrical circuit. Before the film becomes continuous, during the gradual reduction in surface temperature droplets may form and although condensation has occurred, it will not be recorded by the conductivity meter till more condensate appears to provide a pathway for the electric current. It is possible therefore to obtain low values of the acid dew point. Fig. 17.25 gives a schematic diagram of the test system.

Alexander *et al* [1960] used an air cooled probe to measure acid deposition in oil fired boilers, with the air discharge directly into the flue gas stream. The probe was 2.5 *m* long x 25 *mm* in diameter. In a particular test the temperature along the probe wall varied from 80 to 170°C. At temperatures below 94°C the probe surface was damp and analysis of the deposit formed showed that it consisted of metal sulphates suggesting that corrosion of the test surface had occurred.

A probe for the measurement of salt condensation in high temperature gaseous environments has been designed and is being developed by Jenkins [1994]. Fig. 17.26 is a simplified sketch of the probe assembly. The probe is inserted into the flue gas stream to be tested, and hot gas is drawn through an annular space

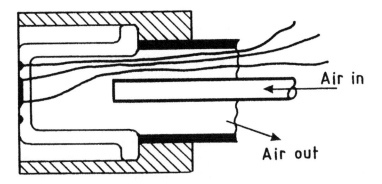

FIGURE 17.25. The BCURA acid dew point meter

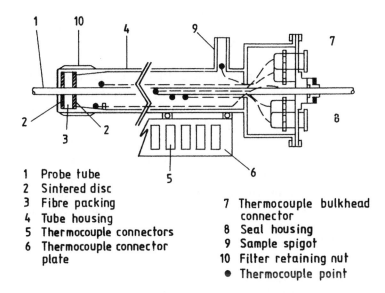

1  Probe tube
2  Sintered disc
3  Fibre packing
4  Tube housing
5  Thermocouple connectors
6  Thermocouple connector plate
7  Thermocouple bulkhead connector
8  Seal housing
9  Sample spigot
10 Filter retaining nut
•  Thermocouple point

FIGURE 17.26. A simplified sketch of a salt deposition probe

between the tube housing and the probe tube. The gas is filtered at its entry to the probe assembly to remove particulate matter that may otherwise interfere with the accuracy of the data obtained. Condensation will occur on the inner tube which is cooler than the outer tube. Salts condense progressively along the tube as they encounter their appropriate dew points with the development of stratified deposits. The bonding of deposit represents a history of condensation as the flue gas cools. The temperature profile within the tube assembly is altered by changing the flow conditions of the gas streams in the probe.

The key variables that are recorded from the probe are the temperature gradient along the probes and the pressure of the gas stream. External to the probe the volume of gas passed is recorded.

Further development work on the probe is required before it can be considered to be sufficiently reliable and robust for industrial applications.

*Optical techniques*

Jenkins [1994] has reviewed optical methods for the detection of deposits in flue gas systems as background to the development of the high temperature salt deposition probe described above. He concludes that the measurement techniques best suited to the test requirements are interferometry and ellipsometry.

The interferometry technique depends on the absorption of monochromatic light by deposition onto a temperature controlled sample piece. According to Jenkins [1994] the technique is most readily applicable to a clean or filtered gas stream. Difficulties are likely to be encountered where the gas is loaded with particulate matter that is likely to affect the light transmission.

Ellipsometry measures the change in phase and amplitude of polarised light. Assessment is by means of the deviation of a beam of light reflected from the deposit surface.

In general the techniques are very much in the development stage and although suitable for laboratory studies under ideal conditions, they present demanding practical requirements for use in full scale industrial locations. Laboratory applications have been reported by Rosener and Kim [1984].

Laser-based, light scattering techniques are available to make on-line measurements on particles in gas streams, e.g. combustion gases. One of the principal advantages is that no sample collection or conditioning is required. No assumptions are required about the shape of the size distribution. According to Insitec [1993] manufacturers of such instruments, the technique can be applied to systems within the following ranges:

Particle size          0.2 - 200 $\mu m$
Concentration
        for sub-micron particles up to $10^7$ particles/$cm^3$
        for super-micron particles up to 100 *ppm* by volume
Velocity               0.1 - 400 *m/s*

The instrument may be used despite extreme environmental conditions, to provide data directly on absolute particle concentration, particle size distribution and particle velocity. Clearly these data would be of considerable help in the assessment of particulate fouling (see Chapter 7).

REFERENCES

Alexander, P.A., Fielder, R.S., Jackson, P. and Raask, E., 1960, An air cooled probe for measuring acid deposition in boiler flue gases. J. Inst. Fuel, 33, 31 - 37.

Anon., 1991, Measuring biological adhesion. Chem. Engr. 505, 17.

Bartlett, G., Santos, R. and Bott, T.R., 1995, Measurement of biofilm development within a flowing water using infra red absorbence. To be published.

Bemrose, C.R. and Bott, T.R., 1983, Theory and practice in gas side particulate fouling of heat exchangers, in: Bryers, R.W. and Cole, S.S. eds. Fouling of Heat Exchanger Surfaces. Proc. Eng. Foundation Conf. Hershey Pocono Resort, 257 - 275.

Bott, T.R., 1990, Fouling Notebook. Inst. Chem. Engrs. Rugby.

Bott, T.R. and Gudmundsson, J.S., 1976, Rippled silica deposits in heat exchanger tubes. 6th Int. Heat Trans. Conf., Toronto.

Chenoweth, J., 1990, Final report of the HTRI/TEMA joint committee to review the fouling section of TEMA standards. Heat Trans. Eng. 11, 1, 73.

Chenoweth, J.M., 1988, Liquid fouling monitoring equipment, in: Melo, L.F., Bott, T.R. and Bernardo, C.A. Fouling - Science and Technology. Kluwer Academic Publishers, Dordrecht, 49 - 65.

Corrieu, G., Lalande, M. and Ferret, F., 1988, On-line measurement of fouling and cleaning of industrial UHT exchangers, in: Melo, L.F., Bott, T.R. and Bernardo, C.A. eds. Fouling - Science and Technology, Kluwer Academic Publishers, Dordrecht, 575 - 587.

Czolkoss, W., 1991, New techniques for on-line monitoring of heat transfer coefficient of single condenser tubes. Koelwater in de Industrie, STI Symposium, Antwerp.

EPRI, 1985, Biofouling detection monitoring devices: status assessment, Report CS3914, Project 2300 - 1, EPRI Palo Alto.

Flint, D., 1948, The investigation of dew point and related condensation phenomena in flue gases. J. Inst. Fuel, 21, 248 - 253.

Fryer, P.J. and Slater, N.K.H., 1987, A novel fouling monitor. Chem. Eng. Comm. 57, 139 - 152.

Harty, D.W.S. and Bott, T.R., 1981, Deposition and growth of micro-organisms on simulated heat exchanger surfaces, in: Somerscales, E.F.C. and Knudsen, J.G. eds. Fouling of Heat Transfer Equipment. Hemisphere Pub. Corp., Washington, 335 - 344.

Insitec, 1993, Particle counter-sizer-velocimeters (PCSVs). Insitec Measurement Systems, San Ramon, CA 94583.

Isdale, J.D., 1988, Debate on gas side fouling, in: Melo, L.F., Bott, T.R. and Bernardo, C.A. Fouling - Science and Technology. Kluwer Academic Publishers, 731 - 733.

Isdale, J.D., Scott, A.C., Cartwright, G. and Glen, N., 1984, The use of a portable probe to study fouling in exhaust gas streams. Proc. 1st UK Nat. Conf. on Heat Transfer. Inst. Chem. Engrs. Symposium Series 86, 415 - 434.

Jenkins, A., 1994, The measurement of high temperature condensation of salt. MPhil Thesis, University of Birmingham.

Marner, W.J. and Henslee, S.P., 1984, A survey of gas side fouling measuring devices. JPL Publication 84.11, E,. and G. Idaho Inc. Idaho Falls.

Melo, L.F. and Pinheiro, J. de D.R.S., 1984, Fouling tests: Equipment and methods, in: Suitor, J.W. and Pritchard, eds. Fouling in Heat Exchange Equipment. 22nd Nat. Heat Trans. Conf. HTD ASME, Niagara Falls.

Moore, W.E. and Hays, G.F., 1985, Guidelines for selection and use of on-line fouling monitors. Paper TP-85-2 Cooling Tower Inst. Annual Meeting.

Mott, I.E.C., 1987, The adhesion of *Pseudomonas fluorescens* to specific surfaces. MSc Thesis, University of Birmingham.

Mott, I.E.C. and Bott, T.R., 1991, The adhesion of biofilms to selected materials of construction for heat exchangers. Proc. 9th Int. Heat Trans. Conf. Jerusalem vol. 5, 21 - 26.

Neal, S.B.H.C. and Northover, E.W., 1980, Measurement of radiant heat flux in large boiler furnaces, I, Problems of ash deposition relating to heat flux. Int. J. Heat and Mass Trans. 23, 1015 - 1021.

Patel, T.D., 1981, Examination of the changes in microbial distribution with time in cooling water slimes. PhD Thesis, University of Birmingham.

Pujo, M.D., 1993, Effects of hydrodynamic conditions and biocides on biofilm control. PhD Thesis, University of Birmingham.

Reid, D.C., 1992, Biofouling in bioprocess plant. PhD Thesis, University of Birmingham.

Rosener, D.E. and Kim, S., 1984, Optical experiments on thermophoretically augmented submicron particle deposition from "dusty" high temperature gas flows. J. Chem. Eng. 29, 147 - 157.

Tsados, A., 1986, Gas side fouling studies. PhD Thesis, University of Birmingham.

Visser, J., 1970, J. Col. Interf. Sci. 34, 26 - 31.

Williamson, R.D., 1990, The deposition of iron oxide particles on surfaces from turbulent aqueous suspension. PhD Thesis, University of Birmingham.

Yeo, V., Bott, T.R. and Bemrose, C.R., 1994, Flow and distribution across a model finned tube. Proc. 10th Int. Heat Trans. Conf., Brighton.

# INDEX

## A

Acid deposition probes, 506, 511-514
  BCURA design, 511
Additives
  applications in combustion
    systems, 352
  antioxidants, 314
  definition of types, 291
  dosing, 288-291
  combustion systems in, 347-352
    aluminium hydroxide, 350
    aluminium oxide, 350
    borax, 350
    copper oxychloride, 351
    limestone, 349
    magnesium oxide, 349
    manganese compounds, 350
    rare earth compounds, 351
    silica, 350
    vermiculite, 350
  liquid systems in, 291-347
    biocides, 316-347
      bromine, 332-333
      chlorine, 449-450
      glutaraldehyde, 321,
        341-345
      non -oxidising, 326
      oxidising, 325
      ozone, 333-341
    corrosion inhibitors, 306-312
    crystal modifiers, 301-306
    dispersants, 306
    flocculants, 293-295
    film forming, 212-313
    scale inhibitors, 295-299
    surfactants, 306
    threshold compounds, 299-301
Adhesion, 45-53

  van der Waals forces, 46-48
  double layer forces, 49-50
Algae, 234-226
Alkali metals in coal combustion,
  448-449
Anodic protection, 310-312, 376
Antifoulants, 19, 317
Antioxidants, 314
Arrhenius plot
  jet fuel coking, 198
Asphaltenes, 189
Autoganism, 231
Autotrophs, 226
Autoxidation, 191-192

## B

Bacteria, 230-234
Barnacles, 238
Base/acid ratio, 445-446
  modified version, 446-447
Biocides, 316 (see also Additives)
  future use, 319-320
  ideal, 346-347
  system factors, 321-323
Biofilms, 223, 228-230, 239-253
  oxygen penetration, 245
  $pH$ effects, 250-253
  suspended solids and, 250
  temperature effects, 247-250
  velocity effects, 242-247
Biology
  micro-organisms, 224-236
  macro-organisms, 226-239
Biot number, 182
Bridging effects, 45
Bromine, 332-333
Bubble formation, 124-125

# C

Calcium
    bridging, 51
    carbonate, 107, 121, 298, 302-306
    sulphate, 120
Carbon dioxide
    solubility, 109
Catalyst attrition, 453-454
Cationic flocculants, 295 (see also
    Additives)
Cathodic protection, 369-376
    impressed current, 372-373
    sacrificial anode, 371-372
Chelating agent, 291
Chemical cleaning, 400-404
Chemotrophs, 226
Cleaning agents, 401
Cleaning devices, 392
    bullets, scrapers and scrubbers,
        393-395
    hollow drill, 393
Chlorine, 323, 325-332
    coal combustion in, 449-450
    dioxide, 332
    dosing targeted, 330-331
Coal
    coke formation, 190
    combustion, 167-169
    mineral content, 439-441
Combustion systems, 435-464
    coal, 437-439
    gas, 455
    oil, 450-455
    waste, 455-463
    mitigation of fouling, 463
Competition, 231
Cooling water systems, 409-435
    blowdown, 418-420
    cost penalties, 428-435
    cooling towers, 414-415
    design, 412-418
    dosing equipment, 417-418

environmental considerations,
    416-417
management, 421-427
pollution problems, 420-421
safety aspects, 416-417
treatment programmes (see also
    Additives)
    advantages and disadvantages,
        423-435
Corrosion, 149-182
    acid affected metals of, 153
    amphoteric metals of, 153
    cathodic protection, 310-312
    controlling mechanisms, 177
    factors affecting, 171-174
    gas systems in, 173-174
        low temperature, 164-165
        high temperature, 165-170
    general theory, 150-164
    inhibitors (see also Additives),
        309-314
    iron and steel of, 154
    liquid systems in, 171-173
        crevice, 161-162
        fatigue cracking, 163
        galvanic, 156-161
        hot spot, 162
        impingement, 164
        leaching, 163
        pitting, 162
        stress, 163
    noble metals of, 154
    resistant materials, 170-171
Copper oxychloride, 351-352
Costs
    capital, 15-16
    lost production, 18-19
    operating, 16-18
    pumping, 17
    remedial action, 19
Cracking (see Pyrolysis)
Critical shear stress, 489
Crystal
    growth, 110-114

modification, 291, 301-302
systems, 101
Crystalline solids, 100-103
Crystallisation boiling and, 122-127
Crystallisation mechanism, 106-114
Crystallisation of organic materials,
127-133
Cunningham coefficient, 73

**D**

Dehydration (see Pyrolysis)
Deposit (see also Fouling)
    crude oil, 193-197
    diffusion of particles, 61-63
    layers in fired heaters, 205
    mass determination
        direct weighing, 491
        thickness, 492
    removal, 37-43, 390
    thickness change with time, 4
    transformation, 38-39
Deposition rate, 3
Deposition and removal, 39-42
Dispersants, 306
Dissolved oxygen
    crude oil, 202
    cooling water, 372
Dosing techniques, 288-291
DVLO theory, 49-51

**E**

ETDA, 303
Electrochemical corrosion cell, 151
Electrode potentials, 152
Energy/distance profiles, 50
Epichlorhydrin dimethylamine, 295
Experimental techniques, 482-499

**F**

Falling rate fouling, 30-31

Ferric oxide, 150
Ferric hydroxide, 150-159
Ferrous oxide, 150
Fick's Law, 37
Flow fluids of, 33-36
Forces
    double layer, 49-51
    long range, 45
    short range, 45
Food fouling
    assessment, 469-470
    mechanisms of, 466-469
    milk, 466-469
    cleaning techniques, 470-473
Food processing, 465-473
    potato crisp manufacture, 466
    cleaning equipment of, 470-473
Foulants
    inorganic, 1
    organic, 1
    thermal conductivity of, 8
Fouling
    assessment, 409-470
    asymptotic, 26
    basic principles, 7-14
    biofilm formation
        microbial adhesion, 257-262
        microbial transport, 254-262
        *pH* effect of, 250-253
        suspended solids effect of, 250
        surface effect of, 249-250
        temperature effect of, 247-249
        trace elements effect of, 253
        velocity effect of, 247-249
    biological, 224-262
        microbial, 224-236
        macrofouling, 236-239
    chemical reaction, 185-218
        basic chemistry, 187-193
        deposit composition, 193-197
        system variables, 197-201
            flow, 200-201
            pressure, 200
            temperature, 197-199

combustion systems in, 435
  coal from, 448-450
  gas from, 455
  oil from, 450-455
  waste from, 455-463
cooling water systems in, 409-435
  basic concepts, 412-414
  cooling towers, 414-416
  cost penalties, 428-435
  design, 414, 418
  dosing equipment, 417
  environmental considerations,
    414-417
  management, 421-428
  pollution problems, 420-421
  safety aspects, 414-417
corrosion from, 149-182
  factors affecting, 171-174
    gas systems, 173-174
    liquid systems, 171-173
crystallisation from, 97-133
  boiling and, 122-127
  mechanisms, 107-114
  nucleation, 110
cost of, 15-21
  additional operating, 16-18
  cost of mitigation, 19-20
  increased capital, 15-16
  loss of production, 18-19
freezing from, 137-147
  flowing systems, 141-147
  static systems, 137-141
obtaining data of, 479-515
  experience use of, 479-481
  laboratory methods use of,
    481-499
    techniques in, 482-491
    estimation of deposit,
      491-499
  plant data use of, 500-515
    heat exchanger performance
      in, 500-503
    sidestream monitoring in,
      503-505

    use of probes in, 505-515
  particulate from, 55-91
    diffusion, 61-63
    electromagnetic effects, 69
    experimental data, 78-85
    fundamental concepts, 56-58
    gas systems, 85-91
      at low temperature, 85-90
      at high temperature, 91
    impaction, 64
    inertia, 63-64
    liquid systems, 78-85
      non-isothermal, 66-68
    transport
      isothermal, 58-66
      non-isothermal, 66-69
  resistance, 9-14
    choice of, 277-280
Free radicals, 191-192, 314-315
Friction, 36
  velocity, 59
Fungi, 236

**G**

Galvanic corrosion, 156-158
Galvanic protection, 369-376
Galvanic series, 155
Gas combustion, 189
Glutaraldehyde, 321, 341-345
Gum formation, 189

**H**

Haematite deposition, 82-83, 486
Heat exchanger
  cleaning
    off-line
      chemical, 399-404
      drilling and rodding,
        392-395
      explosives use of, 395-396
      manual, 389

mechanical, 389-396
osmotic shock use of,
    398-399
steam soaking use of, 396
thermal shock use of, 398
lances, air, steam and water,
    389-391
on-line, 357-388
    air injection use of, 363
    balls use of, 357-362
    brush and cage use of,
        362-363
    chemicals use of, 387
    galvanic protection, 369-376
    inserts use of, 376-379
    magnetic devices use of,
        363-365
    shot use of, 369
    radiation use of, 388
    sonic energy use of, 367-368
    soot blowers use of,
        365-367
    special design features,
        379-387
    water washing use of,
        368-369
commissioning, 283-285
control, 282-285
design
    concentration of foulant effect
        on, 272
    *pH* effect on, 272-273
    surface condition effect on,
        272-273
    temperature effect on, 270-272
    velocity effect on, 269-270
finned tube, 87-90
material of construction, 273-277
non-metallic, 385
operation, 285
scraped surface, 386
selection, 275-277
performance, 500-503
Heat flux meters, 507-508

Heat transfer coefficients
overall
    changes due to fouling, 24
    clean conditions, 10-11, 480
    fouled conditions, 10-11, 480
Heat transfer measurements, 493-496
Heat transfer area, 10-12
Heparin modified siloxane, 384
Heterotrophs, 226
Hydrazine, 313
Hydroids, 238
Hypobromous acid, 332-333
Hypochlorous acid, 325-328

**I**

Ice formation, 137-141
Incentives financial, 20-21
Incineration
    densified RDF, 461
    deposit composition, 458
    corrosion, 459
Inertia of particles, 63-66
Initiation period, 25
Interception, 72-73
Inverse solubility salts, 104

**K**

Kaolin deposition, 81, 83

**L**

Langelier index, 296
Layout of plant, 281
Lithotrophs, 226
Lubricating oil oxidation, 189

**M**

Macrofouling, 236-239
    barnacles, 238

hydroids, 238
mussels, 237
sea mats, 238
sea squirts, 238
serpulid worms, 238
Magnetite deposition, 78-81
Mass accumulation probes, 506, 509-511
Mass transfer, 36-37
Materials of construction, 170-171
Medium composition, 227
Melting points of solids, 166
Metal deactivators, 315
Microbial biology, 224-228
Micro-organisms, 224-236
    algae, 234
    bacteria, 230
    diatoms, 235
    fungi, 236
    surfaces and 228-230
Milk
    composition, 467
    fouling, 466-469
Mitigation (see also Additives)
    liquid system fouling
        biofouling, 316-347
        chemical reaction, 314-316
            use of inserts, 376-379
        corrosion, 306-314, 369-376
        magnetic devices, 363-365
        particle deposition, 292-295
        scale prevention, 295-306
        sponge ball cleaning, 357-362
    gas system fouling, 347-352
        combustion systems, 463-464
            shot cleaning, 369
            sonic techniques, 367-368
            soot blowers, 365-367
            water washing, 368-369
Models of fouling
    biofouling, 253-262
    chemical reaction, 201-218
    corrosion, 174-182
    crystallisation, 114-119

freezing, 137-147
general, 23-32
    Kern and Seaton, 26-30
particulate, 70-78
simple, 26-30
Mussels, 237
Mutualism, 231

**N**

Normal solubility salts, 105

**O**

Oil
    cationic components, 451
    combustion, 169-170, 450-464
        vanadium in, 451-453
    deposit composition, 193-197, 453
    sludge, 190
Optical measurement techniques, 514-515
Orgotrophs, 226
Oxygen
    chemical reaction fouling and, 201
    penetration of biofilms, 245

**P**

Particle deposition (see also Fouling)
    data, 86
    relaxation time, 60-61
Phase equilibria in slagging, 447
Phenol formaldehyde coatings, 383
Phenol epoxide coatings, 383
Phototrophs, 226
Phosphate esters, 301
Phosphonates, 301
Phosphoncarboxylic acids, 301
Phosphonic acid
    amino, 303
    acetodi, 303

Phosphoric acid
  meta, 300
  pyro, 300
  tripoly, 300
Plant data, 500-515
Polyacrylate, 303
Polyamines, 294
Polydiallyl-dimethyl ammonium
  chloride, 294
Polyethylene imine, 384
Polymaleic acid, 303
Polyphosphates, 301
Polymerisation, 192-193
Polyquaternaries, 294
Power station condenser fouling
  costs, 20-21
Pradtl number, 90
Pressure drop
  fouling and, 17, 496-498
  measurements, 496-498
Protein denaturation, 468
Pyrite, 167-168
Pyrolysis, 190-191

**R**

Radial flow growth chamber,
  489-490
Ramsbottom carbon test, 454
Removal rate, 3
Residence time, 274-275, 298, 320
Ryznar index, 298

**S**

Scale composition, 123
Scaling (see Fouling crystallisation)
Serpulid worms, 238
Schmidt number, 62
Sea mats, 238
Sea squirts, 238
Sherwood number, 62
Sidestream monitoring, 503-505

Silica ratio, 445
Silica scale, 121
Slagging
  assessment, 442-448
    temperature indicators,
      442-447
  coal combustion in, 442-448
  definition, 435
  index, 444
Solid/solid interactions, 46-48, 53
Solidification (see Fouling freezing)
Solubility
  solids of, 103-106
  $CO_2$ of, 109
Stanton number, 87
Steam chimney, 126
Stefan number, 144-145
Sticking probability, 57
Straight line fouling, 25
Stress corrosion, 163
Sulphur
  dioxide, 165
  trioxide, 165
Supersaturation, 107
Surfactants, 306, 315-316
Surface treatment
  mitigation for, 379-385
    plastic coatings, 381-382
    thin coatings, 302-385
    vitreous enamels, 380-381
Synergy, 231

**T**

Taprogge system, 357-362
Temperature
  distribution across fouled
    surfaces, 7
  driving force, 9
    mean, 11
  temperature indicators, 442-445
Thermophoresis, 66-68
Transport of particles, 58-61

Tube
  plate fouling and, 2
  tapered, 488
Turbulent bursts, 40-43

# V

Vanadium compounds, 169-170
Velocity profiles, 35
Viscosity index, 447

# W

Waste
  agricultural, 456
  combustion, 455-463
  fouling assessment, 462
  industrial, 456
  non-specific, 457
  specific, 457
  urban, 456
Water
  cooling, 409-435
  composition of
    geothermal, 99
    river, 98
    typical, 411
  deposits, 123
  utilisation, 421-427
Wax crystallisation, 128-133

# Y

Yield stress, 123

Printed and bound by CPI Group (UK) Ltd, Croydon, CR0 4YY

03/10/2024

01040329-0010